UNIVERSITY OF GL▒▒▒RGAN
LE▒ ▒▒ING RF▒ ▒ CENTRE
▒ont▒ ▒DL
443) 48▒▒▒▒

THE PERGAMON TEXTBOOK
INSPECTION COPY SERVICE

An inspection copy of any book published in the Pergamon International Library will gladly be sent to academic staff without obligation for their consideration for course adoption or recommendation. Copies may be retained for a period of 60 days from receipt and returned if not suitable. When a particular title is adopted or recommended for adoption for class use and the recommendation results in a sale of 12 or more copies, the inspection copy may be retained with our compliments. The Publishers will be pleased to receive suggestions for revised editions and new titles to be published in this important International Library.

International Series on
MATERIALS SCIENCE AND TECHNOLOGY

Volume 30—Editor: D. W. HOPKINS, M.Sc.

NOTICE TO READERS

Dear Reader

If your library is not already a standing order customer or subscriber to this series, may we recommend that you place a standing or subscription order to receive immediately upon publication all new issues and volumes published in this valuable series. Should you find that these volumes no longer serve your needs your order can be cancelled at any time without notice.

The Editors and the Publisher will be glad to receive suggestions or outlines of suitable titles, reviews or symposia for consideration for rapid publication in this series.

ROBERT MAXWELL
Publisher at Pergamon Press

EXTRACTION METALLURGY

by

J. D. GILCHRIST, BSc, PhD, ARCST, FIM

Department of Metallurgy and Engineering Materials
The University of Newcastle upon Tyne

SECOND EDITION

PERGAMON PRESS

OXFORD · NEW YORK · BEIJING · FRANKFURT
SÃO PAULO · SYDNEY · TOKYO · TORONTO

U.K.	Pergamon Press, Headington Hill Hall, Oxford OX3 0BW, England
U.S.A.	Pergamon Press, Maxwell House, Fairview Park, Elmsford, New York 10523, U.S.A.
PEOPLE'S REPUBLIC OF CHINA	Pergamon Press, Qianmen Hotel, Beijing, People's Republic of China
FEDERAL REPUBLIC OF GERMANY	Pergamon Press, Hammerweg 6, D-6242 Kronberg, Federal Republic of Germany
BRAZIL	Pergamon Editora, Rua Eça de Queiros, 346, CEP 04011, São Paulo, Brazil
AUSTRALIA	Pergamon Press Australia, P.O. Box 544, Potts Point, N.S.W. 2011, Australia
JAPAN	Pergamon Press, 8th Floor, Matsuoka Central Building, 1-7-1 Nishishinjuku, Shinjuku-ku, Tokyo 160, Japan
CANADA	Pergamon Press Canada, Suite 104, 150 Consumers Road, Willowdale, Ontario M2J 1P9, Canada

First edition published 1967
Reprinted 1969
Second edition 1980
Reprinted 1982, 1986

British Library Cataloguing in Publication Data

Gilchrist, James Duncan
Extraction metallurgy. — 2nd ed. — (Pergamon international library: international series on materials science and technology; vol. 30).
1. Metallurgy
I. Title
669   TN665   78-40820

ISBN 0-08-021711-7 (Hardcover)
ISBN 0-08-021712-5 (Flexicover)

Printed in Great Britain by A. Wheaton & Co. Ltd., Exeter

CONTENTS

PREFACE

THE aims of this second edition of *Extraction Metallurgy* are to provide the student of metallurgy in universities and colleges with a comprehensive preparatory textbook on the production of metals from their ores. It is presumed that the reader will have a working knowledge of mathematics, physics and chemistry up to "A level" and that he will be familiar with phase diagrams, as all metallurgists should be. It will be helpful also if he has made a separate formal study of basic thermodynamics, as the treatment of that subject here is concise rather than rigorous with the emphasis on its applications to industrial problems.

The traditional distinction between "ferrous" and "non-ferrous" extraction metallurgy is now meaningless and has been ignored. The more recent distinction between aspects of the subject which are of a chemical nature and those of a physical or engineering nature is also artificial and is to be regretted. In the development of an extraction process every fact and idea that is relevant is equally important and the process should be developed as a whole. It must be acknowledged, however, that this book has been written with a bias toward chemical rather than engineering aspects of the total subject, a bias which, even in this enlarged edition, it has not been found practicable to redress.

On revision, the basic structure of the book has not been altered. After a general introduction there are four chapters on ores and their mainly physical preliminary treatments called "ore-dressing". These are followed by four chapters on chemical theory – on thermodynamics, on reaction kinetics and mechanisms and on slag structures. Two new chapters have then been introduced on fuel technology and process control. In the twelfth chapter, different types of extraction processes are discussed and an attempt is made in each case to demonstrate the relevance of theory presented in earlier chapters. Finally, brief outlines are given of extraction procedures for most metals now being produced commercially. Through-out the book, the opportunity has been taken on revision to bring the subject matter up to date and to amplify and expand the treatment of many topics which have increased in importance during the past decade. In addition, many numerical examples have been incorporated into the text, complete with solutions, and a short Appendix has been added containing worked examples of the applications of thermodynamics, which could not easily be fitted into the text. It is hoped that these examples will help the student to a good understanding of how to employ thermodynamics for the solution of metallurgical problems.

The student whose main interest is in extraction metallurgy will wish to read much more deeply into at least some aspects of the subject. To help him, a selection of books is assembled in a bibliography which includes most of the small number of general textbooks currently available, along with a number of monographs and several reports on symposia which usually provide a good introduction to related original papers. There is also a short list of references, but it is felt that at the stage of study for which the book is meant to cater, only occasional recourse to original sources should be necessary.

As far as was practicable, the book has been written in narrative form but it is in the nature of the subject matter that the reader will wish to refer backwards and forwards between related topics. Numerous cross references have been made in the text and a copious index has been prepared to help him in this. It is recommended that full use of this index should be made to get the greatest benefit from the book.

ACKNOWLEDGEMENTS

IT WOULD scarcely be possible, even if it were desirable, to record the historical origin of every piece of information incorporated in a book of this kind which is to a large extent a new arrangement of well-established theoretical and factual knowledge, much of it of long standing. The author has made no attempt to make such a record but acknowledges the debt which he shares with his readers to countless workers, writers, and thinkers who have developed the subject to its present status and he hopes that any who feel that more specific mention of their achievements should have been made will forgive him his omissions.

For kind permission to use matter of which they hold the copyright, in the preparation of diagrams and tables, the author is most grateful to the following:

The American Institute of Mining, Metallurgical and Petroleum Engineers (Figs. 61, 63 and 69a).
The American Society for Metals (Figs. 43 and 60).
British Standards Institute (Appendix 3).
Butterworth and Co. (Publishers) Ltd. and Dr. C. J. Smithells (Table 3).
The Cambridge University Press (Fig. 35).
The Faraday Society (Chemical Society) (Figs. 53 and 56 and Table A.1).
Gesellschaft Deutscher Metallhütten und Bergleute e.v. (Fig. 64).
Gordon & Breach, Science Publishers Inc. (Fig. 68).
D. W. Hopkins, M.Sc., F.I.M. (Fig. 70).
The International Nickel Company of Canada Ltd. (Figs. 84 and 85).
Interscience Publishers (Figs. 57, 59, 60).
The McGraw-Hill Book Company (Fig. 6).
The Metals Society (Figs. 45, 46, 62 and 65).
La Revue de Métallurgie (Fig. 58).

The Society of Chemical Industry (Figs. 45, 46, 51, 52, 54, 55 and 80).

Thanks are also due to Professor F. D. Richardson for supplying the most recent information available for incorporation into Fig. 45.

CHAPTER 1

INTRODUCTION

The Beginnings

THE history of extraction metallurgy can be divided into two periods — the ancient period, when only a small number of metals was known, and the recent period, of about the last two centuries, when almost all of the rest have become available commercially.

The discovery of metals by early man depended on their being easily found, easily recognized as ores and easily reduced from these ores. Table 1 in Chapter 2 shows that only iron among the common metals is very abundant in the earth's crust, but the ensuing discussion in that chapter shows how a considerable degree of segregation and local concentration of most metals has occurred naturally. Some metals were concentrated more effectively than others, however, and are therefore more obvious on the ground. The relative ease of reduction of metals from their ores can be read off Fig. 45 in Chapter 7 in which those metals appearing at the top of the diagram are very easily reduced, sometimes simply by heating the ore minerals in air, and in a few cases these metals are found "native", or chemically uncombined. Metals appearing further down Fig. 45 are progressively more difficult to reduce, some requiring extremely high temperatures or electrolytic techniques which have themselves been developed only in the recent period.

. The most favourable metals for easy discovery were gold, silver, copper and mercury. Gold with its distinctive and attractive colour is found native in many parts of the world and was used by man in most of the early civilized societies. Silver is not so distinctive. It too occurs native, however, or as a readily decomposed sulphide and also in a native alloy with gold which was early recognized as being different and was called "electrum". Copper was also produced at a very early date and its brightly coloured

1

minerals are readily recognized. Its production must originally have been by accident in domestic hearths, but it would seem that such accidents would not occur frequently and a sharp mind here and there must have realized the significance of the conditions under which the interesting product was to be found among the ashes. Most of the copper artefacts surviving from these early days are actually bronze which may have been smelted from mixed copper–tin ores in the first instance. Tin conferred not only strength but fusibility which facilitated production of the metal. Its importance as an alloying element was recognized before the beginning of the Christian era and it was as an exporter of tin that Britain first entered the world scene. Later Britain also supplied Rome with large amounts of lead, another readily reduced metal whose principal ore mineral has a very distinctive appearance, and which had been known to the Egyptians and Greeks. Antimony had also been produced by the Greeks but seems to have been confused with lead. Pure zinc was not produced until the recent period, but the Romans made brass by smelting together zinc and copper ores. Many minerals were recognized as distinctive substances with particular uses as pigments, colouring agents for glasses and enamels and in medicine, although the presence of a metal within was not suspected. The use of minerals for alloying may have been primarily as a means of reducing the melting point of the product, as even the 1100°C needed to melt copper must have been near the upper limit of what was possible under primitive conditions. Platinum (with the platinum metals) occurs, like gold, in river sands and as occasional nuggets in the Urals, the Rockies and the Andes, but is not found in quantity in Mediterranean or Near Eastern lands. The Greeks knew of it, but found little and had no use for it. The Spaniards found crude artefacts of platinum in Colombia presumably produced by hammering together grains of the metal which were found in gravels along with gold. The Spaniards found more in Mexican silver mines, but could do no better with it than "debase" gold coinage. It was not melted until about 1847 and only then could its real value be appreciated. Mercury, with its colourful minerals and extremely easy reduction, was known from early times and used by the Romans for the extraction of gold by an amalgamation technique. The Mayas in Central America collected mercury and used it as an object for worship.

At a rather later stage than gold and copper, iron came to be produced by primitive peoples. Its use seems to have been restricted at first to certain tribes and certain localities. The method of production may have been kept secret because an iron weapon could easily break one of bronze. The secret would be easily kept, however, because, although iron ores were to be found almost everywhere, their reduction was much more difficult than that of copper and the yields lower. The iron produced was in the form of a "sponge" of partially reduced ore with a slag mixed through it consisting of unreduced oxide and siliceous impurities. The iron was consolidated and most of the slag squeezed out by hammering the hot bloom while the slag was still molten. The iron was never molten and large pieces had to be built up from smaller pieces by hammering them together while hot. The product resembled the wrought iron of more modern times. In China a product more like present-day cast iron was produced under conditions so reducing as to be carburizing. The alloy, with high carbon and probably high phosphorus and silicon, could be melted easily and cast to complicated shapes. It was not until the Middle Ages that cast iron was "discovered" in Europe. The modern process of making the high carbon alloy first and then decarburizing it to the low carbon alloy called steel was developed only during the last 200 years. Nevertheless, the Romans made their steel by carburization of their iron and despite their ignorance of chemistry as we know it, were able to produce very satisfactory compositions and structures. It should be appreciated that early operators had to select their raw materials entirely by appearance and products passed or failed according to how they behaved in use. There was no chemical analysis and no mechanical testing. Many failures and disappointments must have been suffered, but the large number of beautiful and metallurgically satisfactory artefacts which survive from many parts of the world show that the art of making metals was well developed and practised from very early times and by most of the settled peoples.

The recent period of development of extraction metallurgy begins perhaps about A.D. 1700 when zinc was recognized as a metal. An era was beginning in which chemists would isolate and classify all the elements that could be found on earth, and in which engineers would find uses for most of them. That era continues. The production of aluminium on a large scale dates only from 1886 and that of magnesium in quantity only from the mid-thirties in the present century. Since the Second World War the demands for uranium, beryllium, niobium, titanium, zirconium and other metals hitherto considered rather rare has increased considerably.

This has led to their production in fairly large quantities. Even the transuranic element plutonium can now be added to this list although it has not yet been found occurring naturally on earth. It can now be fairly claimed that any metal could be produced to a high degree of purity — at a price — should it be required.

Extraction Metallurgy Today

Extraction metallurgy has in recent years emerged from the embryonic stage of being an art to become a fully fledged technology. Many of today's processes appear not very different except in scale from those of 50 or even 100 years ago which themselves evolved from even older methods. Today, however, the metallurgist understands his processes much better and can control them better to produce metal of regular analysis. He can also modify his processes to keep costs down as he is forced to fall back on leaner and more refractory ores. Advances in engineering technology have enabled the scale of production to be increased enormously. This has been mainly a materials handling and disposal problem. Advances in electrical technology have made possible the reduction of the most refractory oxides and have contributed to the development of processes requiring particularly high temperatures. Electronics and computers are now leading to new control systems which may soon eliminate the skill and intuition of the operator and replace them with rigidly imposed conditions dictated by complex mathematical equations of his choosing.

The extraction metallurgist takes over the ore as it leaves the mine. In the general case the ore must be *concentrated* to free it of minerals of no value called *gangue*. This is done by operations known collectively as *ore dressing* which involves *comminution* or fragmentation to small sizes to permit easy separation of the different kinds of mineral, followed by one or more *sorting* operations designed to distinguish and separate valuable mineral particles from the rest on the basis of some physical property such as density, magnetism or surface energy. Concentration is followed by *extraction*, to produce the metal itself, and this may be carried out by *pyrometallurgy* or *hydrometallurgy*. Pyrometallurgy involves heating operations including *calcination* and *roasting* at temperatures below the melting point; *liquation* in which the metal melts and runs clear of other materials present; and full *smelting* in which the ore is all melted and separates out into two or more liquid layers one of which contains all the metal (either as metal or sulphide). Further pyrometallurgical operations may be necessary, such as the *conversion* of molten sulphide to metal or of pig iron to steel by blowing air through it to oxidize out sulphur and carbon respectively, and *distillation* processes which are found in the extraction of zinc and mercury. Hydrometallurgical extraction is effected by *leaching* the valuable mineral out of the ore with an aqueous solvent and finally precipitating the metal chemically or electrolytically.

Extraction may be followed by *refining*. This is the adjustment of the amount of impurities to low or controlled levels and may be carried out by pyrometallurgy, or electrolytically. The pyrometallurgical refinement is usually an oxidation of impurities followed by a careful deoxidation of the otherwise pure metal, but, as will be shown, a really high degree of purity is difficult to attain in this way. A final step using a high vacuum technique may be necessary where extreme purity is demanded. Refining by distillation is sometimes appropriate.

An alternative means of obtaining pure metal is by *prerefinement*. The metal is purified while in a state of chemical combination, usually by hydrometallurgical means, sometimes using fractional distillation. A very pure compound is then reduced by one of the standard methods – chemically, by thermal decomposition or by electrolysis, often from a mixture of fused salts containing the metal in solution.

It will be noticed that pyro-, hydro- and electro-metallurgical methods may be used one after another in the extraction of a particular metal. Any extraction process is a sequence of operations each of which has the purpose of preparing the material as well and as cheaply as possible for the next operation. The sequence as a whole should be the one which will convert the ore available to marketable metal and show the largest possible profit margin. There are no "rules" about how these sequences may be made up in terms of classes of operation. The metallurgist must be ready to seize on any new idea and incorporate any new advances in other technologies which may reduce his production costs. He must also be willing to reject a new idea however elegant it may appear if it will not do so.

Extraction metallurgy is not confined to the production of a single metal from each ore. Metals, and sometimes non-metals, present in ores in very small proportions may be concentrated into slags or flue dusts and used to produce small quantities of by-products which are sometimes of great value. The recovery and repurification of metals from the scrap heaps of civilized society is another application of the same techniques which is likely to become more and more important as the need for conservation of metal resources becomes more obvious. In the steel industry the circulating load of scrap is at present about 50%, much higher than in other industries mainly because steel scrap is so easily collected.

The recovery of copper and its alloys, lead, aluminium and the precious metals is now well organized. Process scrap of all metals is, of course, recirculated within industrial plants but when a metal is dispersed in use, whether as large numbers of small artefacts like nuts and bolts or as thin metallic coatings whether of tin or zinc or even gold, collection for re-cycling is comparatively difficult and costly to organize. Other metals are consumed chemically in use like steel by rusting or photographic silver, while others are used in compounded forms like mercury, much of which is scattered widespread for agricultural purposes. A proportion of metal will always be lost to the system. For all metals the levels of recovery for re-use must depend on the cost of collection and refinement compared with its market value. Even tin from tinplate, photographic silver and gold in electronic scrap can be recovered profitably in favourable circumstances.

During his career an extraction metallurgist may find himself engaged in investigations into the fundamentals of process mechanisms, in modifying processes or in designing new large-scale plant or processes, in operating large plant and perhaps on committees or boards making decisions about new projects and how they should be organized and financed. To prepare him for such a career a young extraction metallurgist must be well trained in basic chemistry, particularly physical chemistry, physics and mathematics and in the application of the sciences conjointly to the understanding of industrial process involving heat and mass transfer – in other words, to chemical engineering.| It is of advantage also if he can acquire some knowledge of geology and mineralogy, and enough of mechanical and electrical engineering to enable him to discuss his problems easily with colleagues expert in these fields. He should have studied fuel technology and should be conversant with the techniques available for utilizing energy with the highest possible efficiency. He must learn about economics, finance and markets. He must be able to handle men – both labourers and shareholders. He must also have a detailed knowledge of the basic principles underlying the processes with which he operates and it is with this aspect that the rest of this book is mainly concerned.

Units

The units in the first edition of this book were mainly metric, of the c.g.s. variety such as has been found satisfactory and had been adopted by research workers and technologists in this field in all countries during the first 70 years of the twentieth century. Industrial usage varied, however, with Imperial units being employed not only in countries which had associations with Britain but also in the United States of America and countries receiving technical assistance from her. Since the first edition was published, however, Le Système International d'Unités — SI — has been imposed on the educational system in Britain and less rapidly on her manufacturing industries. In conformity with this move this second edition uses principally SI units but the reader will realize that metallurgists must retain the capability to convert from one set of units to another in order to be able to read the literature of past decades with understanding and in order to be able to communicate with people in other countries who prefer not to change. The change from the calorie to the joule as the unit of energy in thermochemistry and thermodynamics is particularly inconvenient in view of the large amount of data which has been accumulated and recorded in calories (much of it by Americans) — apart from the need to use larger numbers at all stages. The author also finds it difficult to convert such colloquial terms as "a few feet" without straining his style and hopes that the 1.6% error involved in using ton rather than tonne in descriptive passages will be forgiven.

A table of Conversion Factors is printed on page 431 for the convenience of readers.

CHAPTER 2

ORES

The Origins of Ores

ALMOST all metals are derived from ores – concentrations of appropriate minerals accessibly situated at or near the earth's surface. Two exceptions are magnesium, which may also be won from sea water, and plutonium, which is produced in atomic reactors. It is likely that manganese, copper and nickel will in the future be produced from "nodules" swept from the ocean floor. Some preliminary consideration should be given to the question of how ore deposits have come to be where they are.

Some 3000–4000 million years ago the body of matter which was becoming the earth must have cooled to a stage at which most of the vapours had condensed to liquids and some solidification was beginning. At this stage considerable segregation had occurred under the action of gravitational and chemical potentials. Since there was a large excess of metals over non-metals the net conditions were reducing and the planet resembled the products of a gigantic smelting operation with the excess of metals having sunk to the centre of the planet where they were covered first with a layer of heavy sulphide material (matte) and then a layer of lighter oxides (slag). These were enveloped by an atmosphere of permanent gases, halogens, and some of the more stable gaseous compounds like methane, ammonia and the oxides of carbon, much of which has been either dissipated into space or condensed at a later stage.

Further cooling brought about the formation of solidified slag to form a continuous crust over the whole globe. This is now about 50 km thick. It is the only part of the planet to which we have as yet access and from it we must obtain our metals.

The rest of the planet has probably not altered very much in its general description since the crust began to form apart from some loss of temperature. The metallic core or siderosphere is about 7000 km in diameter. The sulphide-rich chalcosphere is 1500 km thick, and the oxide layer, the lithosphere, is a further 1400 km thick. Together the chalcosphere and lithosphere are called the mantle, so that we have crust, mantle and core. The physical distinction between the crust and the upper mantle is based on seismological evidence and qualitatively is that the crust is relatively hard and brittle while the mantle, though solid, is plastic. The crust floats upon the mantle and can be pressed into it or may rise out of it as it is loaded, or unloaded, with, for example, ice or sediments.

Chemically, the partition of the elements among the three major layers was determined by the thermodynamic properties of the elements and the compounds which might be formed from them under the combined effects of the high temperatures and high pressures which prevailed. Modern thermodynamic data should provide a useful guide to the reactions which led to the distribution as it now is. The values of the free energies of formation and of solution at the temperatures and pressures ruling at the great depths under discussion are not the familiar values measured at atmospheric pressure. Equilibrium conditions cannot readily be deduced by extrapolation from known values to corresponding values appropriate to high pressures but the effect of the pressure can be estimated and the salient features of the high-pressure chemistry deduced. The broad picture is as follows. The core is mainly iron, with nickel as the main alloying element (as in meteorites). Also alloyed into this layer are elements which are more "noble" than iron towards possible reactants (mainly oxygen, sulphur and halogens) such as cobalt, platinum metals and gold, along with non-metals which can attain low free energy in combination in this phase—notably carbon and phosphorus.

Owing to the great abundance of iron, it is a principal constituent also of the sulphide layer along with all those elements whose sulphides are more readily formed than that of iron. These include many of the common base metals (see Fig. 46). Sulphur is accompanied by the similar elements selenium and tellurium.

Finally the oxide layer contains preferentially those elements whose affinity for oxygen is high — silicon, aluminium, titanium, the alkalis and alkaline earths and heavy metals which tend to produce oxy-salts (like chromium and manganese) together with large amounts of iron. Of other elements there are only mere traces as is shown in Table 1.

TABLE 1. THE RELATIVE ABUNDANCE OF THE ELEMENTS IN THE EARTH'S CRUST (BRACKETED NUMBERS ARE WEIGHT PER CENT)

Abundance range	
Over 10%	O (46.6); Si (27.7)
1–10%	Al (8.1); Fe (5.0); Ca (3.6); K (2.6); Na (2.8); Mg (2.1)
0.1–1%	C; H; Mn; P; Ti
0.01–0.1%	Ba; Cl; Cr; F; Rb; S; Sr; V; Zr.
0.001–0.01% or 10–100 ppm	Cu; Ce; Co; Ga; La; Li; Nb; Ni; Pb; Sn; Th; Zn; Yt
1–10 ppm	As; B; Br; Cs; Ge; Hf; Mo; Sb; Ta; U; W; and most of the rare earths
0.1–1 ppm	Bi; Cd; I; In; Tl
0.01–0.01 ppm	Ag; Pd; Se
0.001–0.01 ppm	Au; Ir; Os; Pt; Re; Rh; Ru

Compiled from figures given in *Geochemistry* by V. M. Goldschmidt, Clarendon Press, 1954.

Within the crust there has been further general segregation by gravitation into less dense granitic rocks, mainly alumino-silicates of sodium and calcium, predominantly in the continental regions, and the denser basalts with a higher content of iron oxide and magnesia lying underneath and forming the floor of the oceans. The upper mantle below that contains even higher concentrations of the heavier oxides.

Fortunately the separation of the elements into distinct layers is not complete. Even under equilibrium conditions it would be expected that each element would be present at some small concentration in each phase. So far as the crust is concerned — and that is the only part for which direct evidence is available — almost all the elements are known to be present. Their distribution is, however, astonishing. As shown in Table 1 nearly half of the weight of all rocks is oxygen (of which there is much more in the crust than in the atmosphere). Sixty per cent of all the crust rock is composed of silica. Only eight elements are more abundant than 1% and together these account for 98.5% of all the material in the crust. Only five more elements are abundant beyond 0.1% and these first thirteen account for 99.5% of all the rocks. Of these, only Fe, Mn, Al and Mg are familiar metals and only iron has been in the service of man for more than a century. All the common base metals like Cu, Pb and Zn are found to be present only to the extent of 10^{-3} to $10^{-2}\%$, while the more precious metals like gold are present at about $10^{-6}\%$.

The concentration and extraction of metals from rocks containing them in such low percentages would, even today, be impossible and metallurgy could never have developed if geological processes had not effected substantial preliminary concentration of nearly all metals into ore bodies in a manner that will now be discussed.

Magmatic Segregation

In the lower reaches of the crust there existed zones in which the rocks were molten. These large blobs of slaggy material are called "magmas" and that they still exist can be deduced from the manifestations of volcanic activity in various parts of the world. A magma may be primary molten material which has never been solid and has its roots deep in the mantle, or it may be remelted crustal and upper mantle rock the temperature of which has been raised by adiabatic compression during the cooling of the globe and the shrinking of its outer layers. This has been associated with mountain building processes and evidence of magmas is found in the roots of old mountain ranges. The secondary magmas may be acid (granitic) or basic (basaltic) depending on the part of the crust involved so that some earlier segregation may already have occurred before these rocks melted.

The cause of the formation of magmas is not of importance to the metallurgist, but the sites where they occurred are very important because the segregation of metals into ore bodies took place while the magmas were cooling down. The exact mechanism of the process of magmatic segregation is not fully understood — indeed there is probably a variety of detail that might be followed — but the general picture must have been roughly as described below. While magmas are analogous with slags they are much more complex in that they contain nearly all the elements. Being formed from rocks which solidified under a primordial atmosphere which contained not only H_2, O_2 and N_2 but chlorine, sulphur and carbon compounds, magmas contain a few per cent of potentially volatile components. They are initially under enormous pressure, however, and the volatiles remain in solution, their vapour pressures being relatively low.

Solidification of magma is a complicated process but Niggli[1] has considered, as an extreme simplification, that it is essentially the same as the solidification of a solution of about 5% of H_2O in molten SiO_2. Four stages can be distinguished, merging into each other. In the first, SiO_2 crystallizes as from an igneous melt with a progressive lowering of the freezing point as the concentration of water in the residual liquor increases. In the second stage the water content of the liquor is so high that further crystallization of SiO_2 is essentially from an aqueous solution rather than an igneous melt. The crystals are larger and may contain inclusions of liquid. As the proportion of water in the residual liquor increases, its vapour pressure rises until it exceeds the pressure of the system. At this point the third stage has been reached in which deposition is from the vapour phase. If the total pressure is high enough the system may go supercritical. (Deposition in this stage is slight in the SiO_2-H_2O system.) In the final stage the water has cooled and recondensed and the last traces of silica precipitate slowly, probably as a hydrate, as the solution cools down.

These four stages are called magmatic, pegmatitic, pneumatolytic and hydrothermal, respectively. A similar sequence can be used to describe the solidification of a magma in which the SiO_2 is accompanied by Al_2O_3 and oxides of the alkalis and alkaline earths, FeO, MnO and lesser proportions of the oxides of most other metals; and in which the volatile part is mainly H_2O but with H_2S, HCl, HF and CO_2 also present.

"Rock-forming" minerals can be distinguished from "ore-forming" minerals, in that they account for a very high proportion of the total mass of the magma and they solidify according to a pattern like that outlined above, being reduced to quite small proportions, however, by the end of the magmatic stage. The primary minerals crystallized are complex anhydrous silicates of Fe, Ca, Mg, K and Al giving way to hydrated silicates and to SiO_2 in stages two and three.

Simultaneously, and not necessarily independently, the ore-forming minerals also solidify in a fairly well-established sequence. In the magmatic stage, and possibly before the silicates, some sulphides, and particularly those of Ni and Co, precipitate as a liquid phase and may settle out, sinking in the melt and carrying with them noble metals of the platinum group, and copper, gold, etc., in small quantities. Accumulations of these minerals, when discovered, are rich ores indeed. Spinel-type minerals also separate at an early stage – chromite ($FeO \cdot Cr_2O_3$), associated with hercynite ($FeO \cdot Al_2O_3$) and spinel ($MgO \cdot Al_2O_3$), and, if the iron has been sufficiently oxidized, magnetite (Fe_3O_4 or $FeO \cdot Fe_2O_3$) may form also. Ilmenite ($FeO \cdot TiO_2$), though not a spinel, also crystallizes early. It is, of course, necessary that these minerals shall somehow be separated from the bulk of the melt if they are to be found as ore bodies. The most likely mechanisms are that they sink on to the floor of the magmatic chamber (if it has a floor) or that the early forming solids are left behind as the still molten part drains away or is extruded by pressure to another place. Valuable deposits of this kind occur containing Ni, Cr, Fe and Ti, notable among them being the Kiruna magnetite in Sweden and the Sudbury deposits of nickel sulphide in Ontario, Canada. Though large, these deposits are not numerous on the surface though many others may lie as yet undetected deep in the crust. Meanwhile the residual liquor has been enriched in other metals as well as in volatiles (see Fig. 1).

In the pegmatitic stage SiO_2 and silicates are precipitated along with other minerals which are essentially oxy-salts—phosphates, titanates, niobates, tantalates and sulphates with zircon ($ZrSiO_4$). The most important depositions of ore occur, however, in the two final stages. In the pneumatolytic stage the volume of gas involved is immense with the metals probably present as volatile chlorides (see page 330) along with H_2O, H_2S and CO_2. They are usually deposited as sulphides, though wolfram $(FeMn)WO_4$, scheelite ($CaWO_4$) and cassiterite (SnO_2) are deposited at this stage and gold occurs native with quartz and pyrites (FeS_2). The important sulphides are those of Cu, Pb, Zn, Co, Mo and Ag with, of

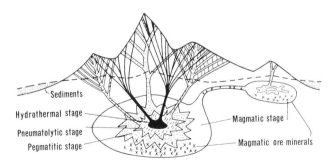

FIG. 1. Schematic representation of a magma and the distribution of minerals from it during the magmatic, pegmatitic, pneumatolytic and hydrothermal stages of cooling. The broken line represents the present horizon. (Not in any sense to scale.)

course, the ubiquitous FeS_2 which is of little value. In the hydrothermal stage further depositions of sulphides take place, especially those of Cu, Pb, Zn, Sb and Hg, along with carbonates of the alkaline earths and fluorspar (CaF_2).

At every stage, transportation must accompany deposition if differentiation is to be effected. The nature of ores formed from a particular magma depends on the stages in its history at which transport becomes possible. The enriched fluid – gas, liquid or supercritical liquid – must flow out of the magma and traverse fissures in the overlying rocks, fill its cavities, and even seep through its intergranular pores. The formation of suitable cracks during "earthquakes" inevitably accompanies the mountain formation processes of which the magma is itself a symptom. Wide cracks and faults are obviously most easily traversed by mineralizing fluids, but wide areas of finely fissured rock present greater opportunity for reaction with the fluids and the intergranular porosity of sedimentaries like limestone provides the greatest opportunity of all – provided the permeability is high enough. The temperature of the magmatic fluids may be 500° or 600°C initially, but will become lower as they pass further from the magma. If the process continues for long, however, the rocks traversed may rise to this temperature. At the same time the pattern of deposition may alter and early deposits may be reabsorbed and carried further out.

The reactions involved may be simple precipitations filling up spaces in the rock, but more important deposits are produced by replacement of the country rock itself on a massive scale. Some rocks are, of course, unaffected, but sedimentaries and particularly limestones are readily dissolved away by the magmatic fluids and replaced according to reactions such as

$$SiF_4 + 2CaCO_3 = SiO_2 + 2CaF_2 + 2CO_2$$
$$2FeCl_3 + 3CaCO_3 = Fe_2O_3 + 3CaCl_2 + 3CO_2$$

this possibly being followed by

$$Fe_2O_3 + 4H_2S = 2FeS_2 + 3H_2O + H_2$$

which provides hydrogen for

$$2AuCl_3 + 3H_2 = 2Au + 6HCl$$

The CO_2 would react with alkaline earths at a cooler stage and the CaF_2 would be deposited as crystals.

These replacements are not just surface effects. They corrode the walls of fissures in depth; advance through the rock via tiny cracks possibly generated by thermal shock; and they may consume massive zones on either side of major fissures until the new minerals are almost continuous over great distances. Massive deposits like these contain several minerals sometimes in zoned or banded formations. Some of the minerals may be worthless, while others even if in quite small percentages may render a deposit very valuable.

The degree to which the country rock is consumed depends on a number of physical and chemical factors. The rock must be soluble in the magmatic fluid. It must be well fissured by tectonic pressures – many fine cracks having greater effect than a few big ones. The pressure distribution must be such that the fluid flows through all the fissures and the viscosity must be low enough to permit this. If the nucleation of the new mineral is difficult, large nodules may grow well separated. If nucleation is easy but growth inhibited, the mineral may appear as myriads of tiny specks. A mineral may replace only certain constituents of a heterogeneous country rock again producing a disseminated ore such as the massive porphyritic copper ores in Zambia. Large areas and great thicknesses of rock may be affected by mineralization or it may be restricted to a few isolated veins or pipes. It is rather unusual, however, to find only one deposit in a particular geographic region.

Mountainous regions almost certainly have ore deposits somewhere in their roots. These are often not accessible in "new" mountain ranges, but become exposed when the mountains are middle-aged or old, though with the passage of time some of the valuable deposits are dispersed by weathering and either accumulate elsewhere as sedimentary deposits, or are lost to the sea. The most interesting and valuable regions of mineralization are the great "shield" areas, particularly those in Canada and Russia where pre-Cambrian rocks (more than 500 million years old) are exposed. These rocks may have been involved in several mountain building operations with as many opportunities for the accumulation of ore – and perhaps some ores have had double treatment. Wherever minerals are found, one metal or a well-defined group of metals predominates, sometimes over vast areas like Central Africa (Cu) or Scandinavia (Fe) or Ontario (Ni and associates). Whether this is simply a matter of reaching a stage of weathering such that the available cross-section is likely to yield a particular group, or whether in each area some earlier differentiation has eliminated other groups and so determined that a particular metal shall predominate, is not known.

Weathering

Weathering in all its aspects must have destroyed many valuable deposits, but has also brought many others within reach by denudation of overlying strata. Weathering can also bring about secondary concentration of valuable minerals in magmatic deposits. Percolating water may oxidize and leach minerals from the upper parts of an ore body and redeposit its metallic content in some other form lower down. This is an advantage if the upper strata are later removed by weathering so that the primary and secondary deposits become available together. As mountains are worn down some of their minerals are transferred into the beds of streams and lakes to form sedimentary deposits possibly with secondary concentration. Tin and gold are both found as "placer" deposits in river beds, often under circumstances which invite prospecting of the head waters for the primary source of the minerals. Titanium as ilmenite associated with magnetite and other dense, hard and chemically resistant minerals form "black sands" on beaches where they are washed free of grains of less dense material by tidal action.

The most important sedimentary deposits are those of iron which are so vast that they must have come from the leaching of basalts and other basic rocks rather than from ore bodies. The solvent would be carbonic acid and the precipitation of a hydrated oxide or carbonate would occur whenever the pH rose above about 6 in lakes and shallow seas. Different conditions yielded different types of ore. The oolites of France and Luxembourg were formed in shallow tidal water where continual motion rolled the sediment into tiny pellets which later consolidated like sandstone. Elsewhere iron replaced calcium in shell sands or simply settled as mud under stagnant conditions, mingled with clays and carbonates brought down from the hills by the same rivers. Such ores are found in Northamptonshire and Lincolnshire in England. In other places iron from solutions has replaced calcium from limestones or has percolated limestones and filled up caves and fissures to form nodular-type deposits such as the kidney ore of Cumberland, England.

Another mode of concentration involves leaching away the gangue materials to leave behind the valuable ones. The nickel silicate ore, garnierite, in New Caledonia is a residual deposit arising from the weathering of a serpentine rock which contained nickel in proportions far too low to be worked. Aluminium is extracted from bauxite which is a residual deposit from the decomposition of alumina rich feldspar type rocks of low quartz content, under rather special tropical conditions which ensure that silica is dissolved away rather than alumina. Some of the richest iron ores are residual deposits derived from taconite ores laid down in pre-Cambrian times. These are vast sedimentary beds of haematite or sometimes of magnetite interlayered with siliceous sediments. The average iron content of taconites is only about 30%, but in many places, in India, West Africa and in the U.S.A., the silica has been leached out, probably by hot magmatic solutions, leaving behind mountainous heaps of iron oxide ready to be shovelled away. Most of these rich ores are very soft and even powdery, reflecting the degree of dissemination in the parent taconite. When the rich residual ore has been used up — as has already happened in some places in the U.S.A. — the lean taconites are available as the greatest reservoirs of iron in the earth's crust, but extensive dressing is necessary before they can be smelted.

Even after ore bodies have been laid down their character may be altered by earth movements or the proximity of later igneous intrusions. In particular faulting and folding of the strata may break the ore body into a large number of small sections which makes mining difficult and expensive; faulting may be along the plane of a vein, breaking up the ore and mingling it with country rock and allowing further dilution by the deposition of more minerals; and the heat from such further intrusions can have interesting chemical effects on the kind of minerals found in the ore which may affect the choice of treatment.

The Winning of Ores

The first stage in the winning of ores is to find them. Traditionally prospectors were hardy enterprising miners seeking their fortunes, with a clever eye for the tell-tale signs of valuable minerals — the "gossan", an iron stain marking the outcrops of veins altered by weather and leached downward to produce secondary concentration; or minerals visible where they outcrop or in gravels downstream in the rivers; or even the colour of the vegetation. Modern prospecting is organized by large mining companies and carried out by geologists and geophysicists who add to the old skills sophisticated techniques – electrical, magnetic, chemical, gravitational and seismic. Electrical methods involve the measurement of soil resistance, or of natural currents in soil surrounding ore deposits due to their slow chemical interaction with the rocks round about. Magnetic determinations of declination and inclination over wide areas reveal large irregularities in the vicinity of magnetic ore bodies in particular and lesser variations in the presence of other kinds of ore. Magnetic readings may be taken from the air and may reveal geological formations thousands of feet below the surface. Dense rocks affect the gravitational pull of the earth' and variations from "normal" for a region can be detected either by sensitive torsion balances or by simple pendulums or spring balances which can be made to show changes in g down to one part in 10^7. Seismic methods are rather like echo-sounding for changes in strata. Geochemical methods include the analyses of vegetation and sands from rivers to trace sources of minerals, and radiometric techniques are used to identify rocks by the emissions of their radioactive content.

By whatever methods a deposit is discovered, a full geological survey must be made to determine the extent of the deposit and the geological complexity of the region before mining operations are commenced.

An ore body is mined by a method appropriate to its size, shape and depth below the surface. The simplest methods apply to deposits which outcrop over large areas. Then large "open pits" are excavated, the ore being broken with explosive and shovelled into trucks. The area is cleared systematically in a series of terraces designed to facilitate transportation. The pit may be worked into a mountainside or downward depending on the configuration of the ore body *vis-à-vis* the local topography. If the ore lies deeper below the surface but is extensive horizontally, open cast mining proceeds by first clearing the over-burden to form a wide trench at one edge of the deposit. The exposed ore is then dug out and over-burden moved from one side of the trench to the other so exposing more ore for stripping. In this way the trench traverses the countryside as the ore beneath is removed. This method can be used where over-burden is several hundreds of feet thick if the value of the ore justifies it (see Fig. 2).

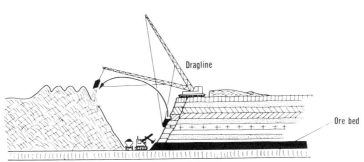

FIG. 2. Diagrammatic representation of open cast mining. The dragline uncovers the ore bed while the shovel loads the ore into trucks.

Other surface operations include dredging and hydraulicking, usually for tin and gold placer ores (sands and gravels). Dredgers are built in gravel beds which can be flooded with water from a nearby river at the same level. They float in ponds of their own making and dig up the gravel from beneath them, stacking worthless gravel to one side or carrying out simple sorting on board when the gravel is thought to contain valuable mineral. Dredgers can reach down 50 m below the water line and dig their way systematically through the deposit (see Fig. 3). In hydraulicking, powerful jets of water are used to "blow away" the barren gravels and expose the mineral-bearing beds which are then washed into sluice boxes for concentration (see page 68).

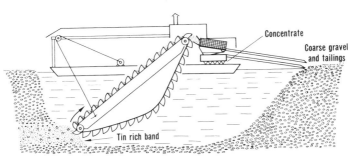

FIG. 3. Schematic dredger working pond in gravel swamp. The dredger is working toward the left depositing a new bank astern all the way. A simple screen and classifier are indicated aft. This unit must generate its own power requirement.

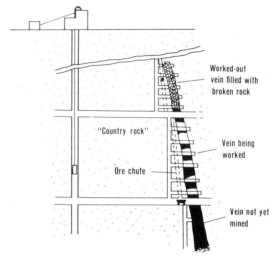

FIG. 4.   Deep mining of a vein deposit. All mining operations are essentially transportation processes for men, materials and air carried out in three dimensions. They are not easily represented on paper and this sketch is very much over-simplified.

Most mining operations, however, involve going underground. Again there are a number of methods available depending on the size and shape of the deposit. Massive deposits extending horizontally would be mined by the "room and pillar" method, the pillars amounting to 25% of the ore body being left to hold up the roof while excavations continue. If the deposit extends vertically, as most veins do, a vertical shaft is cut alongside it and horizontal galleries are driven across at various levels to the vein. The mineral content of the vein above each access point and up to the one above it is blasted down and taken out to the shaft (see Fig. 4). Large but irregular deposits are mined by caving. The deposit as a whole is undercut and blasted down toward a prepared area from which it can be withdrawn, rather like drawing coal from the bottom of a big bunker, and transported to the surface.

In particularly favourable conditions where an ore body extends horizontally the ore might be taken out by retreat mining. A cave is prepared remote from the pit shaft. When it has been cleared it is extended backwards toward the shaft, allowing the roof to collapse behind the current workings. The proportion of ore which can be extracted by this

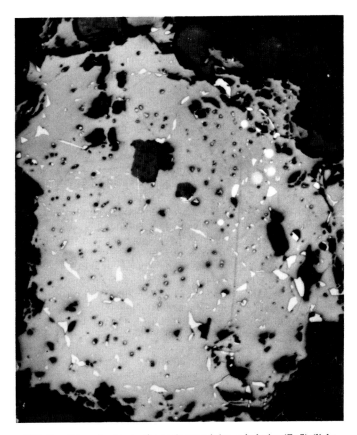

FIG. 5. Microstructure of a rock containing sphalerite (ZnS) (light grey) associated with garnet (dark grey) as the gangue and containing inclusions of chalcocite (Cu_2S) (white). The magnification is 100 so that the grain shown is about 1 mm across and might be released by moderate grinding. The copper mineral grains are so small, however, that grinding far beyond 300 mesh would be needed to release them so that chemical methods would be necessary for the separation of the copper from the zinc.

method is much higher than if room and pillar mining is used but the surface above the workings is liable to severe subsidence. This restricts the use of the technique to regions which are virtually uninhabited.

The Physical and Chemical Characteristics of Ore

Ore as mined may be in lumps up to several feet wide but the maximum size of deep-mined ore must be made much smaller to reduce handling problems. Ore should consist of valuable minerals and gangue with a minimum of "country rock". Often the rock can be eliminated by hand picking at the mine head. The gangue minerals inevitably accompany the valuable minerals, being either partly replaced country rock or an agglomeration of other minerals deposited simultaneously with the mineral of major interest.

In some cases (e.g. magmatic magnetite) the ore is almost a single pure crystalline mineral in very extensive deposits. This case is rare though important. In the general case a vein contains several minerals in banded array, the bands of the valuable minerals being sandwiched between layers of, for example, quartz, pyrites, or fluorspar. In other cases the valuable mineral occupies fissures in another mineral which may be the original country rock or an earlier filling of the vein. The valuable mineral may be deposited as tiny separate crystals simultaneously with another mineral or as tiny specks replacing another dispersion of crystals as in the porphyry coppers, where copper sulphide replaces feldspar. Such minerals may be so dispersed as to appear to have been laid down as eutectics. The type of dispersion is important as it determines the size to which grinding must be carried out to expose the mineral particles to chemical action or so that they can readily be distinguished by physical characteristics as predominantly valuable mineral or gangue (see Fig. 5).

Chemically, an ore may contain three groups of minerals — (1) valuable minerals of the metal which is being sought; (2) compounds of associated metals which may be of secondary value; and (3) gangue minerals of no value. The valuable minerals fall into one of the following groups:

(a) Native Metal

Apart from the precious metals gold, silver and the platinum metals, mercury and a little copper, occurrences of native metal are rare. Some native iron of magmatic origin has been found, the largest deposit being in Greenland. Native copper occurring with sulphides in Zambia is most unwelcome because it will not crush and damages the milling equipment. Native metal is seldom pure, but occurs rather as an alloy. It is more likely to be found as filaments or flakes than massive, though the findings of large nuggets of gold and platinum have attracted much attention.

(b) Oxides

These are often the result of the oxidation of sulphides and if found in veins may gradually give way to sulphide as the veins are worked downwards. A few oxides do occur as primary deposits, notably Fe_3O_4 in the magmatic stage and SnO_2 in the pneumatolytic. TiO_2 is common in igneous rocks and is occasionally found in pegmatitic dykes. Complex double and multiple oxides like niobite and tantalite are also found in pegmatitic veins. These are all resistant to chemical attack and are frequently recovered from beach sands and river gravels.

Other important oxides are the result of chemical action in other ways—for example, the formation of iron oxides already discussed. It is obvious that in the aqueous "hydrosphere", and in the very wet atmospheric conditions which have persisted on earth for many geological aeons, oxidation is the likely fate of any reactive species under the influence of weathering. Oxides and hydrated oxides are the natural end products. Sulphides are oxidized initially to sulphates and from these hydroxides are precipitated at high pH, but under other conditions the sulphate may be converted to carbonate, or may even persist without further change. Such forms would, however, normally be found to be subsidiary to a principal deposit (if accessible) probably of sulphide.

Strongly electronegative metals like aluminium and magnesium occur as oxides rather than as sulphides in primary igneous rocks. Aluminium associates with silicon in rocks like granite and it occurs as an ore where the silica and other associated minerals have been leached out by, say, carbonic acid of volcanic origin, leaving a residual deposit of bauxite $(Al_2O_3 2H_2O)$ contaminated with components like oxides of iron which are not so readily removed by leaching. The still more electronegative alkali and alkaline earth metals tend to form halide salts, most of which are water soluble and are washed away to the oceans. These are concentrated naturally only when they accumulate in enclosed seas with high evaporation rates like the Dead Sea. Eventually in such a sea the salts will crystallize out yielding beds of salt which in rare cases are stratified by differential solubility to provide rich deposits of the less common alkali halides as in the Stassfurt deposits in Germany.

(c) Oxy-salts

Secondary sulphates and carbonates are obviously members of this group, but it contains also many compounds of magmatic origin — silicates, titanates, spinels, etc., some of which might be described as mixed oxides. Some of these, like zircon ($ZrSiO_4$) and the complex rare earth phosphates known as monazite, occur finely disseminated in granites and similar rocks, but are naturally concentrated in placer deposits and in beach sands with other similarly chemically unreactive minerals. Other minerals, like beryl, a complex alumino-silicate, occur rather in the coarsely crystalline pegmatite veins from which they must be won by deep mining methods. Some minerals in this group are associated in families like the spinels which are intersoluble at least at high temperatures. Reduced solubility at lower temperatures may lead to the precipitation of useful minerals in grain boundaries and cleavage planes of the major minerals. In other cases salt-type minerals may be formed when igneous rocks are metamorphosed or altered by heat or pressure. Recrystallizations may produce minerals which are easily separated where, from unaltered rock of the same composition, extraction would not be an attractive proposition.

(d) Sulphides

The last major class of ore mineral is probably the most common and is usually found on the sites at which it was deposited. Again, minerals in this group associate in families with iron almost always present — thus lead and zinc are often found together, and lead is seldom found without some silver. Cadmium is found associated with zinc, iron with copper, copper and iron with nickel and nickel with cobalt. Platinum occurs commonly with copper and nickel.

(e) *Arsenides*

This group is of minor significance except for its nuisance value. Arsenides or more often sulpharsenides occur in association with the corresponding sulphides in some cobalt, nickel, copper and lead ores and arsenates may occur among their oxidized deposits. The presence of these obviously affects the value of the deposit — usually adversely. The most important arsenide is sperrylite ($PtAs_2$), the form in which platinum occurs when magmatically deposited with nickel sulphide.

Gangue minerals are themselves sometimes of value. Veins of sulphide minerals often contain fluorspar which is useful as a flux in steelmaking or as a source of fluorine, or barytes which also has some commercial value in, for example, papermaking. Pyrites is also found in plentiful supply in heavily mineralized areas. It can be used as a fuel in smelting operations and at the same time as a source of sulphur or sulphuric acid. Sulphur must, of course, be fixed and not released to the atmosphere, so it will be counted a value only if it can be marketed locally. Ores found in adjacent deposits sometimes have different gangue minerals each of which can be used as a flux for the other. Some very lean ores become valuable because they are in effect metalliferous fluxes for another ore producing, when mixed with the other in suitable proportions, a "self-fluxing" burden.

The Evaluation of Ore Deposits

Unless a mineral deposit can be worked with profit it would not be designated an ore by miners, but a deposit which can be worked with profit at one period in history need not be profitable at another period.

The credits associated with a mineral deposit are the sum of the "values" of metals and non-metals (e.g. sulphur) which can be extracted from it and sold. The "costs" involved are those of exploration, mining, concentration, transportation, waste disposal, extraction, refining, and marketing, research and administration. If the values exceed the costs by a reasonable amount the deposit is a workable ore.

The situation is, however, fraught with uncertainties and the stock exchange quotations for the shares of mining companies are traditionally liable to wide fluctuations. Owing to the strategic importance of metals, prices are sensitive to the international political climate and in recent years technological advances have also brought about quite sudden changes in the demand for individual metals. The case of lead is outstanding as its principal traditional uses as a roofing material, in paints, for plumbing and for gun shot were all drastically curtailed within a few years. Other uses were developed, however, and the lead industry thrives. On the other hand, the exploitation of nuclear power has created in the past 20 years demands for several metals like zirconium, niobium, beryllium and uranium itself which were not hitherto produced in quantity.

To some extent the fluctuations in demand due to normal trading cycles have been controlled by international agreements of "cartels" fixing prices and sharing available markets among signatories. Such arrangements would be expected to maintain prices high, but should ensure continuity of supply and continuance of activity in all major producing centres even during economic recessions, whereas unrestricted competition could put marginally profitable orefields out of production except spasmodically during periods of boom, with unhappy social consequences locally.

A typical mining venture, particularly in an undeveloped country, involves great capital investment which may extend beyond the cost of mining and extraction plant to the costs of building complete towns, railways and ports and instituting educational and welfare services for considerable populations. To justify such massive investment proven reserves must be adequate to guarantee many years of profitable production and exploration rights assured with a view to extending operations as far into the future as possible. Small though very rich deposits may not attract much attention if the total values present are insufficient to justify the cost of the plant to handle them.

Mining costs are likely to increase as an orefield is developed and less accessible veins are being worked. Geological uncertainties become greater, ore quality may deteriorate, and transportation to the surface may take longer and cost more. On the other hand, the available labour force may become more stable and more skilful.

The decision whether to extract the metal at the mines or ship a concentrate for treatment elsewhere depends mainly on transportation costs and the availability of fuel and power. The very lean gold ores are never shipped. The extraction of gold requires little fuel and only small quantities of chemicals, and the product is compact. Rich iron ores are usually carried to the coal fields for smelting while lean iron ores are smelted beside the mines, the coal or coke being carried if necessary. Aluminium ore is always transported because cheap energy, often hydro-electric energy, is needed for its reduction – but it is calcined first to reduce its weight and volume. Other factors are occasionally important. Technologically sophisticated processes are often completed in industrialized countries where specialized knowledge is available. The market for sulphuric acid brings large quantities of zinc sulphide ores to Europe for smelting, ore being cheaper to carry than acid.

The cost of extraction depends on many factors. The mineralogy of the ore determines the extent of the crushing and sorting operations. The richness determines the cost of handling and the physical characteristics the cost of the crushing and sorting plant. The chemical nature of the ore determines the most suitable process to be applied. The impurities present, valuable or otherwise, determine details of procedure particularly in refining, and hence the cost of refining. Presuming the best process to have been chosen, the cost is then made up of the costs of capital, labour, services and consumables (of which the most important is often fuel). A high degree of concentration obviously reduces extraction costs. The need for a high degree of refining must increase refining costs. All costs must refer to a standard quantity of pure metal.

Capital, fully utilized, costs less per ton of product than if working below capacity. In modern processes the increasing complexity of equipment tends to increase the proportion of total cost due to capital. If this is offset by increases in production rate, so that the charge for capital per ton is reduced, it is justifiable, as it is also if it results in an adequate reduction in labour costs. It seems likely, however, that many industries will soon have to study the efficient use of capital in the same way as they have already made a close examination of the efficient use of fuel to their great benefit.

The amount of energy needed is the sum of the free energy of the reduction of the compound of the metal as found in the ore to the metallic state (which is an inescapable minimum) and the energy required for all mechanical and heating processes which it may be necessary to carry out. This latter depends on the choice of procedure adopted and on the efficiency with which the operations are carried out. A smelting process would usually need more energy than a leaching process but the cost of the energy saved by leaching might well be lost in extra costs incurred in other directions.

The cost of unskilled labour varies widely, but to some extent the variations are offset by differences in efficiency and productivity. It seems probable that these differences, not only in costs but in abilities and attitudes, will steadily decrease in the next decades – a change which may make mining and extraction relatively less profitable in some countries where the cheapness of labour has for long been taken for granted.

The price commanded by a metal at any time is determined by the commodity market on the basis of the current relative levels of supply and demand and it is subject to substantial fluctuations arising from political as well as economic and commercial considerations. These fluctuations are mitigated to some extent where cartel arrangements are in force. The price is unlikely to fall below the average cost of production, but there are periods during which producers whose costs are above average find it difficult to show profit. In areas where the production costs are inherently high production may be forced to cease, sometimes for many years until the economic climate becomes more favourable. The revival of Cornish tin mining in recent years is a case in point. Another case is the South African goldfields where only the richest ores were being mined while the price of gold was being held artificially low prior to 1973, these being the only ones that could be worked with profit. Since that year leaner ore bodies are being worked and the richer ones held in reserve to some extent against harder times.

Price reflects the quality and form in which a metal is marketed. A standard quality, say 99.9% pure, would usually cost less to produce than a special grade at 99.99% (unless it were obtained from a different ore by a different method). It is sometimes appropriate to produce a useful alloy such as a steel rather than go first for a purified metal. There would be no point in removing the copper impurity from nickel which was destined to go to the production of cupro-nickel. Some metals are marketed in part as salts – chromates, for example, for the plating industry – postponing the expensive reduction stage until after the product has been sold. Metals are sold in standard forms – ingots, billets, wire bars – again at prices, per ton, which take account of the extra work done. Wire, sheet and foil require most work to be done in their production and are the most costly forms.

Mining operations in underdeveloped countries have had very important political and social consequences, creating enclaves of European civilization in regions previously almost uninhabited, opening up communications and causing mass movements of the indigenous populations. They have brought a degree of prosperity and some education and political awareness to the native peoples and have interfered with their traditional ways of life. Whether this is good or bad is irrelevant – it has happened and it will continue. It will also come to an end eventually, since every mine approaches the same fate – to become an empty hole in the ground. At this point the miners will, as they have always done, move on to look for other orefields, but the societies which they have created will have to create for themselves some other reasons for staying on.

CHAPTER 3

ORE DRESSING – CRUSHING AND GRINDING

ORES are prepared for the principal chemical treatment by a series of relatively cheap processes, mainly physical rather than chemical in nature, designed to effect concentration of the valuable minerals and to render the enriched material into the most suitable physical condition for the subsequent operations. In general these processes involve: first, comminution to such a size as will release or at least expose all the valuable mineral particles; secondly, a sorting operation to separate particles of ore minerals from those of gangue minerals, and sometimes distinguish several ore minerals one from another; and thirdly, when necessary, a re-agglomeration which may be at the same time a roasting operation (see page 100). Rich ores or homogeneous ores cannot be improved in this way, but they may be crushed, sized and blended with other ores to prepare a uniform feed for, say, a blast furnace, or to produce a suitable specific surface or permeability in a reaction bed.

Roasting and calcining (see page 275) are sometimes considered as part of ore dressing or "beneficiation" probably because they have been used, particularly in the iron industry, to effect preliminary concentration with respect to H_2O and CO_2 and at the same time render the ore lumps more permeable and hence more reactive to the blast furnace atmosphere. Today most of these functions are carried out in sintering or equivalent operations (see Chapter 5). A magnetizing roast is a chemical process which brings about an important change in physical properties and is by virtue of its purpose an ore-dressing operation, but is more conveniently dealt with under roasting (see page 275).

A neat classification of processes is neither necessary nor important. Each extraction sequence must eventually be considered as a whole in which each individual operation (except the last) has to produce the best possible starting material for the next in the sequence.

Comminution

The size to which ore particles must be reduced is dictated by the requirements of the next stage of the process, these in turn depending on the particle size of the grains of the several minerals in the untreated ore. Ideally the ore would be broken down until every particle was either entirely valuable mineral or entirely gangue. This could occur only if fractures were always to pass along the interfaces between grains of different kinds and not across them. In this case, the ore and gangue minerals would be said to be "non-coherent" and it would suffice, to release the valuable particles, if the ore were reduced to a size similar to that of these particles.

If fractures run indiscriminately across one kind of crystal and then another, a high proportion of particles is likely to be "mixed" no matter how small they are made. Gaudin[2] demonstrated this point first with a two-dimensional "chess-board" model of an ore which he "broke" into smaller squares only half the linear size of the chequers, the "fractures" being all transgranular (see Fig. 6). By inspection of this model it can be seen that one-eighth of the "fragments" are all black, one-eighth are all white and three-quarters are "mixed". This clearly shows that only one-eighth of the area could be rejected as being of no value and indicates that

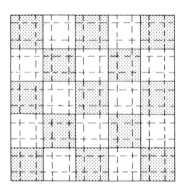

FIG. 6.   Two-dimensional representation of an ore containing ore (dark) and gangue (light) particles, all as equally sized squares in regular array.

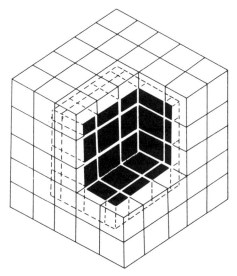

FIG. 7. Three-dimensional representation of a cubic particle of ore mineral (dark) associated with gangue mineral (light) and sectioned into equally sized cubes of side one-third that of the particle shown.

a greater size reduction would have to be effected if a useful separation were to be made possible.

A development of this model into three dimensions can be made using Fig. 7. Suppose that an ore contains $x\%$ by volume of an ore mineral entirely in the form of cubes of side a and that these are "broken" by slicing, making parallel cuts a distance a/n apart in each of the three directions parallel to the cube sides but with the proviso that the plane of no slice should coincide with a cube face. This means that the ore–gangue interfaces are "coherent" so that fractures do not run along the particle boundaries. In these circumstances the volume of each ore particle is a^3. After breaking (slicing), each particle of ore mineral is in the centre of a slightly larger cube as illustrated, the volume of which is $(n + 1)^3 (a/n)^3$ and which comprises $(n - 1)^3$ small cubes of ore mineral, each of volume $(a/n)^3$, and $6n^2 + 2$ similarly sized particles made up partly of ore mineral and partly of gangue mineral coming from the faces, edges and corners of the original ore mineral particle under consideration. Gangue

TABLE 2. VOLUME CONCENTRATION POSSIBLE
AT DIFFERENT VALUES OF THE REDUCTION
FACTOR n

n	Volume concentration = $n^3/(n + 1)^3 \times 100$
0.5	1.2
1.0	12.5
2.0	29.6
3.0	42.1
4.0	51.1
5.0	57.9
10.0	75.1
20.0	86.5

not so closely associated with the ore mineral would be similarly chopped into cubes of side a/n.

If in a subsequent sorting operation it were possible to collect all the little cubes containing any ore mineral and reject all the others the volume percentage of ore mineral in the concentrate would be $100n^3/(n + 1)^3$ which is independent of the original concentration $x\%$. The concentration levels which can be reached in this way are shown in Table 2 for several values of n. It is obvious that a satisfactory level of concentration can be achieved only if the ore is broken down so that ore mineral particles are reduced to a quarter or less of their original size. Grinding to still smaller sizes would make even better concentration possible but grinding costs money so that appropriate level of size reduction must be decided on economic grounds.

In practice mineral particles are not all the same size nor are they all cubes. The shape is not important because an appropriate characteristic dimension can always be recognized to replace the length of the cube edge whatever the shape might be. If the ore mineral is present in a wide range of sizes the question must be asked, how small a particle is it worth the cost of recovering? If, for example, only 0.5% of the value is found in particles smaller than 100 μm but 5% of the value is in particles smaller than 500 μm there would seem to be a case for grinding to about 25-μm size to recover the values in the 100-μm size but little case for grinding further. Such a decision would not be taken casually, however. The actual value of the last 0.5% would have to be compared with the properly calculated cost of recovering some or all of it and indeed the calculations should be continued to determine the optimum cut-off size for the particular ore.

The particles produced by grinding are not of the same size but cover a wide range of sizes. To grind until every particle reaches some predetermined maximum would mean making most of the material very much finer. Some compromise is needed such as to grind until, say, 80% of the material has reached the required size and assume that most of the rest will not be very much over that size. Obviously, larger ore mineral particles will be concentrated to a higher degree and smaller particles to a lower degree than the average so that the theory outlined above provides only a guide to the concentration that will actually be achieved. If the particle size distribution after grinding is known, however, a closer estimate can be made by calculating the probable concentrations for quartiles or, better, deciles of the distribution separately and then finding the weighted mean of the values so obtained.

There are many practical matters to be taken into account. The size to which it seems desirable to grind may be smaller than can be reached economically or may be such as cannot easily be used in later stages of the treatment. Minerals present may grind differently, with one possibly forming "slimes", the particles of which are smaller than about 50 μm. Slimes are difficult to handle in pulp and are easily lost. In the subsequent sorting processes it may be difficult to collect all the values or to reject all the 100% gangue particles so that the predicted level of concentration may not be attainable.

A well-equipped mineral dressing laboratory is an essential adjunct to any mill. Petrographic examination of ores must be made in order to show what kind of treatment is desirable as discussed above. This should be followed by crushing and grinding to the predicted desirable size and careful size analysis of the product. Each size fraction should then be assayed (analysed chemically) to show what has been achieved. Further microscopy on the particles would also show how far separation had been effected. Small-scale flotation, leaching or classification equipment should be available for sorting out the minerals in each size fraction and the products should again be assayed. From the results of this kind of experimentation the most efficient conditions for separating the minerals should be deduced and translated to large-scale operations. In general there would be several minerals in an ore so that the necessary investigation is likely to be rather more complex than has been suggested above. The examination of the ore from each new vein or seam should be in effect a research project which, if successful, should demonstrate the most profitable way to treat the material.

There are two principal kinds of comminution — crushing, down to about 7 mm (max.), and grinding to smaller sizes. Large lumps may have been weakened by cracks introduced during mining or by the existence of bedding planes, and these may make the ore appear to break more easily at the crushing stage than during grinding, but the mechanics of fracture is essentially the same in each stage.

When a prism of isotropic brittle (rather than plastic) material is compressed between two plane surfaces it will first deform elastically and then fracture suddenly along the planes of principal shear stress. If the load is withdrawn immediately the fracture occurs then the prism will be found to have broken into a small number — three or four, say — of major fragments and a much larger number of small pieces, but amounting together only to a few per cent of the original weight. Where there are well-defined cleavage planes, whether of a crystallographic nature or arising from the sedimentary nature of an ore, the fracture will occur along such of these as are nearest to the planes of principal shear stress, i.e. those whose shear stress is first exceeded by the component of the applied stress in the direction in which they lie. Otherwise the fracture pattern would be similar to that for an isotropic material (see Fig. 8).

If, say, a hundred similar pieces were treated like that described above, i.e. one "nip" each, the product would consist of some 300–400 "major" fragments accounting for about 97% of the mass with the remainder in smaller pieces down to dust. After a second similar treatment of the major fragments only, there might be about 1500 "major" fragments (smaller ones now) and some 10–15% of fines. Successive repetitions of the treatment would diminish the size difference between the major fragments and

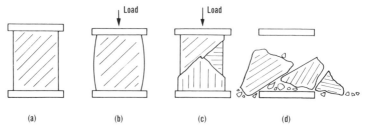

FIG. 8. The fracture of a prism of ore under compressive loading —(a) unloaded; (b) under elastic strain; (c) fracture occurs; and (d) the fragments.

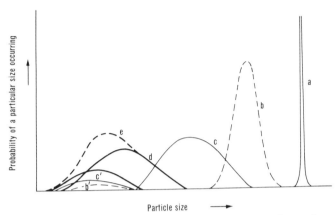

FIG. 9.   The size distribution of ore in several stages of comminution – (a) the unbroken ore of uniform particle size; (b, b′) the
bimodal distribution after each particle has been broken once; (c, c′)
the same, broken twice; (d, d′) the same again, broken three times.
The broken line (e) is the sum of (d) and (d′) indicating a skew distribution of some complexity.

the upper end of the fines range until they overlapped and the size
distribution would change as in Fig. 9 from being distinctly bimodal to
being apparently monomodal but asymmetric – obviously a superimposi-
tion of two simpler distributions.

When the sizes of the major fragments become comparable with those
of the fines the two groups cannot be distinguished so that the former
cannot continue to receive preferential treatment. One distinction between
the crushing and grinding processes is that in the former most of the work
is done preferentially on the "major" fragments, whereas grinding is
usually less selective.

Obviously pressures are not normally applied in as symmetrical a
manner as is shown in Fig. 8. Ore lumps are of irregular shape and are most
likely to have forces applied at edges, corners and projections making it
very probable that small fragments will shear off before major ruptures
occur. This leads to the production of more middle-sized particles in the
early stages and probably more fines at all stages than would be obtained
from the cylindrical model.

A second but related mechanism is sometimes referred to as "attrition".

If two surfaces are nominally "flat" and one is laid upon the other they are in actual contact only at a small number of "high spots". If one of these surfaces is moved across the other, some of the "high spots" will be sheared off producing dust, if the material is brittle. If the surfaces are not very flat similar action may tear off rather larger fragments. This tangential rubbing occurs in most comminution processes along with simple compression or squeezing and indeed some processes are quite clearly designed to include both mechanisms. Obviously this tangential rubbing action produces only small fragments.

A third, somewhat different fracture mechanism is to be found in impact breaking. If an unconstrained particle is struck by a fast-moving object such as a swinging hammer in a hammer mill (see page 44) or another particle, some of the kinetic energy will be converted into strain energy as a compression wave traverses the particle. Should the associated stress exceed the shear strength locally the particle will break in a brittle manner. This fracture is likely to be initiated at cracks or other surface defects already present and it has been suggested that the mechanism is unlikely to produce fresh cracks, so that the product of successful impacts tends to resist further breakdown better than material broken by nipping. It has also been suggested that crack propagation round grain boundaries is very probable under impact fracturing, so that early release of ore minerals can be expected.

Size Distributions

Size distributions observed in crushed materials may be expressed graphically as in Fig. 9 or as cumulative undersize or oversize curves (Fig. 10). Distributions obtained by different methods are most readily compared using these types of curves as they are not amenable to mathematical analysis. The most successful relationship for describing apparently monomodal distributions in broken solids is the Rosin–Rammler equation (as modified by Bennett)

$$R = 100 \exp - (x/\bar{x})^n \qquad (3.1)$$

where R is the weight percentage over size x and $\bar{x}$ is a "mean" particle size corresponding to $R = 100/e = 36.8\%$; n is a constant whose value is near unity and is a measure of the "width" of the distribution. The

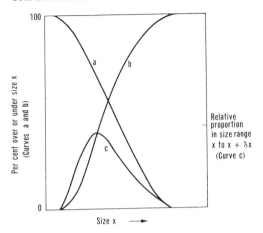

FIG. 10. Size distribution curves. (a) Cumulative oversize curve; (b) cumulative undersize curve; (c) differential curve showing the probability that a particle shall have any size x. "Per cent" usually refers to weight per cent rather than to the number of particles per 100 particles.

equation may be rewritten

$$\log\cdot\log 100/R = n \log x + \text{a constant} \tag{3.2}$$

If the distribution obeys this equation, a plot of $\log\cdot\log 100/R$ versus $\log x$ will be linear. Many crushed and ground products do come near to satisfying this equation and it has been particularly useful to the coal industry. It is, however, rather an insensitive plot and for many purposes the plot of $\log(100 - R)$ versus $\log x$ which corresponds to the differentiated form

$$\frac{\Delta R}{\Delta x} = Ax^{n} \tag{3.3}$$

is just as linear. Carey and Stairmand[3] designate (3.3) a "natural" distribution and postulate that it will occur when a single particle is "free crushed" and that a natural distribution remains natural if the largest particles present are successively "free crushed".

This is similar to an earlier and simpler relationship usually attributed to Schuhmann in the form $R = \exp(-(x/\bar{x})^{n})$ which was itself related to

the equation used by Gaudin in 1926, $(100 - R) = (x/D)^n$ where D was the
sieve size of the largest particle whereas $\bar{x}$ is another characteristic size,
more in the nature of a statistically derived "mean". This seems to ignore
the essentially bimodal nature of the size distributions observed when
single particles are free crushed as described on page 34. There is no
convincing reason why either a natural or a Rosin—Rammler distribution is
more likely to be obtained than the other in any particular case. Each
should be treated as an empirical formula which fits better in some cases
than in others. A comparison suggests that the Rosin—Rammler equation
fits best when the skewness toward a high proportion of the coarsest sizes
and a low proportion of the finest sizes is rather pronounced. In all cases
the equations seem to fit best in the mid-range of sizes with discrepancies
in the top 20% and in the fines. The use of any of the equations to predict
surface areas of broken solids is not practicable because the fine fractions
which contribute a high proportion of the surface are so badly represented
by them. Usually the formulae underestimate the proportion of fines but
they also indicate a gradual fall in the proportion present at any size down
to zero whereas there is probably a cut-off size characteristic of any
particular mineral. This cut-off size is smaller than can easily be measured
but its very existence makes the use of the distribution equations, which
do not take it into account, inappropriate for use in estimating areas. In
practice, comminution is not as simple as has been described. First, lumps
are irregular and pressures cannot be applied uniformly as in Fig. 8, but
concentrate on corners and narrow sections so that more minor fragments
are likely to break off with associated fines. Secondly, crushing is not
usually "free". The lump being broken is likely to be contained by other
pieces pressing round it, or even by virtue of its own geometry. Thirdly,
the crushing tool has a fixed travel and will probably follow through after
fracture commences. These factors will cause much secondary fracturing at
corners and edges by mutual attrition as the particles scrub against each
other. This is usually considered desirable as more work is being done in
every cycle of operation than if the lumps were being treated individually
and the wear on the hard metal linings of the crusher, per ton of ore
processed, is smaller. A fourth consideration is that in some stages medium
sized lumps, being sheltered by larger pieces, may avoid primary fracture
almost entirely. Taking these points together it seems most fortuitous that
the distribution laws mentioned are even approximately valid over a wide

range of materials and breaking methods. In the later grinding stages there is much interaction of particles and "mutual attrition" occurs in which soft minerals are reduced faster than hard ones even to the point of forming some slimes. In grinding there is always a high proportion of attrition by tangential forces.

The actual line of a fracture is probably determined by the presence of small cracks in the surface of the ore particle, which form during the mining operations or during the earlier stages of comminution. Fractures are extensions of these cracks which develop when the local stress at the crack tip exceeds the shear strength of the material. This occurs generally at a lower overall stress than would be needed if the material were free of defects. Large lumps of ore probably always have some cracks present but their density need not be high and small particles may be found without any. These would be more resistant to fracture. An obvious example would be sandstone which can readily be crushed to grains of sand but these are more difficult to grind down to a silica flour. Even glassy types of mineral, however, would be expected to show a similar increase in apparent strength as the size is reduced to a level at which cracks become rare. This is why the size distribution curves probably ought to have a sharp cut-off at small sizes, as mentioned earlier.

Crushing

Primary crushing of ore as mined is carried out in jaw crushers, gyratory crushers or rolls (Figs. 11, 12, 13). Jaws and gyratories accept ore into a wedge-shaped space between a fixed crushing plate and a moving one. In jaw crushers the ore is squeezed until it breaks and the fragments fall down to a narrower part of the wedge to be squeezed again, repeatedly, until they can escape through the minimum gap at the bottom. The jaw may be hinged at the top as shown in Fig. 11 or at the bottom, in which case the maximum size in the product will be more clearly defined. Variations in toggle-plate arrangements give different kinds of jaw movements, some producing a slight gyratory motion which is said to improve production rates and also add some scrubbing action to simple compression with a consequent increase in the proportion of fines produced. Jaw crushers are made in a wide range of sizes from laboratory models up to units which accept lumps of ore up to 3 m across, which gives them an unassailable

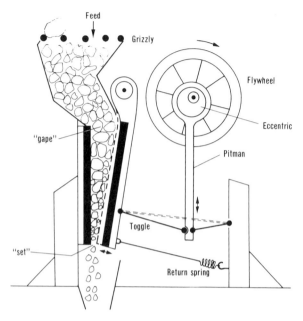

FIG. 11.   Schematic jaw crusher. Hard facings are shown in black.
Forward travel of jaw is indicated by the broken line. Details of
extremely heavy housing have been omitted.

advantage over all other types. They are capable of processing up to 500 tons of ore per hour and reduction ratios range from 4 up to 10 in the biggest machines. In operation the "throw" of the movable jaw is only a few centimetres and the frequency is between 1 and 6 Hz — 60 to 360 strokes per minute. The angle between the jaws is about 20° and the jaws are faced with 13% Mn "Hadfield" steel or chilled iron. Design should be such that the toggle plates should break first on overload, before heavier and more expensive components.

The gyratory crusher is similar in effect, but the relative motion of the crushing faces is due to the gyration of the eccentrically mounted cone and some tangential force is applied in this case as well as simple squeeze. Interaction between particles occurs so that ore is broken against ore producing some fines and reducing wear of the expensive hard metal facings, commonly made of Hadfield steel. Gyratories are built to handle

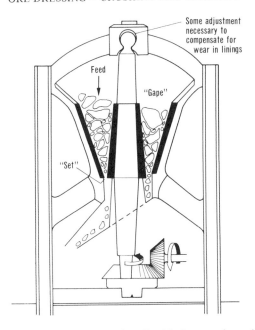

FIG. 12.   Schematic gyratory crusher. Hard facings are shown in black.

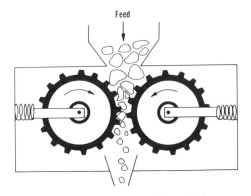

FIG. 13.   Schematic roll crusher with sluggers.

up to 5000 tons of ore per hour which is about ten times as much as the big jaw crushers but the maximum size of lump accepted would never exceed 2 m across and would usually be very much smaller. At the same acceptance size they have much higher capacity than jaw crushers and at the same capacity they are much more compact. Gyratories operate best on hard, brittle materials and do not work well on soft or plastic minerals like clays which tend to pack together and hinder the free flow of ore through the machine. Jaw crushers deal with these materials much better, provided they are set for straight compression without any gyratory action. Gyratory crushers can be adjusted to give a range of product size and to allow for wear in linings. They are considered to be compact and economical in head-room. They are often run with the head buried in ore delivered from hopper bottomed wagons through grizzlies, the top being integral with a large bin.

Rolls draw the ore lumps down into the gap between them and, after one "nip", discharge the product. With a given roll diameter and "set" there is a maximum size of lump that will be drawn in. With smooth rolls this limits the reduction ratio to about 3 : 1, but using "slugger" rolls, which have ridges or knobs on the roll faces, the reduction ratio can be raised above 4 : 1. These are used on iron ores which are usually cohesive but not very hard. Hard ores are likely to wear rolls rather badly.

Fracture of ore between rolls is again mainly by squeezing and must occur by stages as the material is drawn into the narrowing space. Interaction between ore particles is much less than in jaw and gyratory crushers and the proportion of fines produced is smaller. They are often arranged to work in series, each pair of rolls being set closer than the previous pair. If the reduction ratios are low and undersize at each stage is removed by screening between rolls a very close size distribution can be obtained. Rolls are most appropriately used on ores which are "sticky" — usually due to high moisture content or clayey nature. They are self-cleaning, whereas jaws and particularly gyratories can become choked up with this kind of ore.

The choice between jaws and gyratories is an economic one. Jaws can be built to handle larger material — up to 3 m across. At the same gape however gyratories have a higher rate of throughput. If a high production rate is required a jaw crusher would be used for the largest material only — as determined by a very coarse screen — while the bulk of the ore

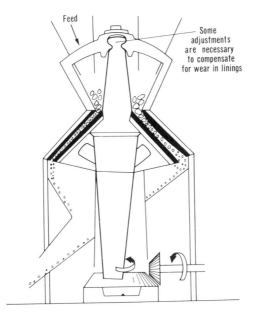

FIG. 14.   Schematic cone crusher for secondary reduction. Hard
facings are in black and details of heavy housing are omitted.

passing this screen would be fed to gyratories, to reduce it to about
50 mm. If only one unit could be justified, a jaw crusher would usually be
considered more flexible and probably cheaper to maintain and run.

Secondary reduction down to −7 mm would be carried out in a cone
crusher, by rolls, or in a hammer mill (Figs. 14, 15). Of these the cone
crusher is the most commonly used on ores. It is similar to the gyratory,
but is self-cleaning because the gap widens to let the product fall through
after the nip has been made. The inner cone sometimes rotates on an
eccentric axis instead of gyrating. These crushers have very high capacity—
much better than rolls at the same setting. Rolls could be employed where
excessive fines were undesirable, but are in fact seldom used at this stage.
Hammer mills are used mainly for breaking weak, brittle materials like
coal. Most ores would cause excessive wear, but soft sticky ores can
sometimes be handled in this type of mill because there is no packing
together of the product (see page 49). The material is fed through a

hopper into the path of the flying hammers which are freely swinging in
smaller machines but fixed rigid with the rotor in the largest sizes. The ore
lumps are broken by the impact fracture mechanism either when hit by
the hammer, or on subsequent collisions with the impact plates enclosing
the working space, with other lumps or with another hammer. All material
must remain in the working space until small enough to escape through the
narrow gap between the hammers and the final breaking plate. In some
cases the material must escape through a metal grid, oversize to the grid
being swept round the mill for re-treatment until small enough to pass.
The largest mills can accept lumps of mineral about 1 m across. Capacities
range up to 2000 tons per hour and the reduction ratio can be as high as
40 : 1, which is exceeded only by "Aerofall" and cascade autogenous mills.
The products are said to be "strain free", a virtue valued where the rock is
to be used for road building. Wear on the hard facings is always heavy and
provision should be made for rapid replacement of these parts (see
Fig. 15).

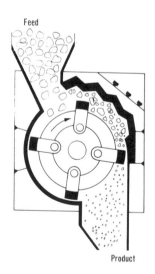

FIG. 15.   Schematic hammer mill.

Grinding

Grinding is mainly carried out in ball mills or in similar equipment (tube, pebble, and rod mills). The typical ball mill is a barrel-shaped vessel rotating on its horizontal axis. It has special replaceable cylindrical liners, lifters, and end plates and is loaded with balls of steel or cast iron. A "pulp" of ore with 20–50% of water is fed into it and discharged axially, the overflow carrying the ground product. In some designs a grate is provided with holes between 3 and 10 mm in diameter to control the size in the overflow, preventing the early escape of coarser particles.

The linings and end walls have traditionally been made of high alloyed steels including 13% Mn "Hadfield" steel or special irons, the sections being designed for easy replacement in so far as is possible. There is now a strong trend toward the use of rubber for linings and end plates. This is said to wear well and to allow mills to be driven harder to give higher productivity. Repair and replacement is faster and cheaper because the rubber can be replaced in small areas by vulcanizing on to the shell without first emptying the mill. For successful operations the lifters must be specially designed for use with the rubber. Steel liners may be used smooth, corrugated or with lifters or "shelves" running parallel with the axis of the mill to control movement of the load. Rubber liners require substantial lifters closely set together to minimize slip between the load and the liners. Rubber cannot be used in larger mills or where the balls are bigger than 85 mm in diameter and in any case the thickness of the rubber must be adequate to withstand damage by its being cut through by the balls. A suitable thickness is usually less than that of a steel liner, however, which gives a mill of any external size a higher capacity when lined with rubber. Rubber is also used in tube and rod mills and in cyclones and pumps. It withstands both abrasion and corrosion and has the additional advantage of being rather quieter in action than steel. The size distribution of the balls should be near to a close packing grading, but in practice replacement balls are fed in at a single size and allowed to degrade to extinction.

The choice between special steel balls and cheap cast iron ones is almost entirely an economic one. Steel would last longer, but seldom long enough to justify the additional cost except where the costs are equalized by, say, high transport charges. There is an occasional chance that contamination by alloying elements in special steel balls might be injurious to the later parts of the process.

In action the balls are lifted up the rising wall of the mill and rolled down or thrown over into the pool on the other side, nipping ore particles against other balls, the linings, or other pieces of ore, and the effectiveness of the operation depends upon there being for each particle a very large number of opportunities for fracture to occur. Attrition also occurs between balls as they are half dragged, half rolled back up through the pulp under continuous impact from above. The rotation speed should be sufficient to raise the balls to about two-thirds of the height of the interior of the mill and throw them across to the "toe" of the load where the coarser material tends to accumulate. With the mill about half filled with grinding media and pulp this speed would be about 75% of the "critical speed" n_c, at which a simple calculation shows that the load will centrifuge − when, of course, no grinding would be performed. The critical speed is given by

$$n_c = 423/\sqrt{D}$$

where D is the diameter of the mill in centimetres. It has been shown, however, that with smooth linings, slip occurs[4] between load and liner over a wide range of speeds so that, particularly at low loadings, centrifuging does not occur until the speed is several times n_c. Under these "super-critical" conditions there is much more attrition between balls, ore

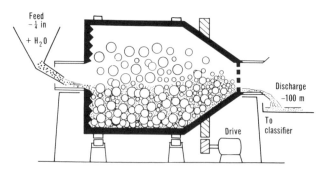

FIG. 16.   Schematic representation of ball mill with conical out-flow end. Hard linings are shown black. The screen at the out-flow is for retaining balls rather than for sizing the product.

and linings and greatly improved efficiencies are claimed, but the wear on linings is excessive. Super-critical grinding is not yet commonly practised, but means of working it with "autogenous" linings have been discussed – that is with ore lumps fixed to or embedded in the lining in such a way as to protect it from wear.

Some ball mills are specially shaped to be narrower at the discharge end so that balls segregate by sizes and the mill is effectively sectioned for coarse and fine grinding – the probability that a fine particle will be nipped being increased by this device (see Fig. 16).

Tube mills are more positively divided by grids into compartments containing different sizes of balls so that the load is ground in, say, three stages. This should be more efficient provided the work is properly apportioned, but it needs more attention to keep it efficient. Tube mills are longer and narrower than ball mills and the term is sometimes used interchangeably with pebble mill which is similarly shaped.

Where contamination of the charge by iron from the balls has to be avoided pebble mills are used. These have porcelain or rubber linings and are filled with porcelain balls or suitable flint pebbles. These mills are narrower than ball mills to avoid excessive breakage of the pebbles and are made longer to compensate for the reduced diameter of the mill and lower density of the balls. Pebble mills are used in the ceramics industries and in the preparation of gold ores, because iron or its salts inhibit amalgamation of gold particles in one process and use up the cyanide reagent which dissolves the gold in the other major process for extracting that metal.

Another variant is the rod mill, which is a similar barrel-shaped vessel, perhaps a little longer than a ball mill of the same diameter, filled with steel bars, 75–100 mm in diameter, laid parallel with the axis. These bars must be the whole length of the mill, which is more troublesome to operate than a ball mill because rods wear thin and break. The broken pieces interfere with the performance of the mill so that thin rods must be removed. The great advantage of rod mills is that they act preferentially on coarser particles at all stages and the production of extreme fines is minimized. This is because rods are held apart by the largest ore particles, the spacing being greatest at the feed end. Secondly, transport is by means of water which flows along the spaces between the rods carrying undersize particles with it toward the discharge end. Once a particle is broken its fragments are carried along the rod slot to the appropriate places for further action — or out of the mill. The process is therefore self-sizing; continuing attrition of fines is minimized; over-grinding, sliming, and heavy oxidation of exposed mineral can, where necessary, be avoided.

Rod mills can be operated at high rotational speeds, with deep lifters for crushing from about 40 mm, or slowly with shallow lifters for controlled grinding as described above.[5] Fracture is mainly by nipping in the former case (i.e. fast squeezing) and by slower squeezing action (with more tangential action) in the latter, but compression between lifters and rods, shearing in the scissors action as tangled rods are straightened and shearing between rods as they are lined up by the end plates, all contribute to the reduction in size. The use of a tapering mill to accommodate the splay of rods toward the feed end has been advocated as giving higher operating efficiency.

Rod mills are not popular except where extreme fines or excessively rounded grains are not desired, although it has been shown that they have a very high capacity for a given size and are very efficient even when heavily loaded.

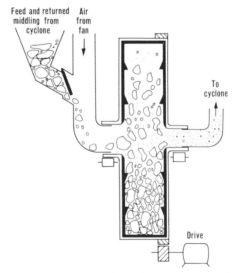

FIG. 17. Schematic representation of autogenous grinding in an air-swept mill coupled in closed circuit with a cyclone for classification and recirculation of insufficiently reduced material.

Another type of tumbling mill is the "cascade" or "Aerofall" mill (see Fig. 17). There are no balls or rods in these, the grinding being autogenous but hard rock may be added if ore itself is relatively soft. Large pieces of ore act like balls on the smaller pieces but break themselves against the linings in the process. The diameters of these mills are very large compared with their length (e.g. 7 m x 2 m long) so that the lumps fall through a great distance to make up for their low density. The mills are swept with a current of air which carries off the fines which are usually classified at once in cyclones (see page 70), the coarse fraction being returned for further milling. The feed may be ore.as mined or it may be prepared to a more suitable size distribution. This type of equipment cuts right across the distinction between crushing and grinding and seems to offer a very cheap, one-step, size reduction process capable of producing a very evenly graded product with little contamination. The reduction ratio can be very large. It is essential that the feed should maintain a large population of heavy particles in the load, that is particles of about 25 cm across, to act as a grinding medium and the product would typically be less than 1 mm and could be even smaller than that. Not every material responds well to this kind of milling however. Brittle materials with a granular texture seem to do best. Under favourable conditions these mills are economical of space, are cheap to build and to operate and give good control over the size distribution of the product.

In favourable conditions, single-stage autogenous comminution is carried out in very large conventionally proportioned mills run in conjunction with hydrocyclones which size the product and return oversize for further treatment. Not all qualities of ore are amenable to such treatment. Ideally large lumps fed into the mill would abrade rather than shatter, so preserving for as long as possible a capacity for damaging smaller lumps.

Hammer mills as already described on page 43 can be used to make a fine product though not as fine as would be produced in a ball mill. Wear on the hammers and linings is very heavy with consequential contamination of the product. Fracture in these mills is thought to be preferentially intergranular, which may sometimes be useful in affording an earlier release of the values than is possible in alternative equipment (see Fig. 15).

In a stamp mill, each of a group of about five heavy "tups" weighing about half a ton is lifted in turn about 25 cm by means of a cam, and dropped on to an anvil. This is repeated about once per second and ore particles in the "mortar" are reduced from about 25 mm down to such a size as can escape through a screen across the discharge opening set at perhaps 25 mesh (see page 58). Stamp mills operate on pulp and the water helps to carry to crushed ore through the screen. They are widely used for crushing gold ore prior to amalgamation, but seldom for any other purpose, and can now be considered obsolescent. Action is preferentially on the larger particles so that sliming is avoided.

Miscellaneous Comminution Techniques

Brief mention should be made of several other types of equipment used for breaking minerals. Not all of these are suitable for use with heavy ores but any of them might find occasional applications in the dressing of lighter minerals.

The disc crusher has been well described and illustrated in older text-books. Like the stamp mill it is little used today. Crushing was between two shallow concave saucer-shaped discs which faced each other in a vertical plane and rotated in opposite directions with some gyration which provided the necessary periodic nipping action and permitted periodic clearance of broken material. The feed was through the centre of one of the discs. These crushers were self-classifying and produced few fines but they had low capacity and were superseded by cone crushers for secondary reduction many years ago.

Pan mills or edge-runner mills, in which two large, heavy, stone or hard metal wheels mounted on a common axle run round a circular track, are simple compression crushers. They are used as secondary crushers but not often on ores. They are more commonly found in brickmaking where batch crushing to provide a product with a particular bulk density or degree of close packing is being practised.

Ring roll and ring ball mills are like giant roller or ball races into which material is fed to be ground down to $-100\ \mu$m between the rollers or balls and the race case. Air streams are used to remove material as it becomes small enough. They are used for pulverizing coal. Their capacity is rather lower than would normally be useful for ores.

Vibration mills are cylindrical drums mounted horizontally on springs. The drum is loaded with hard balls and is made to vibrate so that they rise clear of the lining and fall back with high frequency, nipping particles of minerals charged for grinding between them as they do so. These mills can be operated either wet or dry and can be run under controlled atmosphere or even under vacuum. Alternatively they can be air swept to make them self-classifying. Unfortunately engineering difficulties restrict their size to about 5 tons per hour capacity which is too small for ore handling. They cannot be used on sticky materials which are liable to compaction.

Unorthodox means of reducing the size of materials include weathering, which relies on natural agencies like frost and the chemical effects of rainwater, carbon dioxide or even bacteria to weaken and fragment the material. Roasting can be used on materials which decrepitate or spall due to allotropic changes on heating. Some minerals break with heating because of their coefficient of thermal expansion. Heating with torches playing on to ore travelling along belts, followed by water sprays can bring about release of minerals in suitable conditions.

Electrical methods include induction shock heating and the production of shock waves by the discharge of condensers within a fluid within which solids are in suspension. The solids can be broken as the waves pass through them, as in impact fracturing in a hammer mill. Many devices of these kinds have been shown to be technically feasible in the laboratory but they cannot be developed to be as cheap to operate as the traditional methods.

Fluid energy mills use high-velocity jets of air or steam in which fine solids are suspended. Two or more jets are made to collide and the particles in collision break by impact fracturing as in a hammer mill. More sophisticated equipment uses the jets only to induce violent turbulence. The solids are recirculated through a cyclone to remove undersize. This method is usually used for fine grinding of weak materials. The feed is typically -150 μm and the energy consumption is high. An application of the same principle to coarser materials is the Snyder process[6] which is reported to have reached the pilot plant stage of development. The solids are charged into a magazine with compressed air or steam at about 30 atm pressure. This pressure is released explosively through a special fast-release valve. The solids emerge at high speed to collide either with impact plates or with similar particles released simultaneously from a similar gun facing in the opposite direction. Fracture may be partly due to the sudden decompression of gas within the porosity of the particles but there is certainly impact fracturing also, as the particles collide. It is difficult to envisage conditions in which such complicated equipment would compete with mills of traditional design, but development is said to be continuing.

Efficiency in Comminution Operations

Much effort has been directed toward measuring the efficiency of energy utilization in crushing and grinding processes — that is the ratio of the energy theoretically required to effect a particular degree of size reduction to that actually expended. The latter is readily measured as the power input to the mill and is the sum of the energy needed to move the working parts of the equipment and where appropriate the load, to overcome friction in the bearings, and to grind away the metal from the working faces, along with that needed for size reduction. Most of this energy would normally be dissipated as heat and none of the separate items could easily be evaluated. An accurate energy balance on a comminution process would be a rare achievement.

There has been considerable controversy over the most appropriate measure of the minimum energy required to effect reduction. Rittinger (1867) proposed that it should be proportional to the area of new surface formed. In other words, the minimum energy requirement is the surface energy of the new surface created. This is the net increase in the free energy of the material being treated and as such is irrefutably the minimum energy which could bring about the change observed. A rival theory of Kick (1885), however, suggested that the energy requirement would be proportional to the logarithm of the reduction ratio. This was based on the assumption that geometrically similar particles would always break in a geometrically similar manner, irrespective of their size, when the strain energy rose to a critical value, the strain energy being proportional to volume. Thus if a cube of a hypothetical mineral fractured in compression into eight cubes of exactly half the (linear) size of the original, the energy required to break a lump initially one cubic metre in volume and one metre side into eight cubes each of 0.5 m side would be exactly the same as that required to reduce the same volume of cubes of, say, 1 mm side down to 0.5 mm side, or, indeed, from 1 μm to 0.5 μm. The energy required per unit volume would therefore be proportional to the number of reduction stages — to n — where reduction was continued through n stages until the reduction ratio was (in the above example) $(\frac{1}{2})^n$.

There are obvious differences between the two approaches to the problem. Rittinger is considering the energy in the product of a particular size distribution while Kick is concerned with the energy required to break the original particle irrespective of how many pieces are produced. Rittinger is measuring the net free energy of the change of state while Kick is measuring the activation energy of the process. While Rittinger's energy is an irreducible minimum, Kick's might be modified by altering the mechanism of the process.

Kick's theory is now regarded as being oversimplified. Since fracture is generally brought about by the extension of cracks already present in the surface of the particle it is not necessary to raise the strain energy to the critical value throughout the whole volume of the particle before rupture occurs. It is necessary to reach the critical condition only at the tip or leading edge of the crack to induce its propagation and a small excess of energy would usually lead to fracture. This is because the crack acts as a stress intensifier. The activation energy of fracture is therefore much lower because of such cracks. It is also not proportional to the volume of the particle but rather to the length of crack which has been so charged with energy. The Bond theory[7] follows this line of argument. It puts the crack length at $z^{-1/2}$ where z is a characteristic size of particles in a distribution (the screen size which 80% will pass). The specific surface (surface per unit area) is z^{-1} and the crack length (per unit volume) is the square root of that area. The energy to reduce a material from size z_1 to size z_2 is said to be $W = K (z_2^{-1/2} - z_1^{-1/2})$ where K is a constant for any material which appears to vary over a rather small range for a wide variety of materials. Holmes[8] suggests that the index $-\frac{1}{2}$ is too specific and proposes replacing it with a variable $-r$ which has values between 0.25 and 0.75 depending on the material.

None of these theories can convincingly be confirmed by experiment. This is not surprising. The total energy consumed in any mill far exceeds the fracture energy which is being considered. The fracture energy might be estimated by comparing the power needed to run the mill empty and loaded but a small difference between two large quantities is inherently difficult to measure with accuracy. On the other side, to relate energy used to any parameter of size or surface area requires an extremely accurate determination of size distribution to be carried out, especially in the range of the smallest particles which make a contribution to area quite out of proportion to their weight. Surface area measurements can be made using gas adsorption or dissolution rate techniques — where suitable methods can be devised for the mineral under examination — but these methods are rather specialized and must account for the surfaces of cracks left in the particles without producing new cracks or propagating old ones. The specific surface energies of minerals are not always known with accuracy and these values depend on the medium in which the mineral is immersed, being different for air and water and different again when surface active agents are in solution in the water whether by accident or by design. Indeed, considering the difficulties involved, it is remarkable that so much evidence has been produced in support of one or other of these theories.

Kick's theory has been supported by tests carried out on prepared cubes tested in compression in the manner suggested by Fig. 8 and it may well be substantially valid where fracture is essentially of that type as in jaw crushers and between rolls. The difference between the Kick and Bond theories is probably marginal in these conditions. The strain energy at fracture per unit volume might well be virtually independent of particle size over a wide size range even though the critical strain is reached only locally at crack tips, but obviously in the Bond model the amount of energy stored per unit volume would be less than in the Kick model. If we consider a case in which there is primary crushing from 1 m to 60 mm, secondary crushing to 4 mm, and grinding down to 8 μm, the reduction ratios could be expressed as 2^4, 2^4, and 2^8, respectively. Kick's law would predict energy requirements in the ratios 4 : 4 : 8 or 1 : 1 : 2 in the three stages. In practice the first two stages might require similar amounts of energy but the third grinding stage might need 10 to 30 times as much in a ball mill. This, it could be argued, is because the smaller particles have progressively fewer cracks present so that the model reverts from a Bond/Kick hybrid to a true Kick type. Alternatively it could be claimed that Rittinger's law becomes dominant at small sizes and that the energy is required to make the creation of the very large new surface possible. It is perhaps more likely, however, that it is the mechanical inefficiency of the ball mill that is at fault. A particle entering a jaw crusher cannot escape being crushed and passed through in the first three or four bites but in a ball mill there is a high probability, which increases as the size diminishes, that any particle will escape damage in the next turn of the mill so that the mill must work longer for each particle than if its fracture at each turn were assured.

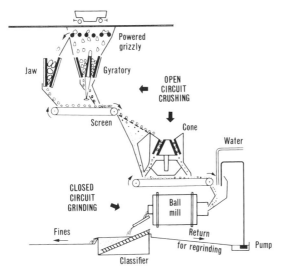

FIG. 18.   Section of an ore-dressing plant showing both open-circuit and closed-circuit operation.

One condition in which the activation energy of fracture is probably low is in impact fracturing as in a hammer mill. Here a shock wave traverses the particle raising the strain energy very locally as it passes through. As this wave passes a crack the associated stress is intensified and may lead to fracture without the total strain energy in the particle ever having approached the level envisaged by Kick or Bond. It is interesting to note that the energy recorded as being consumed by hammer mills is generally somewhat less than that required by other mills doing similar work.

Efficiencies of mills have been quoted in the range 0.02 to 50%. The lower values are based on surface energies. The highest values can be presumed to be based on Kick's criteria but even so are unrealistically optimistic. Enough has been said to show that further discussion can be of little value.

The problem of efficiency is really an economic one. The cost of the necessary fracture energy is small beside the other costs – the costs of other power, capital, steel linings, balls, etc. Efficiency is most likely to be improved in any plant by modifying dimensions and operating conditions within the limits available to find the optimum working arrangements, and is best measured in terms of total operational costs per ton of ore.

Crushing and grinding operations may be conducted in open circuit or in closed circuit (see Fig. 18). In the former the product is all passed to the next stage; in the latter the product is sized or classified in some appropriate manner and the portion over some particular size is returned for retreatment. In the second case the unit can be run faster, a less thorough treatment being accepted. In ball milling this would minimize the formation of slimes and give better control over the size distribution in the product. Where control is inherent in the process as in the rod mill and in jaw crushers, open-circuit operation is more common.

CHAPTER 4

ORE DRESSING – SIZING AND SORTING

THE distinction and separation of large objects from small, heavy from light, cubic from spherical, or magnetic from non-magnetic is very easy provided the objects are reasonable in size and small in number. When there are millions of them and all are very tiny they cannot be handled and examined one at a time and so conditions must be arranged to favour the transport in a particular direction of particles exhibiting one quality rather than another. The winnowing of chaff from threshed corn is a primitive example of this kind of process.

In principle, any physical property might be used as a basis on which to distinguish and separate particles of ore mineral from gangue. In practice most processes fall into one of four classes. *Screening* groups particles according to size – and may or may not effect concentration of the ore minerals. *Classification* processes depend on differences in density, size and shape as these properties jointly affect the motion of particles in fluid media. *Flotation* uses differences in surface properties and *magnetic methods* distinguish magnetic from non-magnetic minerals. Classification, flotation and magnetic separation are all sorting operations.

These processes are complicated by the fact that the distinctions being employed are not clear distinctions. Most particles (see page 31) have mixed characteristics, intermediate between the two extremes of the pure ore and the pure gangue minerals. Somehow a line has to be drawn so that not too much valuable mineral is thrown away and yet not too much gangue retained. Proper comminution can minimize this dilemma, but cannot resolve it. Regrinding of a middling grade comprising the "doubtful" fraction provides a partial solution.

Screening

The simplest, most direct sizing process is screening or sieving. In principle, each particle is tested to find whether or not it will pass an aperture of a particular size and shape so that a number of particles can be separated into two groups – an oversize and an undersize. If the process is repeated using a number of differently sized apertures a size distribution can be determined. A screen is an assemblage of such apertures designed to test many particles simultaneously. Aperture sizes range from no upper limit down to about 40 μm, below which fabrication of screens to a reasonable accuracy becomes extremely difficult. Screens may be made of sheet metal with either round or square holes, but more usually they are made from rods or wires either welded or woven to give square apertures. The maximum size of particle which will pass depends upon the shape and design of the apertures, but this is not likely to be important except in test sieving.

Down to $\frac{1}{16}$ in (1.6 mm), screen sizes have traditionally been designated by the diameter of a round aperture or the side of a square one, in inches in Britain and the U.S.A. or in millimetres in Europe. At smaller sizes European systems continue to use mm or μm while British and American practice is to use a scale of "mesh" sizes. These are closely related to Standard Wire Gauges in the different countries, the precise wire sizes being necessary for the manufacture of accurate screens. The mesh number of a screen is the number of wires in the screen in each direction per linear inch, so that n mesh means that the screen has n wires and n apertures per inch. The width of the aperture is similar to the gauge of the wire so that n mesh means an aperture of about $1/(2n)$ inches. The aperture sizes in the standard mesh series vary from one to the next in geometric progression with a common ratio of very nearly $\sqrt{2}$ which is convenient when the size distribution of the material being sized is of the "natural" type described on page 37 as there is then a similar weight of material left on each screen.

British Standard Sieves have now been brought into line with the European practice (B.S. 410 – 1969), all sizes being designated in mm down to 1 mm and in μm at smaller sizes down to 38 μm which is the old 400 mesh. The opportunity was taken to modify aperture sizes slightly when the change was made but for every standard size in the traditional system there is a corresponding size in the new series which is very little different. Screens having apertures smaller than about 1 cm are woven from wire to give square apertures. Wire may be of steel in coarser sieves, brass or in the finest sizes phosphor bronze. The use of the mesh-size terminology persists in industry, particularly in the U.S.A., and still appears in technical literature. There are differences among even the American systems and there is still one British system not converted to metric units. To demonstrate the differences and to facilitate conversion from one set of sizes to another a table of sieve sizes is included as Appendix 3 at the end of this book.

The other major specifications for standard sieves are those of the American Society for Testing Materials (A.S.T.M.), the Institution of Mining and Metallurgy (I.M.M.) (London), Tyler (U.S.A.) and DIN (German). There is also a French specification which is identical with the German one except that it does not cover the larger sizes. These series differ with respect to the size range covered, the number of sizes in the series and the relationship between mesh number, wire gauge and aperture. In the I.M.M. series the wire diameter equals the aperture width so that the area of the apertures is only 25% of the total screen area. The A.S.T.M. wires are thinner giving about 30% and the B.S. wires thinner still over most of the range giving about 40% of the total area as aperture. Obviously the greater the area of aperture the higher the screening rate per unit area of screen but the less robust the screen. The DIN sieves were originally in meshes per centimetre but are now designated by nominal aperture in millimetres.

It should be noted that whereas the number of particles in a given weight of a particular material is inversely proportional to the cube of their diameter the number of apertures in a screen is inversely proportional to the square of the aperture. To provide equal numbers of apertures per particle, the screen area should increase as the reciprocal of the aperture size (assuming similar weights to come on each screen) – a condition not easily met.

Larger sizes of screens may be made of woven wire or rod like the fine sieves, or the rod may be of tapered section to facilitate the passage of particles. Woven "mat" is, however, expensive and there are cheaper alternatives. Wires may be laid equally spaced in two directions at right angles and welded together where they cross to give either square apertures, or slots of equal width. At the largest sizes rods or bars are used lying in one direction only, the main consideration being that the bars will bear the weight of the load likely to be put on them. The spacing of the bars in grizzlies may be up to 2 m. The largest sizes of "grizzlies" are principally used to safeguard primary crushers against overload and consist simply of an assembly of bars, rails or girders laid across the path of the feed, with suitable spacing, before it enters the hopper of the crusher. It is sometimes convenient to use powered rolls instead of static bars to facilitate the removal of oversize and to give some control over the rate of feed. Powered rolls may be patterned to give better grip or may be made of elliptical bars for the same purpose. For screening at smaller sizes between 10 and 100 mm another form of roll screen may be used. A bank of rolls is arranged as a cascade. These have ∨∧∧-profiles and are positioned so that square apertures appear between adjacent rolls. Material fed over the cascade is divided into an undersize passing through the apertures and an oversize passing over the top. This would be used where heavy hard material would quickly wear out ordinary screens. The equipment looks very costly, but cost must be compared with that of alternatives doing the same job. An obvious disadvantage is that the proportion of the total area occupied by aperture is very small.

The other type of screen is made of sheet steel with round or square holes. This works well in square apertures and the initial cost is low. The available area in round holes is rather small and the surface is rather smooth so that particles slide about rather than roll on them. This makes for inefficient screening, but the punched plates last longer, especially when the material being screened is very abrasive. They can be made only with large apertures, greater than about 40 mm. They are particularly suitable for fitting into trommels (see below).

In any screening operation there will be little doubt whether most of the particles will go through the screen or remain on top of it, but there will always be some near-size particles (unless they are all spherical) which will pass only when a certain aspect is presented to the aperture. To facilitate their passage some mechanism is required to turn them over several times so that there is a reasonable probability that they will go through. In test sieving it is particularly important that a reproducible technique be adopted so that comparable results may be obtained from one occasion to another. It is also necessary to provide some means of transporting the oversize off the screen to make way for new feed.

There are two types of equipment. Reels or trommels are cylinders the walls of which are of screen material. They rotate on a slightly inclined axis so that material fed at the top end is rolled about as it moves towards the discharge end and may pass through the screen to undersize. Compound trommels use several sizes of screen. Sometimes there are in line – fine mesh first, coarse mesh last (which is hard on the fine mesh); sometimes there are two or three concentric cylinders, the inner one of coarse mesh and the outer one of fine mesh, material being fed on to the inner screen. In this design it is not easy to replace worn screens.

Other types of screen are usually flat, inclined to facilitate flow and subjected to a periodic motion either parallel to the screen or normal to it to make the particles either roll or bounce on the deck as they simultaneously move down it. Each particle has therefore a number of opportunities of falling through an aperture. The number of chances depends on the length and slope of the screen and also on the amplitude and frequency of the vibration. The amplitude of a normal vibration should be just sufficient to permit the particles to clear the screen and the frequency just low enough to permit them to return within the cycle. This makes the number of chances a maximum for any screen length and slope. As the gradient is increased the time of passage will be reduced as will the number of chances.

The other important variable is the depth of the bed of particles on the deck. If a single layer is spread out, many particles will immediately fall through leaving much of the screen area unused and more occupied by obvious oversize. Optimum screening rates are achieved with beds about five or six particles deep from which the obvious undersize is quickly eliminated, leaving mainly near-size particles on the deck during the rest of the operation.

The greatest difficulty encountered in screening is to overcome the blinding of screens either by hard near-size particles wedging in apertures or by soft sticky ore which fills up the apertures as would a thick paste. Remedies such as the use of roller screens, screen-drying equipment or ore-drying equipment are very expensive. Light loading of screens; occasional severe jolting; mechanical cleaning with brushes; special types of vibrators – these are simpler alternatives which sometimes prove effective. Slots blind less readily than square apertures especially if the slot can be flared to the underside. Rubber-stranded mat has been suggested to combat blinding and rubber balls are sometimes made to bounce on the underside of the screen. It essential that the cloth be mounted taut so that it has ample springiness in use. This helps to minimize blinding.

Wet screening will also reduce the amount of blinding experienced but it is not commonly practised. Corrosion of the screens is difficult to avoid and the power required to vibrate the screens under water is rather high. It is also difficult to maintain the character of a pulp while the water and fine particles are being removed. Even if the water is retained above the screen the extra buoyancy afforded the larger particles in a pulp by the fines in suspension is lost as the fines are lost so that the coarser particles settle out or require extra agitation to keep them in suspension. One device known as a "sieve bend" seems to have achieved some success. Pulp is made to slide down a concave "surface" which consists of a succession of wedge-sectioned stainless-steel rods lying across the direction of flow and equally spaced to act as a screen. These may be set as close as 50 μm apart but probably work better at a somewhat wider setting, say 150 μm.

It should be noted that screening takes account of only the two smallest dimensions of a particle. The second smallest dimension of a thin flat one may nearly equal the diagonal of a square aperture. In this respect screening on square and round apertures can give rather different results. Particles in the same size fractions can have very different volumes – differing by a factor of 100 even in the same material – entirely because of the wide range of shape which is possible. The mean size of particle in a size fraction is obviously somewhere between the sizes of the apertures defining the fraction. It depends to what use the parameter has to be put, whether the mean of these apertures will suffice, or whether a root mean square or even a root mean cube value would be more appropriate. It may also be necessary to decide whether the mean sought should be on the basis of the weight of material involved, or on the basis of the actual number of particles.

The point should be made that there is a big difference between test sieving and industrial screening. In the former the time allowed can be as long as is necessary to reduce the rate of passage of material to a very low value. Owing to attrition of the particles against the screen this may never stop, but a point can be reached at which no measurable increase in the size of particle on any screen is observed. In industrial screening each particle may be presented to the screen surface a maximum of 10 or 12 times before being removed as oversize. The average size in any fraction would be appreciably smaller in industrially sized material than in a sample carefully sized in the laboratory.

Theory of Classification

The term "classification" is usually restricted to processes in which particles are distinguished by their different rates of travel through fluid media. In practice this often means different rates of fall through water and this particular case will be examined. The other important medium is air. These are also called "gravity" separation processes.

The forces acting on a particle falling freely through water are its net weight (its weight minus the upthrust of the water) and the resistance R offered by the water. The containing walls are at present assumed to be so far away as to have no effect. Applying the Third Law of Motion

$$(m - m')g - R = mf \tag{4.1}$$

where m is the mass of the particle; m' is that of the fluid displaced by it; f is the acceleration of the particle; and g is the acceleration due to gravity.

The resistance R is a complex function of the velocity v of the particle relative to the bulk of the water (which can be assumed at rest). When v is small enough for the flow of water round the particle to be "streamline" or laminar, the resistance is due to shear of the liquid as the particle surface, with water attached to it, moves through the body of otherwise stationary liquid. If a spherical body of diameter d falls freely through a liquid of viscosity η, Stoke's law can be applied as long as the flow remains streamline, i.e.

$$R_s = 2\pi d\eta v. \tag{4.2}$$

When the velocity is high, conditions may be fully turbulent and shearing energy is swamped by energy imparted to the liquid as kinetic energy. This is recoverable only as heat when the particle has moved on and conditions are returning to their initial state. Under these conditions Newton's law applies–

$$R_t = \frac{\pi}{8} \cdot p_l d^2 v^2, \tag{4.3}$$

p_l being the density of the liquid.

Inserting R_s and R_t for R in equation (4.1) and putting $m = \frac{1}{6}\pi \cdot \rho_s d^3$ and $m' = \frac{1}{6}\pi \cdot \rho_l d^3$ where ρ_s is the density of the particle, it follows that—

$$\frac{dv}{dt} = \frac{\rho_s - \rho_l}{\rho_s} g - 18 \frac{\eta v}{\rho_s d^2} \qquad \text{for laminar flow} \qquad (4.4)$$

and

$$\frac{dv}{dt} = \frac{\rho_s - \rho_l}{\rho_s} g - \frac{3}{4} \frac{v^2 \rho_l}{\rho_s d} \qquad \text{for turbulent flow.} \qquad (4.5)$$

When $dv/dt = 0$, v becomes the maximum or terminal velocity v_T, given by

$$v_T = \frac{d^2}{18} \frac{(\rho_s - \rho_l)}{\eta} g \qquad \text{for laminar flow} \qquad (4.6)$$

and

$$v_T = \sqrt{\left(\frac{4}{3} dg \frac{(\rho_s - \rho_l)}{\rho_l} \right)} \qquad \text{for turbulent flow.} \qquad (4.7)$$

It is demonstrable that streamline flow of liquid round a particle is limited to conditions where Reynolds Number $Re < 0.2$ and that flow is not fully turbulent until $Re > 800$. Reynolds Number is a dimensionless ratio equal to $vl\rho_l/\eta$ where l is a parameter characteristic of a linear dimension of the particle — the diameter in the case of spheres, for which $Re = vd\rho_l/\eta$. Since the density of water $\rho_l = 1$ and its viscosity $\eta = 0.01$, the following critical values of v_c can be calculated—

Diameter	Critical velocity v_c cm sec^{-1}	
d cm	$Re = 0.2$	$Re = 800$
1.0	0.002	8
0.1	0.02	80
0.01	0.2	800

This suggests that pure laminar flow around falling particles is likely only when the particles are very small, and fully turbulent flow only with rather large ones. In practice most of the applications involve the intermediate range of Re between 0.2 and 800 for which the theory is not yet fully developed, but Rittinger modified equation (4.7) to

$$v_T = \sqrt{\left(\frac{4}{3} dg \frac{(\rho_s - \rho_l)}{\rho_l} \cdot R \right)} \tag{4.8}$$

where the coefficient R has a value about 2.5 under typical ore-dressing conditions.

The theory as outlined ignores a number of other complications of importance in the applications. First, the particles are never spherical so that the derived equations are in error with respect to the numerical coefficients. A "shape factor" k would have to be incorporated which would be constant for all sizes of any particular mineral but might vary considerably from one mineral to another. The diameter would have to be replaced by some statistical size parameter l, conveniently the diameter of the sphere having the same volume or the same projected area in the direction of motion. Equations (4.2) and (4.3) become $R_s = kl\eta v$ and $R_t = k'\rho_l l^2 v^2$ respectively. Shape affects rate of fall mainly through the flow pattern. Spheres fall faster than most other shapes. Flat plates like mica flakes fall most slowly.

A second complication is that free fall conditions are not approached in practice. Interference by the walls and bottom of the containing vessel and by other particles is significant. The magnitude of these effects can be indicated by reference to empirical corrections which must be applied to equation (4.6) if it is to be used to determine the viscosity of a liquid from the terminal velocity of a sphere falling through it. If the experiment is conducted in a cylinder of radius r and the velocity is measured a distance h from the bottom end, equation (4.6) becomes[9]

$$v_T = \frac{d^2}{18} \frac{(\rho_s - \rho_l)}{\eta} g \cdot \frac{1}{(1 + 1.2d/r)} \cdot \frac{1}{(1 + 0.55d/h)} \tag{4.9}$$

Thus the correction for wall effect alone amounts to 1% when $r = 100d$, increasing to 30% when $r = 3d$ and that due to end effect alone is about 5% when the end is about 10 diameters distant. The effects of the containing walls and other particles cannot be ignored.

Under laminar flow conditions energy dissipation, i.e. resistance to flow, can be explained in terms of a mechanism whereby the liquid "lattice" is subjected to strain which it sustains only temporarily, but which is relieved by spontaneous shear in a "mean relaxation time" τ. Thus as a body passes through a fluid it moves with it an attached film of fluid relative to the stationary fluid beyond. If the body moves a distance δx in time τ the strain energy developed will be proportional to the angle through which the "lattice" has been deflected. If the fluid is stationary at a very distant boundary the angle is small and resistance attributable to the boundary is small. If the fluid is stationary at a nearby surface the angle subtended by the same displacement δx in the same time τ is much greater. Should the boundary be very close a further complication arises that the fluid velocity in the narrow annulus between the particle and the container wall cannot be ignored. The relative velocity becomes much higher and so also the resistance to motion.

When two similar particles fall together with moderate separation they would therefore be expected to fall rather faster than either alone because for each, in one narrow sector, boundary drag is reduced — a part of the boundary in effect moving with the particle. If many similar particles fall all at the same level this effect would be accentuated especially at the centre, i.e. away from the container wall. If many similar particles are falling at all levels, however, each interacts with its neighbours all round and indirectly with the walls and the bottom. Even when the volume concentration of similar spheres is only 1% the average separation is down to 3.74 diameters and the terminal velocity is down to 94% of the free falling value. At 10% volume concentration the separation is 1.74 diameters and the terminal velocity is only half that for free fall.[10]

It is obvious that a large particle would interact with a nearby small one tending to drag it down with it, but itself suffering delay. A small, very dense particle falling past a slower one would have less effect on it than a large one at the same distance (centres) and the same relative speed — the surface—surface distance being greater.

A large particle falling through a suspension of very small particles behaves as if the effective density and viscosity are those of the suspension as a whole rather than those of the true fluid medium.

The general case where there is a wide range of particle sizes and several densities is one of great complexity and the motion of any particle will depend not only on its own size but also on the distribution of sizes and velocities of the particles around it. Particles of similar size and weight will tend to move together through a suspension of the finest particles which may tend to flow upward with the displaced liquid. Particles of any particular size will tend to be dragged down by larger ones, but will be impeded by smaller ones. The effective density and viscosity will be different for different sizes of particle. In a given time a dense particle will probably go further than a lighter one of the same size but under such complex conditions of "hindered settling" it could not be calculated whether adequate separation could be effected between any two classes of particle in a reasonably short distance. The odds, however, would appear to be heavily against.

The effects discussed can, however, be exaggerated by exploitation of "differential acceleration". Equations (4.4) and (4.5) show that the *initial* acceleration (v small) is greatest in the most dense and in the largest particles. If particles can be allowed to fall only short distances, repeated many times for cumulative effect, a better separation is probable.

Such a system would involve checking the fall periodically and this enables a completely different effect to be introduced. After the large particles have come to rest on say a grid, small particles will continue to fall and may trickle through the interstices between the big ones. This "consolidation trickling" can be exaggerated if the water is allowed to drain for a short time through the arrested solid bed. (See "Jigging", p. 73.)

Classifiers

Water is used in several types of equipment to effect separation by size or density.

In the simple *panning* technique, sands or gravels are swirled with water in a shallow conical dish with the effect that the dense particles stratify in the bottom while lighter minerals, being more buoyant, remain partly in suspension and can be decanted with water from time to time. The mechanism of separation is not simple but involves differential acceleration and consolidation trickling coupled with the peculiar flow pattern in the water due to the shape of the pan, which helps to keep the lighter particles in motion. This is most successful when small heavy grains are being separated from relatively coarse gangue – particularly for treating gold-bearing sands. It is essentially a manual operation and is now mainly of historical interest.

Perhaps the modern successor to panning is the Humphreys Spiral – a spiral chute of concave cross-section down which closely sized sands are washed by a shallow stream of water. Light particles, being more buoyant, are washed to the outside. Heavy particles are retarded by friction against the floor and move more slowly down the inside of the spiral whence they are diverted into separate channels.

The simple *sluice box* (Fig. 19) was a tank, fed with pulp at one end from a launder, and with an overflow weir at the other end over which water carried such solids as did not settle below its level during the time of passage. Solids collected between riffles on the floor of the tank which prevented their being stirred up by the agitation of the incoming stream. The water flow was horizontal and the degree of separation depended on the dimensions and densities of the mineral particles and on the dimensions of the sluice which determined the transit time. All dense particles and most of the large ones would be retained and would be dug out of the sluice box periodically. In olden times elaborate systems of sluices were a regular feature of most mining sites, local streams being diverted for the purpose.

This crude type of classifier developed in several directions. Removal of solids is now usually continuous or is at least possible during the operation

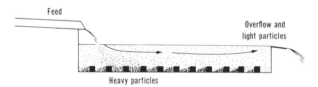

FIG. 19. Simple sluice box probably operated by diverting a stream through it. Development into hopper-bottomed tanks for continuous operation was an obvious early step toward greater efficiency (see Fig. 22).

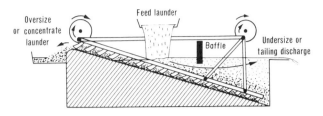

FIG. 20. Rake classifier. A further development from the sluice box. Transport of heavy particles may alternatively be by means of a spiral or a screw on a bottom of circular section.

either through gates at the bottom of conical hutches (Fig. 22) or by means of rakes (Fig. 20) which draw the settled solids up a sloping floor and at the same time agitate the water and release fines which have been trapped at the bottom. The rectangular tank may be replaced by a conical vessel with peripheral overflow (Fig. 21) and the solids may be assisted to the central outlet by rotating rakes. This system – the bowl classifier – has a very large overflow area and can therefore carry very large volumes of water with moderate linear velocity. An extreme example of this feature is found in the Dorr thickener – a very large shallow cone used to settle out the finest of solids and produce clean water for re-use industrially. A long transit time and a minimum of agitation are the desirable conditions achieved in this equipment.

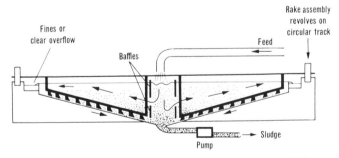

FIG. 21. Diametric section through bowl classifier or thickener. This is similar to the rake classifier but oversize is raked down to the sludge sump. This enables extreme fines to be taken out so that the bowl can be designed to take all solids to oversize. Diameter may be 50 m. Rake assembly rotates very slowly.

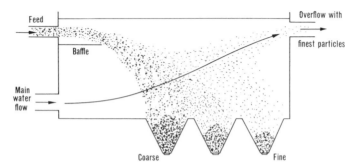

FIG. 22. Modified sluice box with imposed vertical flow of water
gives increased control over products. (After Coulson and
Richardson[11].)

A vertical flow of water may be imposed in any system to reduce the
absolute rates of fall and possibly increase the size of material in the
overflow. Figure 22 shows how it can be applied to produce several grades
in the settled product.

This principle is extended to the point where the upward velocity of
the water controls the characteristics of the settling and overflowing
fractions. A *hydraulic classifier* is a cylindrical (or tapering) vessel in which
water rises at a controlled rate. Pulp is fed centrally near the top. Particles
whose terminal velocities are greater than the upward velocity of the water
go down. The rest flow up and over, possibly into another vessel in which
the water velocity is less, so that a second fraction can be isolated. This
technique is also used for sub-sieve particle size analysis — usually using an
air-classifier.

Cyclones achieve similar results from more compact equipment. They
may operate with gas or liquid, in which latter case the term *hydrocyclone*
is often used. Pulp is injected tangentially with very high velocity at the
top of a conical vessel (Fig. 23). The flow is constrained to follow a spiral
path down toward the apex but as the available cross-section diminishes
an inward flow is superimposed toward a vortex which develops along the
axis and in which the net flow is upwards through a central pipe called the
"vortex finder" which stabilizes the vortex and acts as an overflow. Any
particle in the cyclone is subject to (a) a centrifugal force $m \cdot v^2/r$
characteristic of its tangential velocity v and the radius of curvature r of its
path; and (b) a drag force imposed by the local inward flow of the water

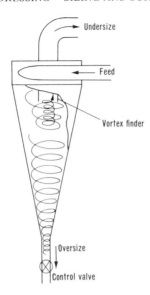

FIG. 23. Schematic cyclone showing typical paths followed by oversize and undersize particles. The general form is common to cyclones for use with gases and liquids but dimensions and proportions vary considerably depending on the purpose for which the unit is designed.

and depending on the relative radial velocity, the viscosity of the water and the dimensions of the particle according to equations (4.2) and (4.3). For each particle there must be a path with a particular radius at which these forces are balanced $(dr/dt = 0)$ and toward which it will move. This radius will be greater for larger and denser particles (those with high terminal velocities in free fall). If the radius of this equilibrium orbit is less than that of the vortex the particle will be carried in the vortex stream to overflow. Otherwise the particle will join other large and dense particles at the wall of the cone and gravitate to the apex discharge aperture.

Cyclones are well established for cleaning gases of dust. Hydrocyclones were first used for desliming water, but they have since been developed as classifiers. Each unit must be designed for its own special purpose though any particular cyclone can be operated to give limited control over its products through variations in the feed velocity and by throttling the apex discharge pipe – but not without affecting the rate of throughput. The major advantage of hydrocyclones is that for a given capacity they are small, economical of floor space and cheap to build but wear and tear on the feed pumps which must carry pulp can be severe and expensive. They are most useful for sizes below about 6 mesh (2.8 mm) and separations can be made down to about 5μ.

The classifiers discussed are usually fed with pulp containing 20–60% solids (by volume) – rather less in cyclones. The underflow may have up to about 83% solids. Except at the lowest densities, considerable interaction occurs between the particles and separation is by size as well as by density – that is the heavy fraction will contain large particles of both ore minerals and gangue as well as the smaller ore mineral particles. This is suitable for open circuit operation in which the product is to be passed to another process such as tabling (see page 74) in which the large gangue particles can be removed later. Low density operation, in for example rake classifiers (Fig. 20), is more appropriate when in closed circuit along with, say, a ball mill where classification by size alone is desired. Low density pulps require bigger plant to handle the large volumes of water involved. Except for the cyclone, feed may contain particles up to 7 mm or more and all the processes can make separations down to sizes of the order of 200 mesh (76 μm).

When a sharper separation is required between ore and gangue minerals other methods must be adopted.

Heavy medium separation, dense medium separation or DMS does not depend on rate of fall or on size but only on density. Since silica has a density of 2.65 and metallic oxides and sulphides about 3.5–4.5, the former will float and the latter will sink in a liquid of density 3·0. No suitable liquid has so high a density, but suspensions of heavy minerals or alloys ground to −100 mesh (−150 μm) are reasonably stable up to densities of about 3.4 – the denser suspensions being prepared with rather coarser solid content. Suitable substances commonly used are galena, magnetite and ferrosilicon. The choice depends principally on cost and the contamination risks involved. Blends of two together have been mentioned as giving a medium of lower viscosity, at the desired density, than could be obtained from one only. Pulp, all > 3 mm, is fed into a conical vessel containing the suspension of suitable density. Gangue is skimmed off into a launder. Concentrate may be pumped by an air lift from the bottom of the cone. The main difficulty and expense is the recovery of the dense medium and its reconstitution for re-use. The density of the suspension must constantly be restored to its correct value and losses of the medium either to the concentrate or to the tailing should be held low. The technique operates best on coarse material and is cheap enough to have been used even on iron ore.

The use of dense media in cyclones has been developed particularly for the separation of diamonds from quartz but applications to ores of uranium and tungsten have also been reported. Quartz is crushed small enough to release the diamonds which are then separated with a very high recovery rate. The cyclone is operated so that the quartz (S.G. 2.5) rises cleanly into the vortex while the diamonds (S.G. 3.5) go down to discharge as "oversize". Separation is obviously by density in a medium of about S.G. 3.0 and not by size which would be very similar for the two minerals. The magnetite would go both ways. It would be separated magnetically from the quartz and the diamonds would be recovered using suitable screens.

As with screening, so with classifying, laboratory sizing to very small sizes using elutriators can produce separations which would not be reproduced in the equivalent industrial equipment. Elutriators are hydraulic classifiers and may use either water or air as the working fluid. The fluid is passed up through a series of tapering tubes, wide end up and of progressively increasing cross-section so that the fluid velocity progressively falls. Particles are classified by their rates of fall which determine which column in the series is the one in which each particle will be retained. This gives some guidance as to the behaviour of similar particles in classifiers but firm predictions of that behaviour cannot be made on the basis of experiments made under such idealized conditions.

Jigging

Another sorting process is called *jigging*. Thick pulp is supported on a grid (often overlaid with heavy ore to yield a permanent bed of low permeability). A pulsator forces water up through the grid with sufficient velocity to bring all the particles momentarily into suspension so that the bed is "fluidized". The water is then allowed to drain back through the grid and the cycle is repeated continuously. Details of the cycle are important and vary from case to case, but would normally include a very

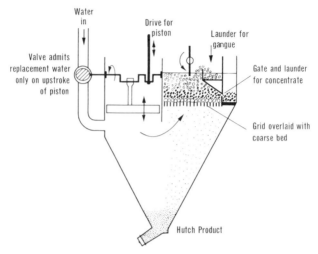

FIG. 24.   Schematic jig.

sudden upward thrust, a spell of free fall (in which hindered settling and differential acceleration bring the dense particles below the less dense) and finally a period of draining or even suction in which small particles (and particularly small dense ones) are drawn low in the bed and may even come through the grid (see Fig. 24).

There are three products — the gangue tailing is skimmed over a weir; the concentrated ore is drawn under a gate which excludes the tailing; and a "hutch product", usually of dense fines, which is drawn from a spigot at the bottom of the cell. The separation is not quite so ideal as is suggested and jigs are operated in series, in batteries, to produce a concentrate and a tailing and a middling which may be re-ground and returned for further sorting.

Tabling

Concentrating tables or *riffle tables* (Fig. 25) employ similar principles in a completely different way. The riffle table has a large flat surface inclined slightly both front to back and from left to right. Its smooth surface is divided into narrow strips by "riffles" of wood which form a

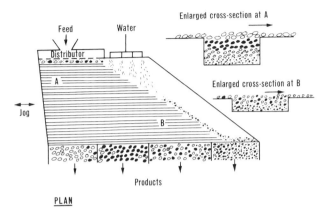

FIG. 25. Representation of a riffle table showing how it distributes particles by size and density (the black particles being the heavy ones).

large number of parallel channels perhaps 1 cm wide and 1 cm deep, but tapering to zero depth at the lower end. Pulp is fed at the top high corner and enters the channels. A stream of water is maintained across the table. The table is vibrated with a horizontal motion having a frequency of the order of 300 cycles per minute, each cycle being composed of a relatively slow forward movement followed by a very rapid return so that particles are induced to creep forward along the channels. The effect of these pulsations also brings the small particles below the larger ones, and heavy ones below light ones of the same size, so that lighter and larger particles are crowded out and washed down the table to lower placed channels. The largest and lightest should proceed almost straight down the table while the smallest and heaviest must follow a diagonal route via the bottoms of the channels. The process is continuous with the idealized effect that particles are segregated into four groups – large light, large heavy, small light and small heavy, but in fact the medium and small gangue is likely to be mixed in with the ore. Tables work best on well-classified material so that the operator can concentrate on separation by density rather than on density and size simultaneously. There are many variations on this theme, one of which is the *pneumatic table* which has a finely perforated surface through which air is blown to keep particles moving down the table while a jogging action keeps them moving across it.

Flotation

Froth flotation is the most specific of the sorting processes as it operates through the sensitive surface properties of the individual minerals. It is readily applied to very fine concentrates (excluding colloidal dimensions or "slimes") and it can distinguish not only ore mineral from gangue but one ore mineral from another. Briefly, conditions are arranged so that when a pulp is agitated and air bubbles are blown through it certain minerals attach themselves to the bubbles and are floated out in a froth which is skimmed off and discharged of its mineral burden.

The surface property of interest is really its chemical or perhaps crystallographic capability of interacting with specific organic ions with such effect that a layer — probably a monomolecular layer of these ions becomes attached to the surface, conferring on it a high effective surface tension. This condition of the surface is manifest in what is more readily recognized as "wettability" and is measured in terms of "contact angle" at an intersection with an air–water interface. If a drop of water is placed on a solid surface it takes up a position as in Fig. 26(a) in which θ is the contact angle. This is not easily measured, being sensitive to contaminants at the interfaces, to surface irregularities, and to the last direction in which the water flowed across the surface (which gives advancing and retarding angles). There is, however, an equilibrium angle for any solid–liquid pair in contact with air at any temperature.

At equilibrium the forces acting at a point of contact of the solid, liquid and gaseous phases are the three surface tensions γ_{s-l}, γ_{l-g} and γ_{s-g} (where s, l and g denote solid, liquid and gas). These are related by the equation

$$\gamma_{s-g} - \gamma_{s-l} = \gamma_{l-g} \cos \theta. \qquad (4.10)$$

If the liquid is pure water γ_{l-g} has a value of about $0.072\ \mathrm{Nm}^{-1}$ depending slightly on temperature. The contact angle then depends on the difference between γ_{s-g} and γ_{s-l} for the solid surface in question. If these are equal, $\theta = 90°$. If $\gamma_{s-g} > \gamma_{s-l}$, $\theta < 90°$ and approaches zero as $(\gamma_{s-g} - \gamma_{s-l})$ approaches γ_{l-g}. This limit is approached by surfaces very easily wetted by water–clays, most oxides and hydrates. Very low values are observed less often than might be expected because as the solid–liquid interfacial energy decreases the probability of inter-solution increases.

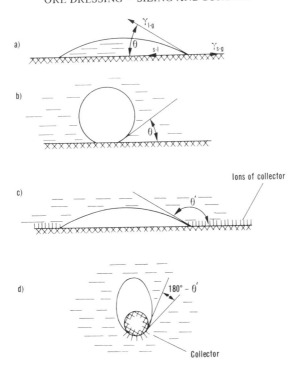

FIG. 26.   The relationship between contact angle and surface tensions and the effect of a collector on the shape of an air bubble at a liquid–solid interface.

When $\gamma_{s-g} < \gamma_{s-l}$, $\theta > 90°$ and would approach $180°$ if $(\gamma_{s-l}-\gamma_{s-g})$ approached γ_{l-g}. The surface would then be completely non-wetting. In fact the largest angles observed, on surfaces like that of paraffin wax, are only about $110°$. The condition for wetting is that the work of adhesion of liquid to solid exceeds the work of cohesion of the liquid which is equal to twice the surface tension of the liquid, i.e.

$$W_{ad} = \gamma_{s-g} + \gamma_{l-g} - \gamma_{s-l} > 2\gamma_{s-l} = W_{Co} = 0.144 \text{ Nm}^{-1} \text{ for water.}$$

Combining this with equation (4.10) we have for wetting,

$$W_{ad} > \gamma_{l-g}\,(1 + \cos\theta)$$

and for non-wetting,

$$W_{ad} < \gamma_{l-g} \, (1 + \cos \theta).$$

Against air and water, for complete wetting with $\theta = 0°$, $W_{ad} \simeq 0.144 \, Jm^{-2}$ or for $\theta = 90°$, $W_{ad} \simeq 0.072 \, Jm^{-2}$. In practice in flotation θ values are found to lie in the range between $25°$ and $80°$ so that the range of W_{ad} will be from about 0.14 to 0.085 Jm^{-2}, or rather less in view of the reduced value of γ_{l-g} appropriate to the less than pure water found in flotation cells. Unfortunately attempts to measure this energy or indeed to apply thermodynamics quantitatively to flotation have not been very successful and the practice remains largely an art.

If an air bubble is placed on a solid surface immersed in water the contact angle should be the same as for a drop of water on the same surface in air. Figure 26b shows that if θ is small the air bubble tends to become detached but if θ' is large the bubble becomes firmly attached to the surface (Fig. 26c). If the surface is part of a small particle the net density of the particle and the bubble together may be less than that of the water so that the two will rise to the surface (Fig. 26d). If the particles are *very* small compared with the bubble the exact position with respect to the surface film (see page 83) is not known with certainty, but the whole surface of the bubble can appear to be covered with particles apparently on the air side.

Surface properties of ore and gangue minerals vary within too narrow a range to be useful for effecting separations directly. Such differences as do exist must be amplified by bringing about selective adsorption of chemisorption of certain organic compounds called "collectors" on to the mineral it is desired to float.

There are three main types of collector — oils, organic acids and their salts, and organic bases.

The oils may be kerosenes, creosotes, diesel or fuel oils. Their main applications are to the flotation of fine coal from shale and they can be used for separating the hard durain from the bright vitrain and clarain fractions. They are not selective enough for ore-dressing today, but they served well until about 1925.

The organic acids and bases are all compounds whose molecules are composed of large non-polar alkyl groups $(R = CH_3(CH_2)_n)$ and a polar group which may be $-COOH$, $-OH$, $-SH$, or $-NH_2$. These compounds all ionize in aqueous solution yielding ions like $R \cdot COO^-$ or $R \cdot NH_3^+$ which should form insoluble compounds with ions in the minerals which they are to float. They are effective by virtue of the fact that the polar group can attach itself to an ion of opposite sign on the mineral surface, the alkyl group being oriented outwards. If the mineral surface is sufficiently covered in this way it assumes the surface properties of the corresponding hydrocarbon particularly in so far as it displays a high contact angle with water.

The organic acids and their salts fall into several groups. The simplest are the fatty acids and soaps like stearic acid, $CH_3(CH_2)_{16} \cdot COOH$ and stearates, and the unsaturated oleic acid, $C_8H_{17} \cdot CH:OH \cdot CH_2 \cdot COOH$ which is the commonest in use. These are seldom obtained pure but are used as commercial products of variable composition in admixture with other homologues and derivatives. A common source of oleic acid is "tall oil", a by-product of the wood-pulp industry. The fatty acid collectors are used mainly on oxide minerals and oxy-salt type minerals like silicates and apatite. They can also be used to collect halide minerals like fluorspar. They have been used elsewhere but are not very selective. Attachment seems to be through the cations present and success seems to depend on the cations being suitable rather than on the anion type.

The most successful group of compounds in the dithiocarbonates or xanthates. Potassium ethyl xanthate, a salt, with the formula

$$S = C \bigg\langle {\begin{array}{l} O-C_2H_5 \\ SK \end{array}}$$

is the most commonly used reagent in this group. Higher homologues (propyl, butyl, amyl, etc., up to hexyl groups replacing C_2H_5) are more powerful collectors, but cost limits their application to specially difficult conditions. Attachment to the mineral is by replacement of the potassium by a metal ion in the lattice surface. The heavy metal salt being insoluble, its molecule in effect precipitates on the surface though the mechanism of its attachment is in some doubt. The main use of these is on sulphide ores but xanthates are very flexible in use and their potential uses are numerous.

The other major group is the dithiophosphates or "aerofloats". These have the general formula

$$\begin{array}{ccc} R_1 & & S \\ & \diagdown \diagup & \\ & P & \\ & \diagup \diagdown & \\ R_2 & & SK \end{array}$$

and there are many of them. They too are used mainly for the flotation of metallic sulphides, but they are weaker collectors than xanthates. They can be used first to pull out minerals which are very easily floated, leaving less floatable minerals to a second treatment with xanthate or xanthates may be used as scavengers after aerofloat treatment. The difference is that in the first case two different minerals may be separated. Aerofloats are mild frothers. This might be convenient in some circumstances but in general it is found that better control can be maintained if the two functions can be exercised separately. (See page 82.)

Organic bases are less commonly used than the acids and salts. They are mainly amines, $R \cdot NH_2$, or quaternary ammonium compounds

$$\begin{array}{ccc} R_1 & & R_3 \\ & \diagdown \diagup & \\ & N & \quad X \\ & \diagup \diagdown & \\ R_2 & & R_4 \end{array}$$

where X is Cl, SO_4, etc. The most commonly used is probably dode-cylamine, $CH_3(CH_2)_{11}NH_2$, which yields a positive ion and must attach itself via a negative ion either in the mineral surface or already adsorbed onto it. A particular application is to the floating of silica — usually con-sidered an "acid" oxide.

In recent years the use of chelates has been investigated[12] for col-lecting. This could open up a large new class of reagents for this purpose. The use of "oxine" (8-hydroxyquinoline) has been shown to be technically possible for the collection of lead and zinc minerals. Chelates form distinct compounds rather than adsorbed layers on mineral surfaces. They are less specific in action than the collectors already mentioned in so far as a suitable chelate will react favourably with lead in ore whether it is present as sulphide, sulphate, carbonate or oxide and it may collect other metals at the same time. This may not always be useful but clearly the right reagent for the job must be selected. Oxine must be added to the pulp in solution in acetone and the chelated mineral must be further collected by attachment of the alkyl radical to droplets of oil dispersed in the pulp. These are incorporated into the froth and removed in the usual manner. At present the cost of using chelates appears to be higher than can normally be accepted. (See page 298.)

The choice of collector for a particular job is usually determined by experience and experiment. It depends on many factors, particularly the nature of the minerals involved, their degree of oxidation, and the presence of even traces of other heavy metals in the pulp.

Additional reagents called "regulators" or "conditioners" are usually mixed into the pulp. The most important of these are to control the pH within narrow limits. Carbonates, bicarbonates, lime and sodium silicate are used as alkalis and sulphuric acid for adjustment to low pH. There is usually a pH above which a mineral will *not* be collected by, say, a xanthate collector. For galena and potassium ethyl xanthate this is 9.5 probably because lead hydroxide is even less soluble than the xanthate in more alkaline solutions so that OH^- ions displace xanthate ions from the mineral surface. Other minerals have different critical pH values for the same collector so that some degree of differentiation between minerals is possible by control of pH alone. Some alkalinity is desirable because it ensures the precipitation of heavy metal ions as hydroxides, preventing their using up valuable reagents to no purpose. Alkalis may also cleanse surfaces and lower the surface tension of the water so assisting froth formation. The choice of pH is important and a search for the right value may lead to a change in the choice of collector if a suitable compromise on the basis of the cheapest collector is not possible.

Two other classes of reagent which may be used are "activators" and "depressors". Activators render a surface more amenable to the action of a collector. Cupric ions deposit from copper sulphate on to sphalerite (ZnS) particles and this copper film in turn adsorbs xanthate so rendering sphalerite floatable. Similarly barium ion adsorbed on to silica would allow it to be floated with oleic acid, but would prevent its being collected with dodecylamine so that toward this latter collector the barium salts acts as a depressor.

Depressors render a surface inactive toward a collector and, as indicated, excess alkalinity alone may be depressive. Sodium sulphide will so coat galena (PbS) that xanthate will find no exposed lead ions on which to attach itself. Activation and depression are effected largely by promoting or preventing the precipitation of a compound of the collector on the surface of the mineral particle — activation by interposing films of more favourable substances, depression by precipitation of something less soluble.

In special cases a wetting agent may be added to ensure that the gangue minerals are properly wetted and do not tend to float. When slimes are present reagents may be added either to disperse them to ensure the individual treatment of each particle or to flocculate them and encourage them either to float or to sink all together.

A "frother" must also be added to ensure the formation of a stable froth with sufficient buoyancy to carry the load of floatable mineral out of the pulp. It should not itself be a strong collector, especially of minerals intended to sink, and the froth formed should not be so persistent as to resist destruction by sprays after separation of the pulp. It should work in the presence of the other essential reagents. Soaps and organic detergents are too vigorous and the usual reagents used are pine oil and cresylic acid, but the use of new synthetic frothers is now increasing. These again contain compounds comprising polar and non-polar groups which are adsorbed strongly at water–air interfaces, the polar groups toward the water. When two bubbles come together their surface films meet with the non-polar groups facing each other and they tend to repel each other rather than coalesce. This gives the froth stability. It is essential that the froth is not too stable, as it must be easily dispersed to release its burden immediately it has been removed from the cell.

Contact angles achieved in flotation are not very large. A mineral fully covered with potassium ethyl xanthate will show a contact angle of about $60°$ with air and water and higher homologues will give a somewhat higher value but in practice coverage is only partial and the ethyl xanthate collector will give a contact angle of only $20–30°$. Fatty acids give rather higher angles than xanthates. In principle these angles can be increased by increasing the concentration of collector in the pulp and so the degree of coverage, but apart from cost the use of too much collector is always found to be counter-productive. It tends to reduce selectivity and can cause the agglomeration of fine particles into flocs which may be too heavy to be floated out. The low angles mentioned are sufficient to effect separation provided, of course, that other minerals are sufficiently wetted to ensure that they sink. The distribution of particles in the interface is not known, nor is the mechanism of attachment. An electrical "double-layer" theory has been used to explain why slimes are difficult to collect.[13] The chemistry of flotation is not as simple as has been suggested in the foregoing discussion. The flotation of sulphides by xanthates, for example, is very dependent on a partial oxidation of the mineral surface. Freshly ground galena is difficult to float but after some exposure it floats satisfactorily. Potentiostatic studies have shown that xanthate will collect galena only if the potential maintained between the mineral and the solution of the collector lies within a narrow critical range. This corresponds to a fairly narrow range of oxygen potential and suggests that either an over-oxidized surface or a fully reduced surface will not adsorb xanthate ions. The role of oxygen is still obscure but it has been suggested that it accepts electrons from the surface, so polarizing it that it becomes more receptive of the xanthate ions. Claims have been made that control of oxygen can improve the selectivity of collection in some conditions.

Minerals prepared for flotation must be ground below about 300 μm depending on density, with slimes < 30 μm held to a minimum. Material coarser than 300 μm is generally too heavy to be carried out of the pulp by the air bubbles. Slimes are also difficult to collect, usually because they fail to enter the bubble interface. This must be a kinetic effect associated with the energy needed by the tiny particles to penetrate the water–air interface. Slimes may also be lost by spontaneous flocculation to a size too heavy to float or they may coat larger particles with the effect of preventing the collector from acting on them satisfactorily. Slimed values and slimed gangue present rather different problems. It is sometimes appropriate to prepare the pulp by classification or by tabling to remove at least the coarser gangue particles. This will reduce the pulp volume and thereby the quantity of reagents required. The pulp density is important and should in general be fairly high. In roughing cells solids may be up to 50% of pulp weight but this will be reduced to about 25% in the cleaning cells. These figures depend on mineral density and vary greatly between one practice and another. Pulp is usually conditioned in a special tank with agitators to allow a uniform addition of each reagent to be made and to allow time for its reaction with the surfaces to be completed. Frothers are usually added after the conditioner, just before the pulp enters the first cell. Sometimes reagents are added in the ball mill immediately preceding the flotation cells. This can have the advantage that the fractured surfaces are still "clean" when collector ions reach them – but as indicated above, this is not always a desirable condition. The quantities of the various reagents added are small – only 10–50 ppm or up to 0.05 kg t^{-1} of pulp of the surface-active agents, which are required to form only a monomolecular layer which is in fact usually far from being complete. About 0.1% of alkalis is required, but these may enter into a number of side reactions.

In flotation cells, pulp is agitated and air either injected or entrained to form many bubbles which rise through the cell, collecting mineral particles as they go. They assemble as a froth on top and this is scraped off into a launder and discharged with water spray. Pulp is continuously pumped into the next of about ten cells in series for repeated treatment. The successive floats become less rich and the later ones are returned as a middling, possibly via the conditioner, for another treatment. The pulp from the last cell is barren and is discarded or directed to some other process.

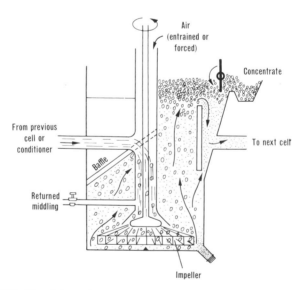

FIG. 27.   Schematic flotation cell. These are operated in batteries as indicated in Fig. 29.

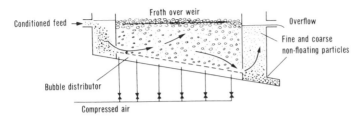

FIG. 28.   Simple flotation cell operated by compressed air. (After Coulson and Richardson.[11]).

Two types of cell are sketched in Figs. 27 and 28 which are self-explanatory. The flow of materials in batteries of flotation cells is shown in Fig. 29.

The distribution of mineral particles and bubbles in the column of froth which collects at the top of the cell is of interest and importance. Small bubbles about 1 mm diameter rise and accumulate in the water surface. These then rise into the froth column above, bursting as they do so. The floatable minerals are redistributed over a diminishingly small bubble surface while at the same time surplus bubble wall solutions drain back into the pulp, carrying back down reagents no longer required and washing out slimes of gangue minerals which may have become entrained. It is essential for clean separations that the froth performs this cleansing function. It is not sufficient to catch the bubbles as they arrive at the surface. Clearly the rate of skimming is critical and must be well matched to the rate of bubble formation. The net effect in the froth bed must be an upward flow which carries out the values and the depth must be sufficient to give time for the separations to take place.

A modified technique described as "flotation separation" has been developed in Russia which uses the froth column as a filter. The same range of reagents can be used but the conditioned pulp is delivered on to the top of a prepared column of froth which may be up to a metre deep. Hydrophobic particles are trapped in the froth while hydrophilic particles are washed through the bubble wall fluid to escape into the water phase below. Froth is skimmed from the top and the bed is continuously regenerated from below as in the conventional cell. The advantage is that coarser material can be retained in the froth bed than can be floated out of a conventional cell. Conversely, however, it is likely that fine sinks will be

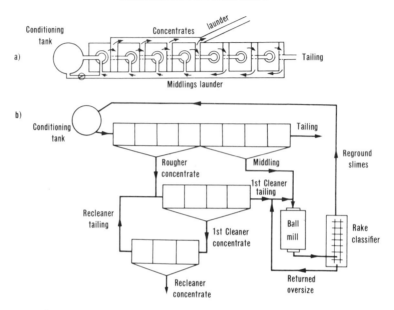

FIG. 29.   (a) Simple battery of flotation cells showing typical flow
of materials. (b) More elaborate arrangement of flotation cells in
closed circuits which include regrinding of middlings.

retained also and it is usual for the new process to be operated as a
complement to conventional working on a coarse fraction separated by
tabling or classification prior to flotation.[14]

Flotation is best applied to pulp which has been classified or tabled free
of at least the coarse gangue. Pulp volume is thereby reduced and reagents
costs diminished. Flotation acts on particles in which enough floatable
mineral is exposed for a bubble to fix on to it. The particle must not be
too heavy to be lifted – and slimes are to be avoided as far as possible but
an advantage of flotation is that it *can* handle finer material than other
sorting techniques if necessary. A new technique is being developed[15]
whereby flocculation of -20 μm slimes can be effected on a selective basis
using small additions of polymers which are adsorbed on to specific
minerals causing them to form large flocs which can readily be separated
from mineral species which remain dispersed. These flocs are larger and
stronger than those obtained by conventional electrochemical coagulation
techniques.

Flotation is a cheap and versatile process but it requires skilful and experienced operators to get the best out of it. In the case of a lean highly disseminated ore there is usually no alternative to flotation concentration unless extraction by leaching is appropriate (see page 293).

Magnetic Separation

Only a few minerals are separated by magnetic means. The most obvious case is that of the ferromagnetic magnetite but it will be immediately apparent that other iron minerals can be chemically altered to produce magnetite. This is usually done using a magnetizing roast (see page 276) which involves heating in a suitable controlled atmosphere to convert at least part of the ore to magnetite. Another possible way of reaching the same condition, which is currently being suggested for the treatment of some carbonate ores, is to attack the mineral with sodium hydroxide to convert the iron to ferrous hydroxide which can then be oxidized at about 50°C to magnetite.[16] Apart from magnetite only pyrrhotite, a nickel-bearing sulphide of iron, is attracted to a bar magnet but there are a number of other minerals with sufficient paramagnetism to make their separation by magnetic means a possibility. These include moderately magnetic minerals like siderite, garnet, chromite, ilmenite, haematite and wolframite and weakly magnetic minerals such as spinels including franklinite and the constituents of monazite sands – rare earth phosphates, thoria, ilmenite, zircon, rutile, along with garnet and magnetite. Magnetic fractionation of these sands is possible.

Magnetic separation can be effected either with low-intensity fields of 500–1200 oersteds or high intensity fields of up to 22,000 oersteds. The former will be used only on magnetite and the latter for all of the others.

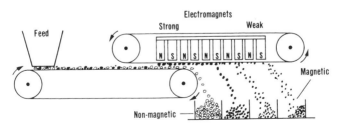

FIG. 30.   Dry magnetic separation of mineral particles.

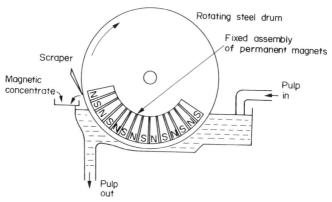

FIG. 31.   Schematic wet magnetic separator.

Separation may be carried out either dry or wet. Dry separation gives a more complete separation of magnetic from non-magnetic materials but where ore has been milled wet it will usually be uneconomic to attempt to dry it before magnetic processing. At sizes >5 mm dry separation will be the rule and at sizes <150 μm separation will always be wet with a choice at all intermediate sizes. Magnetic fields may be obtained using either permanent magnets or electromagnets the latter being necessary only for the most intense fields or where there might be an advantage in being able to vary the field intensity.

Two types of magnetic separation devices are shown in Figs. 30 and 31. When a mixture of magnetic and non-magnetic particles is carried across a magnetic field the magnetic particles tend to remain in the field while the others pass through. Magnetic separators simply provide two paths — a gravitational path for non-magnetic particles and a magnetic path for magnetic material. The diagrams show how this may be done. If necessary the intensity of the magnetic field may be diminished by stages to make possible the separation of minerals of progressively weaker magnetic susceptibility. Where several magnets are assembled in line the polarity facing the mineral stream alternates. This makes the magnetic particles turn over so releasing any non-magnetic particles trapped among them.

Magnetic separators whether wet or dry can work efficiently only if the material is presented in rather a thin layer only a few particles deep. Consequently the design of high capacity plant for use with fine material at reasonable cost is scarcely practicable.

Electrostatic Separation

Minerals have a wide range of electrical conductivity and can be distinguished by that property. If several kinds of particle are given an electrostatic charge and are then brought into contact with an electrical conductor at earth potential the charge will leak away from the good conductors much more rapidly than from the poor conductors. While the charge remains the particle will cling to the earth conductor by electro-static attraction. The weakly conducting minerals will therefore remain attached to the conductor longer than the good conductors so affording a means of separating minerals whose conductivities differ appreciably.

Figure 32 illustrates how this can be applied to the separation of different kinds of minerals in heavy sands from beaches and stream placer deposits where the particle sizes are all very similar. The material must be dry and the atmosphere dry too because under moist conditions conduc-tion is mainly by adsorbed surface films of water and the distinctions between minerals vanish.

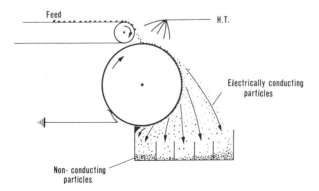

FIG. 32.   Electrostatic separation of mineral particles.

Like the magnetic separators, electrostatic separators operate only on very thin layers of material and there is the same difficulty in designing high capacity into the plant. Attempts are being made to remove this limitation by making the separations within a fluidized bed with a high electrostatic potential imposed across it in such a way that conducting and non-conducting particles move in different directions. Close control of humidity appears to be essential for success. The total removal of solids from gases by electrostatic precipitation is, of course, another application of the same principles. (See page 92.)

Filtration

Solids are recovered after concentration by thickening the pulp and then filtering. Thickening to about 50% water (a sludge) is performed in a very wide bowl classifier (Dorr thickener) or in a hydrocyclone set so that all the solids go in the oversize fraction. The same processes may be used (e.g. after wet cleaning of gases) for producing clean water.

Filtration is carried out by passing the water through a cloth of natural or synthetic fibre which will not permit passage of the solids. The cloth should offer minimal resistance to water flow and act mainly as a support for the deposited solids which form the effective filter bed. The water may be forced through the filter under pressure or drawn through by suction. Pressure methods are economical of space and equipment, but are necessarily discontinuous as filter plates have to be replaced whenever the solid cake builds up to a critical thickness beyond which its resistance to the flow of the water becomes too great. Labour costs are therefore high. The use of suction makes a slower but continuous process possible. The rotary drum filter (Fig. 33) is a cylinder made up of vacuum blocks covered with cloth to support the deposited solids. The water is drawn away through the cloth to suitable traps. As the drum rotates the solid cake can be washed and dried partly by heat, partly by suction. Finally the vacuum can be reversed to dislodge the cake which will probably retain some 5–10% of water depending on particle size, and may require further drying.

Thickening or "de-watering" is necessary before roasting and other pyro-metallurgical processes, but the later stages of drying may be carried out in the preheating stage of the first furnace used and some processes are

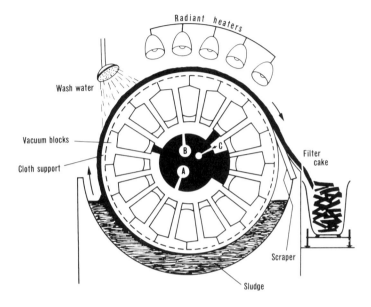

FIG. 33. Cross-section of valve end of rotary filter for de-watering sludge from thickener. The black central section – the valve – remains stationary while the rest of the assembly rotates slowly clockwise bringing vacuum blocks successively into connection with ports A, B, C or closure. A and B lead to suction pumps via traps for concentrated liquor and dilute liquor (washings) respectively. Opposite C compressed air loosens the cake.

designed to accept a fairly wet slurry so that filtering and drying can be eliminated as separate stages. Filtration is necessary in hydro-metallurgy for separating pregnant liquor from gangue prior to, say, electrolysis.

Gas Cleaning

The separation of solids from gases is another important operation which is related to the wet processes just discussed. Gas cleaning follows gas–solid reactions in blast furnaces and fluidized beds and is essential after "fuming" operations to recover valuable volatile oxides (see page 275). It may be employed to recover valuable materials or to cleanse the gas either for utilization or before emission to the atmosphere, following a wide range of metallurgical operations.

Coarse particles can be removed by gravity or momentum separators. A gravity separator is analogous to a sluice box. A momentum separator involves a change (or reversal) of gas velocity at which point the coarse solids tend to maintain their initial velocity and are thrown out of the stream. Simple primary dust catchers using these principles take much of the load off more refined equipment. It is essential that they operate above the dew point of the gas, however, as they do not work well wet.

More complete separation is made in cyclones, bag-houses, by washing or by electrostatic precipitation.

To effect complete separation a gas cyclone (see page 70) must be designed with a high inlet velocity and have a narrow cross-section to maintain a high centrifugal force on every particle throughout its transit of the unit. Bag filters are arranged in batteries (see Fig. 34). Each bag is a long fabric tube — perhaps 5 m long by 150 mm in diameter — through the walls of which the gas must pass. The cloth is woven fine enough to prevent the passage of the coarser dust particles but, as with most filters, a build-up of fine particles on the surface very quickly creates a barrier to the passage of dust of all sizes. The process need not be interrupted to empty the bags. Periodic shaking dislodges most of the clinging dust which falls into a collection area beneath.

Washing involves a counter-current flow of gas versus water, the latter cascading down through a system of hurdles in a tower or being projected through the gas stream as a series of sheets by means of a disintegrator carefully designed to present as large a surface of water as possible to the gas which is itself flowing under very turbulent conditions. This type of gas cleaner may be coupled with a hydrocyclone to clean the washing water.

In electrostatic precipitators gas is passed between two electrodes — a large, earthed, receiving electrode and a smaller, discharge electrode maintained at a potential between −30,000 and −100,000 volts with respect to earth. A coronary discharge from the smaller electrode ionizes the gas and the ions become attached to the dust particles which then move rapidly across to the earthed electrode from which they are removed at intervals either mechanically or with a stream of water. The transit time through the cell must be sufficient to allow each charged particle to travel across

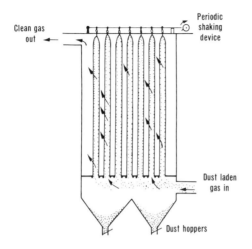

FIG. 34.   Schematic bag-house for cleaning gas.

the width of the cell under the electrostatic force that is applied to it. The electrodes may be in the form of vertical tubes (earthed) with a wire suspended axially within each, or rows of plates (earthed) with series of wires suspended between each pair of plates. The wires should be specially shaped with finely pointed projections branching off at regular intervals to ensure a good distribution of the coronary discharge. The surface of the earthed plates may also be specially formed so that the build up of dust, retaining a substantial negative charge, is not allowed to impair the efficiency of the operation. Apart from the obvious decelerating effect of the negatively charged layer, when it is dislodged (dry) by rapping, the particles tend to fly back on to the plate. Parallel arrays of "chutes" are attached to the receiving plates. Most of the dust attaches itself to the outer edges of these chutes. When dislodged it falls inward on to the chute which is inclined at a sufficient angle to allow the dust to slide down along it to discharge.

The choice of gas cleaner depends on the temperature and moisture content of the gas and on the size of particle which has to be removed. Gravity separators and simple dust catchers should be operated above the dew point of the gas or the dust becomes a mud. Bag filters must also be operated above the dew point but are also limited to a maximum temperature of $180°C$ even with synthetic fabric. Glass or asbestos cloths can, however, be used higher, but not abové $350°C$. Washing cools gases and leaves them saturated with water. Electrostatic precipitators can accept gas at up to $500°C$ or can accept it wet. They may be used on washed gas and indeed saturation with water improves their efficiency.

Simple dust catchers catch particles only down to $100\ \mu m$ and cyclones to $10\ \mu m$ but bags and washers can take out particles down to about $1\ \mu m$. Electrostatic precipitators would remove even smaller particles if required to do so. Their efficiency depends mainly on the residence time of the particle between the electrodes and can be almost 100% if required.

CHAPTER 5

AGGLOMERATION

WHEN the particle size of an ore or concentrate is too small for use in a later stage of treatment (e.g. in the blast furnace) it must be re-formed into lumps of appropriate size and strength. This may be done by briquetting or by pelletizing at ordinary temperatures or by sintering or nodularizing at temperatures so high that some partial fusion may occur which will effect binding. Pelletizing and sintering are the most commonly used processes. Sometimes they are used together, fine ore being pelletized as feed for the sinter plant.

These agglomerates are usually required for blast furnaces and especially in the case of the iron blast furnace the amount of burden preparation has increased greatly during the past 20 to 30 years. This now involves crushing to about 30 mm and screening at 10 mm. The oversize goes directly to the blast furnace but the large quantity of fines produced must first be agglomerated. Some ores are so fine when mined as to require agglomeration. Others need to be agglomerated only after appropriate mineral dressing. Ore from a single source may be fed to the furnace in all three forms — natural, sintered and pelletized. A strong swing toward pelletizing which was observed during the 1960s has stabilized as it has been appreciated that the two agglomeration processes are complementary — sintering being more appropriate on coarser grades of small ore while very fine material is required to make good pellets. Sintering has the advantages that combined water, carbon dioxide and, most important, sulphur can readily be eliminated from the ore simultaneously with agglomeration and that flux can be incorporated into the agglomerate to the point that slag formation may be started. Lime may be added in excess of that equivalent to the gangue content of the agglomerated ore, the surplus becoming available for oxides in the coke ash and other constituents of the burden. Flue dusts can be conveniently reintroduced to the production line at the agglomeration stage.

95

Pelletizing

Pellets are made by rolling critically moist ore around in a drum or on a rotating inclined disc. Some small pressure is necessary to consolidate the pellets as they form but this comes mainly from their own weight applied to each small particle as it is picked up. The particles are bonded together by capillary and surface tension forces in the moisture between them as is explained below. The pellets may be used in the air-dried condition if some suitable bond can be developed to confer adequate strength. Portland cement may be added for this purpose. More usually they are hardened or "indurated" by firing at such a temperature that a good bond is produced either by recrystallization of the minerals present or by the formation of glasses. Flux may be added to promote this vitrification.

The bonding of particles by liquid bridges has been studied by workers in several fields — notably by Fisher[17] and Rumpf.[18] The forces acting on two equal spheres through a drop of liquid at their point of contact (Fig. 35) are (a) that due to surface tension and (b) that due to the deficiency of the hydrostatic pressure within the drop, under atmospheric pressure. These can be shown to be $2\pi b\sigma$ and $\pi b^2 \sigma(1/c - 1/b)$, respectively, where σ is the surface tension of the liquid; the meniscus is assumed to be circular and of radius c; and b is the radius of the lens at the "neck" (Fig. 35). The term $(1/c - 1/b)$ is an estimate of the curvature of the liquid surface (Fisher). The force holding the spheres together is then

$$F = 2\pi b\sigma + \pi b^2 \sigma(1/c - 1/b)$$

$$= \pi b\sigma(b + c)/c. \tag{5.1}$$

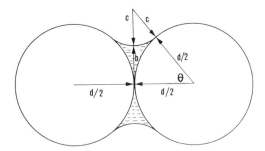

FIG. 35.   Two equal spherical particles held together by a lens of water (see text). (After Fisher[17].)

This is best evaluated in terms of the angle shown as θ which increases with moisture content. Assuming that the liquid wets the particles perfectly (contact angle zero) b and c can be expressed in terms of the particle diameter d and θ thus—

$$b = \tfrac{1}{2}d(1 + \tan \theta - \sec \theta),$$

$$c = \tfrac{1}{2}d(\sec \theta - 1)$$

and $\qquad b + c = \tfrac{1}{2}d \tan \theta$

so that $\qquad F = \pi d\sigma/(1 + \tan \theta/2)$ (5.2)

which is proportional to the particle diameter and surface tension and appears to approach a maximum as θ approaches zero, i.e. when the moisture content is very low. This maximum has a value $\pi d\sigma$ and falls to $2.9d\sigma$ when $\theta = 10°$ and $2.3d\sigma$ when $\theta = 45°$ beyond which the case of two particles cannot be extended for reasons of geometry. If θ is small, however, the displacement possible before rupture is also very small. A weaker bond with more water could stretch before breaking and F would increase initially as it did so.

The strength of aggregates of similar spheres has been considered by Rumpf and clearly depends on the coordination number k, that is the average number of bonds attaching each particle to its neighbours. This depends on the closeness of packing and can range from 6 to 12 but in a naturally shaken down assembly would be about 9 corresponding to a voidage ϵ of about 0.35. Rumpf derives a formula for the tensile strength of an aggregate—

$$T_1 = 1.1 \frac{1 - \epsilon}{nd^2} kF.$$ (5.3)

Using the values of k and ϵ above it is noted that $k\epsilon \simeq \pi$. Combining this with equation (5.3),

$$T_1 = \frac{2\pi\sigma}{d(1 + \tan \theta/2)}$$ (5.4)

which is inversely proportional to the particle diameter, reflecting the number of bonds involved in any cross-section. Like F, the tensile strength appears to rise to a maximum when the liquid lenses are vanishingly small, but again in this case the displacement to cause rupture would be negligible and the aggregate would be brittle. A higher moisture content would make a tougher pellet.

When θ exceeds $45°$ air bubbles become isolated and tend to occupy the largest spaces available. An intermediate range of conditions occurs which cannot be discussed mathematically, but Rumpf has dealt with the case when the voids are completely filled with water. He shows that the tensile strength of the pellet is then due almost entirely to capillary forces, the contribution of surface tension, which now acts only at the surface of the pellet, being negligible. The pressure deficiency in the capillaries is inversely proportional to their "mean hydraulic radius" – or to the diameter of the individual particles. It is approximately equal to the tensile strength of the pellets which is given as—

$$T_2 = 8\left(\frac{1 - \epsilon}{\epsilon}\right)\frac{\sigma}{d}. \tag{5.5}$$

When $\epsilon = 0.35$, $T_2 \simeq 15(\sigma/d)$ which is more than twice the (maximum) value of T_1 when the voidage is the same and $\theta = 0°$.

In practice particles are neither spherical nor of uniform size and a distribution of shapes and sizes in which packing is close would obviously reduce the effective hydraulic radius (replacing d in equation (5.5)) and increase the pellet strength. A moisture content sufficient to fill the voidage completely would appear to offer the greatest strength in a given material, but since even a small excess would confer mobility and rapidly reduce strength a small deficiency in water content below that necessary to fill the pores is desirable. The proportion of water used depends on the voidage ϵ plus available pore space within the particles and is therefore determined by their shape, size, distribution and physical nature. Small particles will readily attach themselves to larger ones and may form a cementing bridge between them. The bond will be strongest when the surface tension is high so that soluble substances in the ore are likely to have a weakening effect on the pellets. Minerals which are thoroughly wetted will form stronger pellets than others which are not readily wetted, unless wetting agents are added where they are needed (but these reduce σ).

Pellet formation involves nucleation and growth. Large particles, picking up small ones, probably act as nuclei. Irregularly shaped particles like fragments of straw have been observed to initiate nucleation and might be added to hasten this phase of formation. Growth rate depends on the flow of young pellets in the machine and size is largely determined by time of residence. It seems to be necessary that each pellet is gently squeezed by its neighbours as it picks up new material so that the loading of the machine is important too. The choice of equipment is between a rotating disc (or a circular tray about 1.5 m in diameter with a raised edge round it several inches high) set at a suitable angle and rotated at a suitable speed, or an inclined drum rotated at a suitable speed. The disc throws off larger particles and is self-sorting. Drums provide a more effective rolling motion. A drum with several baffles has been suggested[19] as a compromise which offers better efficiency and control than either the simple drum or the disc.

Pellets are made from ore ground to -50μm. Pulp must be thickened, filtered and dried to a suitable water content between 5 and 10%. Bentonite may be added up to 1% as a bond. Large-scale production is usually in drums about 2 m in diameter inclined at $5-10°$ from the horizontal and rotated at 15–20 rev/min. The output is screened at 10 mm and undersize is returned while oversize pellets proceed to the hardening kilns. These may be shaft kilns, travelling grates or some hybrid of the two. The pellets must be handled carefully until they are strong. The kiln temperature is about 1100°C and heating should be arranged on the counterflow principle because the pellets cannot be fed to the blast furnace hot. Finished pellets should be strong without having been vitrified and should not exhibit too large an expansion when heated up in reducing gas as this can lead to breakdown in the furnace. Pellets may be made to incorporate some flux but this is not yet commonly practised. Lime cannot be dead burned so that, unless it can be chemically compounded, it will hydrate in the atmosphere and cause the pellets to disrupt. Pellets may also be reduced at least partially during hardening. This policy is not much pursued either, probably because reduction in the blast furnace is more efficient. The loss of weight accompanying this reduction can be advantageous when the pellets have subsequently to be transported over long distances. Complete reduction of pellets is carried out when they are to be used as feed for electric steelmaking furnaces. A high proportion of pelletizing is carried out at the mines where it is not appropriate that the exact requirements of distant ironworks should be anticipated. When large pelletizing plants are built adjacent to blast furnaces greater efforts can be expected to be made to exploit the potential advantages of the process.

Prepared pellets should be strong enough to resist mechanical damage as they pass through the blast furnace — either by fracture or by abrasion. Even if only 5% of dust is ground off 12-mm pellets it can impair the permeability of the burden to a dangerous level. This can be alleviated by using larger pellets but at the cost of reducing the reduction rate and so increasing production costs. Pellets are inherently very reducible but the cost of their production must be met by exploiting that advantage to the full by using as small a pellet as is practicable. A much smaller pellet is sometimes made for use as feed for sintering machines to confer high permeability on the sinter bed. These pellets are not hardened.

Sintering

True sintering involves no fusion, only diffusion. It is applied to the consolidation of metallic and ceramic powder compacts which are heated to temperatures approaching their melting points to allow diffusion to take place at the points of contact of the contiguous particles so that they grow together to produce a rigid entity. The process can be envisaged as a net migration of vacancies into the solid at the highly curved high energy surfaces near to points of contact, and out again at low energy areas away from contact points. The net migration of atoms is in the opposite direction whereby it decreases the free energy of the system as a whole by decreasing the curvature at the contact points, so that a solid bridge is built up at each of these points. As a consequence the aggregate shrinks.

The process known as sintering in ore preparation is also conducted below the melting point, but involves important chemical reactions and in the formation of new chemical species crystallizations occur which provide the main mechanism for the formation of bridges between particles. There is usually some local fluxing also and some sinters have an appreciable amount of glassy bond. In most cases sintering is used for roasting as well as agglomeration, to reduce or eliminate especially sulphide and carbonate by oxidation and dissociation. Sometimes sintering is referred to as "blast roasting". The process may also be used for blending in fluxes needed at a later stage and for absorbing returned sinter fines, flue dust, filter cake, etc.

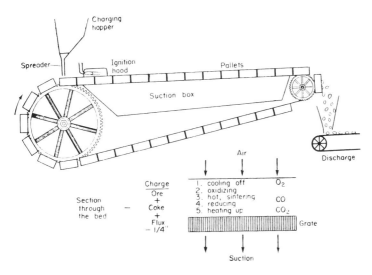

FIG. 36.   Schematic Dwight–Lloyd sintering machine.

Sintering machines are of two types — the Dwight–Lloyd type continuous process (Fig. 36) and the Greenawalt type batch process. The former is more common particularly in large plant but its capital cost is so high that the use of the simple batch process is not unworthy of consideration when steady high production rates cannot be assured. In the batch process each grate sits on its own suction box. In the Dwight–Lloyd machine the grates are in sections or pallets which travel in an endless belt over a series of suction boxes which can be separately controlled. The first of these is under the ignition hood and the last just short of the discharge point. The most important operational control is the permeability of the bed which must be maintained high during the *whole* process. The preparation of the feed and particularly its moisture content are very critical. When the materials available are very fine they may be prepared as small pellets on a disc or in a drum to be fed green on to the sintering grate. These pellets might contain ore, fuel and other additions.

The components of the charge are much coarser than those used for pellet manufacture. Ore should be -15 mm but very little should be $-150\,\mu$m. Coke breeze and limestone should be reduced to -3 mm but rod milling is advocated for the coke to minimize production of dust.

Sulphide ores are sintered pyritically – that is without fuel being added other than the sulphur in the ore. Indeed in these cases the recycling of a large amount of returned fines is essential to keep the temperature on the strand from rising too far. The charge must be mixed in a drum with about 5% of water. The opportunity is taken here to achieve a crumby consistency compatible with a high permeability to the passage of air through the sinter bed. The bed has a depth of 400 to 500 mm. It may be laid down in two layers of different characteristics, e.g. less fuel at the bottom. Ignition is by gas or oil, fired under the hood over the first suction box. In modern plant for sintering iron ores the strand may be up to 5 m wide and 100 m long. It moves at about 7 m/min so that the time taken for material to pass along the strand is about 15 minutes. Outputs of the order of 20,000 tons per day are being achieved. The scale of operation is considerably smaller in non-ferrous practices. After the ignition of the surface a combustion front passes slowly downwards through the bed accompanied by a series of physical and chemical changes which are described below. The temperature reached in the bed is transiently about 1250°C and the sinter cake leaves the grate with the bottom still at about 1000°C. It is usually cooled – in air – crushed if necessary, screened and sent to the blast furnace. Sinter is fed cold to the iron blast furnace and to others where the top is run at a low temperature but the zinc blast furnace is operated with a hot top and requires that sinter is fed at about 800°C. In that case steps are taken to retain as much of the process heat as possible. Gases drawn through the bed must also be cooled and scrubbed free of dust and sulphur dioxide before being released to the atmosphere. These heat losses and the heat loss from the extensive surface of the bed make sintering appear to be a very inefficient process but it should be appreciated that the coke breeze used as fuel is virtually a waste product and is not used in large amount and that the energy used in pyritic sintering is also cheap. The cost of recovering the waste heat would be hard to justify. The sinter-bed temperature is controlled through the fuel content of the mix and the suction rate or in pyritic sintering by dilution of the charge with inactive returned fines which also help to maintain high permeability in the bed. Updraught sintering, using an ignited bottom layer covered by 20–30 cm of sinter mix, is applied in the special case of materials rich in lead minerals. Metal produced by interaction of sulphide and oxide is frozen and retained in the sinter.

At any point in the bed the series of events is as follows. First, the charge is dried out by hot gases coming from the reaction front above. Then the temperature rises as hot gases from the approaching reaction zone sweep through. When the temperature reaches the ignition temperature the combustion front has arrived. Up to this time the hot gases have been reducing and, where appropriate, the particles of ore should have been partially reduced, at least on their surface. As the combustion front passes the conditions become oxidizing. The fuel burns or sulphides are oxidized and reduced oxides are re-oxidized. As the oxides re-form the new crystals forming at points of contact between particles bridge over from one particle to the next producing a permanent bond. The first stage of this process may start during the reduction stage, but this is less likely to be effective because of the lower temperature at that stage.

During the re-oxidation stage a large amount of heat is evolved and the temperature rises so that metal oxides and gangue, particularly silica, react to form low melting silicates which contribute to the bonding as the temperature falls. The later stages of the process are under oxidizing conditions and higher oxides continue to form as long as the temperature is sufficiently high. There is a continuous exchange of heat between the hot sinter and the air being drawn through it. The process is complete when the combustion front reaches the grate.

A good sinter should be strong but capable of being crushed to its required size without generating a large quantity of fines. It should also be chemically reactive, permeable to gases and of large specific surface. In sintered iron ore these requirements are incompatible. The highest strength is obtained when the proportion of glassy silicate bond is high, but this bond is mainly iron silicate which is slow to reduce in the blast furnace. Silica is particularly high in lean ores but if lime is incorporated in the mix the iron oxide may be left relatively free and active (see Chapter 9). In sinter made from rich ore the strength depends on the crystalline bridges of magnetite and haematite and is often disappointingly poor. An improved product may be obtained by the expensive technique of circulating a high proportion (up to 40%) of returned fines. Despite the adverse effect on productivity this practice is widely followed to improve sinter quality from both rich and lean ores.

Nodularizing and Briquetting

Fine ore and process fines from dust catchers, etc., may be agglomerated in a rotating kiln if heated to such a temperature that some of the components just begin to melt. The process is probably rather like pelletizing, but with a slag for bond instead of water. The main difficulty is the tendency of the materials to build up on the lining of the kiln. The method was formerly very popular in ironworks in Europe, but in recent years sinter strands have been preferred.

Briquetting is essentially a mechanical process of agglomeration in which the materials, after mixing and tempering with water and after the addition of any necessary bonding agents, are pressed or extruded into brick or block form. These are then dried and hardened. Hardening is usually by heating but the use of a hydraulic cement allows it to be done cold and the cement can be useful as part of the flux requirement of the subsequent smelting operation. Large briquettes are cheaper to make but more difficult to make sufficiently strong. Briquetting is not popular but interest in the process keeps recurring. It would appear that it is not far off being economically attractive even for iron ores. The best known use of briquettes is as feed for vertical retorts used for the reduction of zinc. In that case the reducing agent is incorporated into the briquette (see page 327).

CHAPTER 6

THERMODYNAMICS – THEORY

THE extraction metallurgist uses thermodynamics as an aid in his study of equilibrium conditions in systems of interest to him. A large body of data has been assembled, mainly in the past 30 years, which enables him to calculate or at least estimate the equilibrium positions of hundreds of reactions under specified conditions of temperature, pressure and physical state, including particularly dissolved states, without further recourse to experiment. These data, which are mainly calorimetric, are still being accumulated. Thermochemical data are also needed for estimating the heat requirements of processes and for preparing heat balances.

In this chapter the basic theory of those parts of thermodynamics which are likely to be of use to metallurgists is briefly outlined. The treatment given lacks the detail and mathematical rigour found in numerous excellent textbooks on chemical thermodynamics. The intention has been rather to show in this chapter and in the next how thermodynamics can usefully be applied. (See also Appendix 1.)

Internal Energy

A thermodynamic system is a portion of matter under consideration which may be "isolated", "closed" or "open". An isolated system, being one which can exchange neither matter nor energy with the universe beyond, is a useful but entirely theoretical concept; a closed system can exchange energy but not matter; an open system can exchange both.

An isolated system has an *internal energy U* which may be present in several forms but cannot be altered in its total quantity. This, the First Law of Thermodynamics, ignores the possibility of converting mass into energy.

A closed system can exchange energy with its surroundings by the transfer of heat or by the system "doing work" on the surroundings and thereby losing energy, e.g. by expansion against the pressure exerted by the surroundings (or vice versa). By these means the temperature, pressure and volume of the system may be seen to have been altered and the system may be said to have been brought into a new "thermodynamic state". Its internal energy would assume a new value characteristic of this new state and U is said to be a "thermodynamic function" of the state of the system and depends only on the parameter's temperature, pressure, etc., and not on the mechanism whereby the state was achieved.

An open system can receive matter from the surroundings so altering its state with respect to its material content. If the matter acquired does not alter the composition, temperature or pressure of the system, the internal energy of the system will be increased in proportion to its increase in volume.

The important forms of this energy are *vibrational energy*, and *chemical bonding energy* or *structural energy*. Vibrational energy is the kinetic energy of all the atoms and molecules present as they oscillate on their mean lattice positions in solids or traverse randomly the space they occupy if gaseous, with the case of liquids being similar to that of gases. The chemical bonding energy is the sum of all the potential energies in bonds joining atom with atom and including energy of attraction between atoms which are not actually immediate neighbours. When energy is transferred between one system and another heat is said to pass between them or work is said to be done by one system on the other. The net effect of such a transfer must be that the sum of the vibrational and structural energies of one system is increased while that of the other is reduced by the same amount. This follows from the First Law if the two closed systems are presumed to form together one isolated system. The terms "passage of heat" and "doing of work" suggest mechanisms of transfer of energy and do not necessarily indicate the form in which the energy is distributed after transfer.

Vibrational Energy and Entropy

Vibrational energy depends upon temperature. The higher the temperature of the system the greater the amount of its vibrational energy.

With a given amount of vibrational energy, however, a system comprising species of low heat capacity (specific heat per mole) will be observed to be at a higher temperature than another system containing species of higher heat capacity. Moreover, the vibrational energy in a liquid at its boiling point is less than that of the same mass of its vapour at the same temperature, by the latent heat of vaporization. A similar, but much smaller, difference occurs at the melting point and indeed at any transition temperature.

The *vibrational energy* of a system is expressed as the product $T \times S$ where T is the temperature in $^{\circ}K$ and S is the *entropy* of the system. All forms of energy can be expressed as a product of a potential factor (here T) and a capacity factor (S in this case). For example, mechanical energy can be measured by the product force x distance or pressure x volume change, and electrical energy as the product potential difference x ampere-seconds. If the student of thermodynamics has difficulty in understanding the physical significance of entropy he should reflect for a moment upon the physical significance of the coulomb or ampere-second when completely divorced from potential difference. The coulomb is accepted because it can be counted using the familiar items, an ammeter and a clock, but without a potential gradient down which to flow it is a particularly meaningless quantity. It is merely that which, when multiplied by volts, gives energy. Similarly entropy is that which, multiplied by temperature, gives energy. It cannot be measured with a meter, but the entropy of a system can be deduced, at least in principle, from a study of the distribution of its matter in space and of its energy among the available energy levels, and entropy change can be measured calorimetrically. Entropy is a measure of the dispersion of the system – of its matter in space and of its energy among possible energy levels. The entropy of any solid increases slowly with temperature as its lattice expands and includes more vacant sites. At the same time the number of energy levels which may be occupied by its atoms or molecules increases and contributes further to a rise in entropy. Each atom occupies a little more space as the temperature rises and it vibrates a little faster and with a slightly higher amplitude. The uncertainty as to the exact position and velocity of any atom at any moment increases – or, in the language of thermodynamics, the system experiences an increase in entropy. A small abrupt increase occurs on melting as the structure loosens and the concentration of vacancies increases sharply. With the addition of the latent heat of fusion the average energy per atom increases and so its velocity, momentum and dynamic entropy. When evaporation takes place the volume occupied increases about a thousand-fold and molecules move with very high velocities on random courses. The entropy gain on evaporation is very large and it continues to increase gradually with the temperature in the gaseous state. The less order in a system, the more random the distribution of its molecules and their energy levels and the greater the entropy.

The concept of entropy is probably best seen in terms of statistical mechanics in which the definition is given that the entropy of an assembly of similar molecules forming a thermodynamic system is given by

$$S = k \ln W \qquad (6.1)$$

where W is the number of ways of distributing the molecules among the various positions and energy levels available. W is obviously increased by reducing the pressure or by raising the temperature. Equation (6.1) accounts only for the entropy due to position. It can be extended to account also for the entropy due to the momenta of the molecules making up a system and this makes it possible, at least in principle, to calculate the total entropy of a system if the total kinetic energy of its molecules and the volume they occupy are known. For the purposes of the present study a quantitative examination of the different kinds of entropy is unnecessary and entropy or entropy change will be considered as quantities which

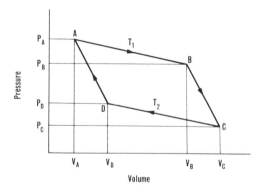

FIG. 37.   The Carnot cycle (see text).

would normally be evaluated from thermal data obtained experimentally (see eqn. (6.23)).

That entropy is a thermodynamic function of state can be confirmed by examining the Carnot cycle (see Fig. 37). One mole of an ideal gas is supposed to form a closed system contained in a very special "engine" in which it is made to go through a four-stage cycle as follows.

A to B	Reversible isothermal expansion	at T_1	Heat supplied $= q_1$	Work done $w_1 = RT_1 \ln (V_B/V_A)$ $= q_1$ (by 1st Law)
B to C	Reversible adiabatic expansion	Cools to T_2	Heat supplied $= 0$	Work done $w_2 = C_V(T_1 - T_2)$
C to D	Reversible isothermal contraction	at T_2	Heat supplied $= -q_2$	Work done $w_3 = RT_2 \ln (V_D/V_C)$ $= -q_2$ (by 1st Law)
D to E	Reversible adiabatic contraction	Heats to T_1	Heat supplied $= 0$	Work done $w_4 = C_V(T_2 - T_1)$ $= -w_2$

C_V is the heat capacity (specific heat per mole) at constant volume and R is the gas constant with the value 8.314 J mol^{-1} K^{-1}. The work done in the isothermal stages, remembering that the working fluid is an ideal gas and that the changes are carried out reversibly, is given by

$$w = \int_{V_A}^{V_B} P\,dV = RT \int_{V_A}^{V_B} \frac{1}{V}\,dV = RT \ln \frac{V_B}{V_A}$$

as shown. This is equal to q since $\Delta U = 0$ for an ideal gas at constant temperature, so that the net work done is given by

$$W = RT_1 \ln (V_B/V_A) + RT_2 \ln (V_D/V_C).$$

Since the cycle returns the gas to its original stage, $\Delta U = 0$ so that $W = q_1 - q_2$. All stages are carried out reversibly which means under potential (i.e. temperatures and pressure) differences which are at all stages infinitesimally small so that the maximum work is done at each stage. The efficiency of working the engine is therefore a maximum, given by

$$W/q_1 = (q_1 - q_2)/q_1.$$

Combining the gas law for adiabatic expansion (PV^γ = a constant) with the equation for Charles' law ($PV = RT$) leads to the equation $V^{\gamma-1}$ = a constant/RT so that

$$V_B^{\gamma-1}/V_C^{\gamma-1} = T_2/T_1 = V_A^{\gamma-1}/V_D^{\gamma-1}$$

and therefore $\quad V_C/V_D = V_B/V_A$.

It follows that $\quad W/q_1 = (q_1 - q_2)/q_1 = (T_1 - T_2)/T_1$

and therefore $\quad q_2/T_2 = q_1/T_1$. $\hfill (6.2)$

Now q/T has the dimensions of entropy and if ΔS is put equal to q_1/T_1 and q_2/T_2 it is clear that the change in entropy from A to C is equal to the change from C to A, so that the total change in the cycle is zero. S like U is returned to its original value on completion of the cycle. It is also an extensive quantity being proportional to the mass of gas present in the system. Entropy has therefore the necessary attributes of a thermodynamic function of state.

It must be noted, however, that q_1 and q_2 are the heats supplied and rejected in the *reversible* changes between A and C and like the work done in the cycle, these are maximum values. A thermodynamic process is conducted reversibly when, at all stages, the system is only infinitesimally different from an equilibrium state. If the temperature is to be raised reversibly from T_2 to T_1 this must be done in an extremely large number of extremely small steps, so that each small addition of heat dq raises the temperature by an increment dT and this change is accompanied by corresponding changes in pressure, volume, etc., before more heat is supplied. Such a process would be infinitely slow and quite impracticable, but the concept is useful and indeed a partial approach toward reversibility in industrial processes can be advantageous. A feature of reversibility is that the process can be reversed at any stage and the system returned to its original state without either the system or its surroundings suffering permanent change. Some processes are inherently irreversible in this respect — diffusion for example being impossible to unscramble without introducing permanent changes in the system or its environment. A simple example of a reversible process is the evaporation of a liquid at its boiling point – i.e. when the pressure over the liquid equals its vapour pressure. Strictly, the partial pressure of the vapour in the atmosphere over a boiling liquid must be equal to its vapour pressure at its boiling point. This condition is reached fairly closely at the surface of a liquid boiling in an enclosed space like a flask or kettle. If the vapour is contained in a cylinder fitted with a piston which can be withdrawn at exactly the correct rate for maintaining the pressure at exactly the vapour pressure of the liquid at the ruling temperature, the evaporation will be reversible. If the vapour pressure is P and a volume ΔV of gas is produced the work done is $P\Delta V$ and this energy is equal to the latent heat of evaporation of the appropriate quantity of the liquid. If the pressure is suddenly reduced to $P - \Delta P$ by withdrawal of the piston through a volume ΔV, and evaporation allowed to continue until the original pressure is restored, the process is no longer reversible and the work done as the same amount of liquid evaporates is clearly less than $P\Delta V$ but greater than $(P - \Delta P)\, \Delta V$. The heat required to evaporate this liquid would also be less than in the reversible case, by the First Law. In general it can be shown that the work done and the heats supplied or rejected are greater during a reversibly conducted process than during the same process conducted in any irreversible manner. Hence it can be written that $q_{rev} > q_{irrev}$ and that the entropy change $dS = q_{rev}/T$ is a limiting value of the quotient q/T which is greater than any other value. Irreversible changes of state between A and C in Fig. 37 are possible involving less heat in which case q/T is not the same as dS but smaller. Irreversible processes are sometimes referred to as "spontaneous" or "observable". In the example given above, when the pressure was reduced, evaporation would obviously proceed spontaneously even without addition of heat from an outside source while the reversible process always depends on an external influence if it is to progress in any particular direction.

The Third Law of Thermodynamics asserts that the entropy of a system comprising a pure perfectly ordered crystalline solid at 0 K is zero. Accepting this assertion it is possible to calculate the entropy per mole of a pure substance at any other temperature from its specific heat and latent heat data—

$$S_{T_n} = \int_0^{T_1} \frac{C_p}{T}\, dT + \frac{L_{T_1}}{T_1} + \int_{T_1}^{T_2} \frac{C_p}{T}\, dT + \ldots \frac{L_{T_{n-1}}}{T_{n-1}} + \int_{T_{n-1}}^{T_n} \frac{C_p}{T}\, dT \quad (6.3)$$

where C_p is the heat capacity (specific heat per mole) at constant pressure and where the integration ranges (except the last) cover temperature ranges between transition temperatures. There is additional entropy, entropy of mixing, in a system containing more than one component, given by equation (6.63) so that even at 0 K a solution cannot achieve zero entropy.

Chemical Bonding Energy

The *bonding energy* is the sum of the potential energies of the constituent atoms in the fields of force imposed by their neighbours or the sum of the energies of the individual bonds between the atoms. Each type of bond (C–C, C–H, H–O, etc.) has a particular bond energy. Among similar types of compound each of these energies has a constant value so that in principle they can be added up throughout a system. These energies are additive so that, in principle, if the structural formula of a compound is known and the energy of each kind of bond involved, the free energy of formation of the molecule can be calculated. This can be demonstrated most effectively in the field of organic chemistry where the free energies of formation of successive members of homologous series are each greater than that of the member below by a similar amount. Resonant bonding and double bonding can sometimes be distinguished by examining the free energies of formation. Vibrational energies are also related to the types and numbers of bonds in the molecules, however, so that similar deductions can be made about structures by examining enthalpies of formation.

In any particular thermodynamic state each chemical species i has a *chemical potential* μ_i (joules/per mole) and the chemical bonding energy or "Gibbs free energy" of a system is calculated as

$$G = \Sigma \mu_i n_i \qquad (6.4)$$

where n_i is the number of moles of the species i present. This free energy is the sum of all the bond energy in the system.

The characteristic value of μ_i may be modified locally by imposed conditions. At boundaries, for example, all valencies are not equally satisfied and this leads to surface tension and surface energy which is a part of the free energy of the system which may be evaluated separately from $\Sigma \mu_i n_i$ for the bulk. Strain energy in metal lattices is rather similar.

Surface energies will be found to have an important influence on many extraction processes – in comminution, agglomeration, flotation and in gas–liquid reactions, to name but a few.

Heat, Work and Enthalpy

The internal energy of a system cannot be evaluated exactly because there is no temperature at which an absolute value can be determined– not even at 0 K. Changes in U can be determined experimentally, however, and values are usually referred to an arbitrary value U^o at some standard state and temperature. Hence

$$U_A = U^o + \Delta U_A \qquad (6.5)$$

where ΔU_A is the change in the internal energy of the system as it is brought from the arbitrary standard state to the state designated A.

Such a system, if closed, may gain (or lose) energy either as *heat* or as *work*. Conventionally heat gained by a system is positive, and work done by a system (on its surroundings) is positive. In a typical process a quantity of heat q might be supplied to a system which would undergo changes which would involve the expenditure of work w in expansion against the resistance offered by the surroundings. The net gain in energy to the system is then expressed as

$$\Delta U = q - w. \qquad (6.6)$$

Work may be electrical or mechanical, the latter being more generally important in chemistry. If, in a reaction, a gas expands against a constant pressure P from a volume V_1 to a volume V_2, the work done is $P(V_2 - V_1)$ or $p\Delta V$. If only this kind of work is being done equation (6.6) can be written

$$\Delta U = q - P\Delta V. \qquad (6.7)$$

It is one of the properties of an ideal gas (to which most real gases approximate especially at high temperatures) that under isothermal conditions its internal energy is independent of the volume it occupies, i.e. $\Delta U = 0$ or $q_T = P\Delta V$. Under adiabatic conditions, $q = 0$ so that $\Delta U = -w$ or $(\Delta U + w) = 0$. Under constant volume conditions $\Delta V = 0$ so that $U = q_V$, and under constant pressure conditions—

$$\Delta U = q_P - P\Delta V \qquad (6.8)$$

or $$q_P = \Delta U + P\Delta V \qquad (6.8a)$$

$$= (U_A + PV_A) - (U_0 + PV_0) \qquad (6.8b)$$

where the initial and final states are designated 0 and A.

Hence $$q_P = H_A - H_0 = \Delta H \qquad (6.8c)$$

where $H = U + PV$ is called the *enthalpy* or *heat content* of the system. Like U, H is a thermodynamic function of state being an extensive quantity and having a unique value for the system in any state irrespective of how the state was attained. This means that the difference in the value of either U or H, i.e. ΔU or ΔH for any change in a system, also has a unique value independent of how the change has been brought about. ΔU and ΔH are called complete or perfect differentials. The quantities q and w, as has already been seen do not have values independent of the route by which the change has been effected. Both values depend on whether a change has been conducted reversibly or not, for example. Therefore dq and dw are not perfect differentials but if all the work done is mechanical work against a constant pressure then it can be evaluated as $P\Delta V$ and with P constant and ΔV a definite quantity. In these particular circumstances $dw = P\Delta V$ is also definite. In the case that the volume is constant $w = 0$ so that $dq = dU$ and dq is also definite.

Enthalpy is important because most chemical reactions are conducted at constant pressure (atmospheric) and the enthalpy change during a reaction is the familiar *heat of reaction*. It will be appreciated that when there is no gaseous component in a system the difference between U and H will be small and that between ΔU and ΔH negligible (unless forms of work other than $P\Delta V$ work are involved).

Like U, H can have no absolute value but must always be compared with an arbitrary value assigned to some standard reference state. At the absolute zero of temperature, where V becomes zero for ideal gases and is very small for other states, U and H tend to become equal in value, though still indeterminate. This basal energy must be principally bonding energy with the atoms at their closest approach but with minimal vibration and the value of $U_0 \simeq H_0$ in any closed system will depend upon which chemical compounds are present at 0 K, the energy of compounds being lower than the sum of that of the constituent elements.

If a quantity of energy q is supplied to such a system so that its temperature is raised to T K reversibly, while the volume increases from 0 to V against an external pressure P, the enthalpy of the system will become

$$H_T = H_0 + q.$$

This will be manifest as vibrational energy and bonding energy so that

$$H_T = TS + \Sigma\mu_i n_i, \qquad (6.9)$$

but as work PV has been done against the surroundings, this has been "banked" outwith the system leaving within it the internal energy U, given by

$$U = TS + \Sigma\mu_i n_i - PV. \qquad (6.10)$$

Free Energy

The *Gibbs free energy* already discussed is now seen to be related to enthalpy by

$$G = \Sigma\mu_i n_i \qquad (6.4)$$

$$= H - TS, \qquad (6.11)$$

being the difference between two thermodynamic functions of state. G is itself a function of state and its differentials are perfect. There is another similar function of state in

$$A = U - TS,$$

$$= G - PV. \qquad (6.12)$$

This is called the *Helmholtz free energy* or *work function* (A for Arbeit). Now "free" energy signifies energy that can be retrieved from the system during an isothermal process and made available for "doing work". If in the course of a particular change the Gibbs free energy of a system is reduced by ΔG we can write

$$w_G = -\Delta G$$

for the energy which becomes available for doing work. At the same time, however, the change in A will be

$$\Delta A = \Delta G - P \Delta V$$

and the energy available for doing work now appears to be

$$w_A = -\Delta A$$
$$= w_G + P \Delta V$$

The explanation of this apparent discrepancy is that A is the total or maximum energy which can be converted into work while the Gibbs free energy is a value which discounts that energy which, under constant pressure conditions, must be absorbed in doing mechanical work against the surroundings, i.e. $P \Delta V$ where the volume change associated with the change in the system is ΔV. This is quite rational because this $P \Delta V$ energy is really no more "available" than additional vibrational energy $T \Delta S$ associated with any increase in volume incurred during the change.

In general the Gibbs free energy G is the more appropriate function to use when a reaction is being carried out at constant pressure, as many chemical reactions are, at one atmosphere, for example. The Helmholtz function A should be used when the reaction being studied is being carried out under constant volume conditions as some important reactions are, in autoclaves. In such a case, of course, $V = 0$ so that $A = \Sigma \mu_i n_i = G$.

The term "work" means in practice mechanical, electrical or even simply thermal work. ΔG is evaluated for a reaction carried out at a particular temperature. This energy is available after allowing for any energy required for expansion of the system. What remains can then be used in an electrolytic cell to deliver a current at the cell voltage round a resistive circuit, as happens in lead–acid accumulators. More often the energy is used in raising the temperature of the system to a higher level as in any furnace. Sometimes this is done with the intention of causing a controlled expansion of the products with the object of developing mechanical energy as in an internal-combustion engine. This discussion has been about isothermal changes without prejudice to what happens to the energy after it has become available. Changes in free energy consequential upon temperature changes will be dealt with later.

Relationships Between the Thermodynamic Functions

Changes in the values of these functions, as a system is brought from one state to another, are more important than their arbitrary values in any one state. These changes are expressed as follows, in equations which are quite independent of any assumed arbitrary values of U or H at 0 K:

$$\Delta H = \Delta U + P\Delta V, \qquad (6.13)$$

$$\Delta G = \Delta H - T\Delta S, \qquad (6.14)$$

$$\Delta A = \Delta U - T\Delta S \qquad (6.15)$$

or in consideration of the fact that each thermodynamic function has a unique value for every possible state, the difference between the values for two states is a perfect differential so that equations (6.14) and (6.15) can be written

$$dG = dH - T\,dS \qquad (6.16)$$

and

$$dA = dU - T\,dS \qquad (6.17)$$

and when no chemical change is involved so that only $P\Delta V$ work is possible,

$$dU_{ni,nj,...} = dH - P\,dV. \qquad (6.18)$$

By implication these equations apply when T and P are both constant.

These functions are evaluated experimentally mainly by calorimetry supplemented by information on bonding energies derived from spectroscopic data or otherwise, and by values derived from chemical equilibrium data. Their values are usually expressed in joules and related to the gramme-molecule (mole) or gramme-atom of pure substances, or, when chemical reactions are involved, may be related to the "gramme-formula weight".

The enthalpy per mole of any single substance is related to its heat capacity at constant pressure, C_p (which can be measured) by Kirchhoff's equation—

$$H_T = H_0 + \int_0^T C_p\,dT \qquad (6.19)$$

provided there is no phase change in the temperature interval $0 - T$. The heat capacity, C_p, is the heat required to raise the temperature of one mole of the substance by one degree Celsius, at constant pressure, and it is related to temperature by an empirical equation having the general form—

$$C_p = a + bT + cT^2 + dT^{1/2} + eT^{-2} \qquad (6.20)$$

of which usually only the first two terms and sometimes one other are found to be necessary. A set of coefficients in this equation is valid only between one transition temperature and the next. At each transition temperature, an increment of enthalpy, equal to the latent heat L_t of the phase change, is added to the substance and above the transition temperature C_p is related to temperature by an equation of the form of (6.20) but with a new set of coefficients. Hence equation (6.19) can be extended thus—

$$H_T = H_0 + \int_0^{T_1} C_p \, dT + L_{T_1} + \int_{T_1}^{T_2} C_p \, dT + \dots \qquad (6.19a)$$

From equation (6.20) it follows that for any reaction or other change in any system at constant temperature, the change in heat capacity is given by—

$$\Delta C_p = a' + b'T + c'T^2 + d'T^{-1/2} + eT^{-2} \qquad (6.21)$$

and by equation (6.19),

$$\Delta H_T = \Delta H_0 + \int \Delta C_p \, dT, \qquad (6.19b)$$

i.e. $\qquad \Delta H_T = \Delta H_0 + \alpha T + \beta T^2 + \gamma T^3 + \delta T^{1/2} + \epsilon T^{-1} \qquad (6.22)$

where a' is the change in a or the difference given by the weighted sum of the a's of the several species present *after* the reaction *minus* the weighted sum of the a's *before* the reaction; b', c', d' and e' are obtained in a similar manner and α, β, etc., are derived from a', b', etc., in the integration. ΔH_0 is a fictitious quantity characteristic of the phases of each species present in the reaction. It is the extrapolated value of ΔH at 0 K assuming that no phase changes have occurred on the way down.

Entropy changes are derived from the same data, using (6.3)

$$\Delta S_T = \Delta S_0 + \int_0^T \frac{\Delta C_p}{T} dT \qquad (6.23)$$

$$= \Delta S_0 + \alpha' \ln T + \beta' T + \gamma' T^2 + \delta' T^{-\frac{1}{2}} + \epsilon' T^{-2}. \qquad (6.23a)$$

From (6.22) it can also be deduced, using the Gibbs–Helmholtz equation (6.43a), that

$$\Delta G_T = \Delta H_0 - \alpha T \ln T - \beta T^2 - \tfrac{1}{2}\gamma T^3 + 2\delta T^{\frac{1}{2}} + \tfrac{1}{2}\epsilon T^{-1} + IT \qquad (6.24)$$

where I is an integration constant which must be evaluated by experiment (see page 157). This apparently complex function, when evaluated at a number of temperatures and plotted against temperature, almost always gives a very close approximation to a straight line and can be condensed graphically to

$$\Delta G_T = A + CT. \qquad (6.25)$$

Any error so introduced is small compared with the uncertainty arising from the basic data used. The constant A is approximately ΔH_T and C is approximately ΔS_T both of which appear to remain fairly constant over temperature ranges which do not include phase changes (see page 162). Actually the small variations in ΔH_T and ΔS_T are of opposite sign and practically cancel out.

It will be obvious that the thermodynamic functions along with P, V and T are very closely interrelated and they give rise to a large number of relationships obtained by applying simple algebraic operations and differentiations – particularly partial differentiations – to the equations already discussed. Keeping the chemical composition constant (as indicated by the suffixes n_i n_j, etc.),

from (6.10),

$$(\partial U)_{n_i, n_j \dots} = T dS - P dV \qquad (6.26)$$

and from (6.9) and (6.26)

$$(\partial H)_{n_i, n_j \dots} = T dS + V dP \qquad (6.27)$$

and from (6.12) and (6.26)

$$(\partial A)_{n_i, n_j \dots} = -P dV - S dT \qquad (6.28)$$

and from (6.11) and (6.27)

$$(\partial G)_{n_i, n_j \ldots} = V \, dP - S \, dT. \tag{6.29}$$

Each of these gives rise to another pair on the addition of a further restriction in each case (again indicated by the suffixes),

$$\left(\frac{\partial U}{\partial S}\right)_{n_i, n_j, \ldots V} = T, \tag{6.30}$$

$$\left(\frac{\partial U}{\partial V}\right)_{n_i, n_j, \ldots S} = -P, \tag{6.31}$$

$$\left(\frac{\partial H}{\partial P}\right)_{n_i, n_j, \ldots S} = V, \tag{6.32}$$

$$\left(\frac{\partial H}{\partial S}\right)_{n_i, n_j, \ldots P} = T, \tag{6.33}$$

$$\left(\frac{\partial A}{\partial T}\right)_{n_i, n_j, \ldots V} = -S, \tag{6.34}$$

$$\left(\frac{\partial A}{\partial V}\right)_{n_i, n_j, \ldots T} = -P, \tag{6.35}$$

$$\left(\frac{\partial G}{\partial T}\right)_{n_i, n_j, \ldots P} = -S, \tag{6.36}$$

$$\left(\frac{\partial G}{\partial P}\right)_{n_i, n_j, \ldots T} = V \tag{6.37}$$

from these in turn can be derived the four Maxwell equations—

$$\left(\frac{\partial P}{\partial T}\right)_V = \left(\frac{\partial S}{\partial V}\right)_T \text{ from (6.34) and (6.35),} \tag{6.38}$$

$$\left(\frac{\partial T}{\partial V}\right)_S = -\left(\frac{\partial P}{\partial S}\right)_V \text{ from (6.30) and (6.31),} \tag{6.39}$$

$$\left(\frac{\partial T}{\partial P}\right)_S = \left(\frac{\partial V}{\partial S}\right)_P \text{ from (6.32) and (6.33)} \tag{6.40}$$

and $\qquad \left(\dfrac{\partial V}{\partial T} \right)_P = -\left(\dfrac{\partial S}{\partial P} \right)_T$ from (6.36) and (6.37). $\qquad$ (6.41)

From these in turn further useful equations can be derived. In particular, combining (6.11) and (6.36) the Gibbs–Helmholtz equation is obtained–

$$H_T = G_T - T \left(\frac{\partial G}{\partial T} \right)_{n_i, n_j, \ldots P} \qquad (6.42)$$

This can be integrated by rewriting it as

$$\frac{H}{T^2} = \frac{G}{T^2} - \frac{1}{T} \left(\frac{\partial G}{\partial T} \right)_{n_i, n_j, \ldots P}$$

and noting that

$$\frac{d(G/T)}{dT} = \frac{1}{T} \left(\frac{dG}{dT} \right) - \frac{G}{T^2}$$

hence

$$\int \frac{H}{T^2} = -\left(\frac{G}{T} \right)$$

or

$$G_T = -T \int \frac{H_T}{T^2} \, dT. \qquad (6.42a)$$

There follow the corresponding equations in ΔG_T and ΔH_r

$$\Delta H_T = \Delta G_T - T \left(\frac{\partial \Delta G}{\partial T} \right)_{n_i, n_j, \ldots P} \qquad (6.43)$$

and

$$\Delta G_T = -T \int \frac{\Delta H_T}{T^2} \, dT. \qquad (6.43a)$$

The importance of this lies in the fact that from a knowledge of the free energy change and its temperature coefficient the heat of a reaction can be calculated. The temperature coefficient can be obtained by differentiating equation (6.24) if it is available. It also means that a heat of reaction can be deduced from an e.m.f. measurement (see page 140) and its temperature coefficient.

Thermodynamic Functions and Chemical Change

Thermodynamic data should provide criteria for equilibrium and should indicate the direction in which spontaneous changes may occur. Spontaneous changes are irreversible changes. In reversible changes (taking place under equilibrium conditions) it has been shown (page 111) that $\Delta S = q_{rev}/T$ whereas for irreversible or "spontaneous" or "observable" changes $\Delta S > q/T$. By combining these with equation (6.14) expressed as $\Delta G = q - T\Delta S$, it can be shown that for equilibrium $\Delta G = 0$ and for spontaneous change $\Delta G < 0$. (These apply under constant pressure conditions. Under constant volume conditions ΔA replaces ΔG.)

This means that the (appropriate) free energy tends to a minimum value at equilibrium, and this is in accord with the Second Law of Thermodynamics which says that the direction of spontaneous change in an inanimate system is such that the system can thereby be made to do work on its surroundings. In other words some free energy must become available as "q" which can be converted to "w".

It may be noted also that while for an irreversible change in a closed system $\Delta S > q/T$ the loss of entropy by the surroundings is likely to be almost equal to q/T because the surroundings lose the q from a vast heat reservoir which suffers an infinitesimal drop in temperature, so that the heat transfer *from* the surroundings is almost reversible. There is therefore a *net increase in entropy*, taking the system and its surroundings together, accompanying all ordinary reactions and a strong implication that the entropy of the universe is for ever increasing.

Consequent upon equation (6.14) either loss of heat content or gain in entropy tends to produce spontaneous reactions. At very low temperatures $T\Delta S$ is so small that only exothermic reactions are spontaneous. At high temperatures even endothermic reactions may occur spontaneously if $T\Delta S$ is large enough to ensure that ΔG is negative. Unless ΔH and ΔS are of opposite sign, the sign of ΔG remains ambiguous and actual values are needed if the direction of a reaction is to be deduced.

At constant temperature equation (6.29) reduces to

$$dG_{ni,nj,...T} = V\,dP \qquad (6.44)$$

and if the system consists of n moles of a single ideal gas i, $V_i = nRT/P_i$ so that the equation (6.44) becomes

$$dG_i = nRT(dP_i/P_i).$$ (6.44a)

Integrating from the lower limit at which the gas is in a standard state at 1 atm pressure,

$$G_i - G_i^\circ = nRT \ln P_i$$ (6.44b)

the affix $^\circ$ indicating the value at the standard state which cannot, of course, be determined. From this it follows that the chemical potential of the gas, or its Gibbs free energy per mole, is—

$$\mu_i = \mu_i^\circ + RT \ln P_i.$$ (6.44c)

$RT \ln P_i = \mu_i - \mu_i^\circ$ is referred to as the chemical potential of i with respect to the prescribed standard state. If the species i is a gas it is customary but not obligatory to refer to it at one standard atmosphere pressure as its standard state − that is, the chemical potential of i at one atmosphere pressure $= \mu_i^\circ$. If i is one of several gases present along with $j, k, \ldots$, etc., and P_i is its partial pressure, equation (6.44c) is still valid for i while $\mu_j, \mu_k, \ldots$, etc., can be calculated in a similar way. ΣP_i is, of course, the total pressure of the system, while the Gibbs free energy is obtained from

$$G = \Sigma n_i \mu_i = \sum \frac{P_i}{\Sigma P_i} \mu_i$$

for ideal gases. If the species i is either solid or liquid and its vapour pressure over this condensed phase is P_i, equation (6.44c) can still be applied and P_i can still be referred to the standard state at one atmosphere pressure. It is often convenient, however, to adopt the pure condensed phase as the standard state − or the vapour in equilibrium with it at pressure P_i, the two being, of course, at exactly the same chemical potential. A change of standard state cannot be made without altering the (indeterminate) value of μ_i°, but this has no effect on μ_i for any other state under consideration. Clearly the second term on the right-hand side of equation (6.44c) must be adjusted to compensate for the change in μ_i°. This is done in principle by altering the lower limit from which the integration of (6.44c) is carried out to the pressure appropriate to the new standard state. This matter is considered in some detail below. The use of the pure solid or liquid as the standard state is particularly convenient when solid or liquid solutions are being considered.

Thermodynamic Functions and Chemical Equilibrium

If several substances are involved in a chemical reaction such as

$$aA + bB = cC + dD$$

$$\Delta G = c\mu_C + d\mu_D - a\mu_A - b\mu_B$$

$$= (c\mu_C^\circ + d\mu_D^\circ - a\mu_A^\circ - b\mu_B^\circ) + RT(\ln p_C^c + \ln p_D^d - \ln p_A^a - \ln p_B^b)$$

$$= \qquad \Delta G^\circ \qquad + RT \ln \frac{p_C^c \cdot p_D^d}{p_A^a \cdot p_B^b}. \qquad (6.45)$$

At equilibrium, $\Delta G = 0$,

therefore $\qquad \Delta G_T^\circ = - RT \ln \dfrac{p_C^c \cdot p_D^d}{p_A^a \cdot p_B^b} = - RT \ln K_T \qquad (6.46)$

where K_T is the equilibrium constant of the reaction.

This relates the standard free energy change – that referring to the condition that all of the reactants and products shall be in their standard states as somewhere defined – with the reaction constant K_T expressed as a function of the vapour pressures of the reactants and products when actually in equilibrium. The standard state in this case is, implicitly one atmosphere for all species.

Combining equation (6.46) with the Gibbs–Helmholtz equation (6.43a) in the form $d/dT(\Delta G^\circ/T) = -\Delta H^\circ/T^2$ leads to the *van't Hoff reaction isochore* –

$$\frac{d \ln K_P}{dT} = \frac{\Delta H^\circ}{RT^2} \qquad (6.47)$$

which shows quantitatively how K_p increases with temperature if the forward reaction is endothermic and vice versa. This is one expression of the Le Chatelier principle which states that if a change occurs in one of the factors such as temperature or pressure under which a system is in equilibrium the system will tend to adjust itself in such a way as will diminish the effect of the change.

ideal gas 0.0224 m³ or 273 k
gives press 1 atm
contains 6.0228 x 10²³ molecules
Non ideal gas, press < 1 atm

Solutions and Activities

The theory developed so far has applications only to reactions occurring between ideal gases and has therefore no direct application to metallurgical problems. When gases being considered are not ideal the values of the pressure p in equations (6.44) to (6.47) must be replaced by the *fugacity* f – an "effective pressure" which can be determined only by experiment. This is due to association between gaseous molecules. If one mole of an ideal gas like H_2 or N_2 is adjusted to a volume of 0.0224 m³ at 0°C its pressure will be found to be one standard atmosphere and we can be certain that there are 6.0228×10^{23} (Avogadro's number) molecules present. If a non-ideal gas like SO_2 were used in the same experiment the pressure would be found to be less than one atmosphere because some of the SO_2 would have associated or polymerized to form S_2O_4, S_3O_6 or even larger molecules. The total number of molecules falls short of Avogadro's number so that although each molecular species may well be behaving ideally the pressure developed falls short of the standard value. This is because there is not in the equilibrium mixture of 64 g of what we recognize as "SO_2", a whole mole of matter. There is generally a high degree of association at temperatures just above the melting point and again at very high pressures but at high temperatures and at low pressures and particularly when these conditions occur together, as they do in a high proportion of cases in which the metallurgist is interested, the behaviour of gases closely approaches that of ideal gases so that fugacity need seldom be considered. Most metallurgical reactions, however, involve condensed phases which appear not to be catered for in the equations so far derived. This difficulty is overcome, in effect, by considering that the reactions occur between the vapour phases in equilibrium with each of the condensed phases and therefore at the same chemical potential as the condensed phases. No particular mechanism is assumed for these reactions. Complete reliance is placed on the fact that phases in equilibrium are at the same chemical potential as each other, so that an equilibrium position worked out for the gaseous phases will also be an equilibrium position for the solid and liquid phases with which the gases are in equilibrium. Since the vapour pressures of the solid and liquid substances of interest in extraction metallurgy are often very low, the further assumption that their gaseous phases behave ideally is usually quite valid.

In practice the vapour pressures of condensed phases are seldom evaluated. Instead, pressure is replaced by an "effective concentration" or *activity* a which, like f, must be evaluated by experiment. The activity of a component i in a solution at a particular temperature is given by

$$a_i = \frac{\text{vapour pressure of component } i \text{ over the solution}}{\text{vapour pressure of component } i \text{ when } i \text{ is in some standard state}}$$

$$_r a_i = \frac{\text{vapour pressure of component } i \text{ over the solution}}{\text{vapour pressure of the pure component } i \text{ at the same temperature}}$$

If the concentration is expressed as a mole fraction x_i and the solution is "ideal" or "Raoultian", i.e. behaves like a mixture of perfect gases, and obeys Raoult's law, $_r a_i = x_i$, which is, of course, a quantitative expression of that law. More often solutions are not ideal because the attractive forces between like and unlike molecules (or atoms) are different. Tendencies toward immiscibility or toward compound formation are reflected in positive and negative "deviations" from ideality (see page 136).

The activity of a species is related to its mole fraction in a binary system $A - B$ by a curve of one of the forms shown in Fig. 38. When B is the solvent the curve always comes close to the ideal curve or Raoult's law line so that as $x_B \to 1$, $_r a_B \to x_B$. In this range and near it, it is convenient to put pure B as the standard state. Then reading the strong curve at $x_B = 0.9$, $_r a_B \cong 0.9$ but at $x_B = 0.5$, $_r a_B \cong 0.2$ and the ratio $_r \gamma_B = _r a_B / x_B = 0.4$ is called the *activity coefficient* at that composition. The prefix r is used here to distinguish Raoultian activity and activity coefficient, that is when the standard state is the pure substance.

When x_B is small and B is the solute, reference to pure B as the standard state becomes inconvenient. As $x_B \to 0$, however the curve is again straight in a short range in which the solution obeys Henry's law. Activity is proportional to mole fraction, i.e. $a_B = \gamma_B^\circ x_B$ where γ_B° is the Henry's law constant (see Fig. 38). If the standard state is deemed to be at point R the activity in that non-physical state being put to unity, then in

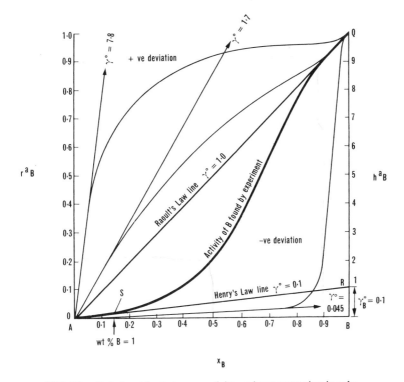

FIG. 38. Relationships between activity and concentration in solutions. The discussion in the text refers to the heavy line. The light lines indicate alternative shapes in which these curves may be found.

the range in which the curve for the activity of B approaches the Henry's law line, the "Henrian" activity $_h a_B \to x_B$ as $x_B \to 0$ and $_h a_B = x_B$ as far as that law is valid. Under this convenient arrangement when $x_B = 0.01$, $_h a_B = 0.01$ in the case shown in Fig. 38 but this correspondence fails at about $x_B = 0.17$; when $x_B = 0.5$, $_h a_B = 2.0$ and when $x_B = 1.0$, $_h a_B = 10$ which would usually be inconvenient. Under this arrangement the activity coefficient is designated $_h \gamma_B = {}_h a_B / x_B$ which at $x_B = 0.5$ has the value 4.0. Note that $_h a_B = {}_r a_B / \gamma_B^\circ$ and $_h \gamma_b = {}_r \gamma_B / \gamma_B^\circ$.

A third possibility is to choose point S as the standard state (which may be physical or non-physical depending on the shape of the curve) where the Henry's law line intersects the ordinate through 1% wt B. Then as $x_B \to 0$, the modified Henrian activity $_{\%}a_B \to$ wt$\% B$ and $_{\%}a_B =$ wt$\% B$ as far as Henry's law is valid or as far as the wt$\% B$ can be accepted as being proportional to x_B, whichever comes first. The values of wt$\% B$ and $_{\%}a_B$ are related to those of $_h a_B$ and $_r a_B$ through the molecular weights of the two components in the following manner. If a binary solution contains x_B moles of B which has a molecular weight M.W.$_B$ and $(1 - x_B)$ moles of a which has a molecular weight M.W.$_A$ then

$$\text{wt}\% B = \frac{x_B \text{M.W.}_B}{x_B \text{M.W.}_B + (1 - x_B)\text{M.W.}_A} \times 100$$

$$= \frac{x_B \text{M.W.}_B}{\text{M.W.}_A + x_B(\text{M.W.}_B - \text{M.W.}_A)} \times 100. \qquad (6.48)$$

If $x_B(\text{M.W.}_B - \text{M.W.}_A)$ is small enough to be neglected,

$$\text{wt}\% B = x_B \frac{100\text{M.W.}_B}{\text{M.W.}_A}. \qquad (6.48a)$$

When x_B is small enough for Henry's law to be valid $x_B = {}_h a_B$ and in a limited range wt$\% B = {}_{\%} a_B$ so that

$$_{\%}a_B = {}_h a_B \frac{100\text{M.W.}_B}{\text{M.W.}_A} = {}_r a_B \frac{100\text{M.W.}_B}{\gamma^\circ \text{M.W.}_A}. \qquad (6.48b)$$

Clearly the approximation at (6.48a) is always justified when x_B is very small but the upper limit of x_B to which it can be applied depends on the ratio $(\text{M.W.}_B - \text{M.W.}_A)/\text{M.W.}_A$ and should be tested with numbers in any particular case. At concentrations of B higher than that at which the approximation can no longer be justified the wt$\% B$ is not proportional to x_B nor to $_h a_B$, so that if $_{\%}a_B$ continues to be put equal to wt$\% B$ it is no longer proportional to the other values of activity. This is not a satisfactory situation and it is necessary to limit the equality of wt$\% B$ to $_{\%}a_B$ to the range of concentration at which the approximation (6.48a) can be justified. If values of $_{\%}a_B$ are calculated from (6.48b) at higher concentrations of B they will not be numerically equal to the wt$\% B$ value even if they are still on the Henry's law line. Much of the advantage of using the 1% wt standard state is lost when activity and wt$\%$ are no longer numerically the same but beyond the range of maximum usefulness an activity coefficient $_\%\gamma_B = {_\%}a_B/\text{wt}\% B$ is available. This coefficient is not, however, simply related to the Henrian activity coefficient because it must account not only for deviations from ideality but also for the non-linear relationship between mole fraction and wt%. The 1% wt standard state is most usefully applied to dilute solutions of industrial significance such as molten steel or other metals at their refining stage. It is used in the case of blast furnace iron also, but the total amount of solute is so great in this case that it is doubtful that the underlying assumptions are all justified. (See page 135.)

Whatever standard states are used, equation (6.46) is valid in the form

$$\Delta G_T^\circ = -RT \ln K_T = -RT \ln \frac{a_C^c a_D^d}{a_A^a a_B^b} \qquad (6.46)$$

but the actual numerical values of K_T and ΔG_T° depend on the standard states adopted, which determine whether a_i is $_r a_i$, $_h a_i$ or $_\% a_i$, or even some other value.

Consider the chemical potential of B in a particular solution in the solvent A and suppose that it has the absolute value μ_B. Then, referring to the pure gaseous B at 1 atm pressure as the standard state, by equation (6.44c)

$$\mu_B = \mu_B^{\circ(\text{gas})} + RT \ln p_{B(\text{over solution})} \qquad (6.49)$$

where $\mu_B^{\circ(\text{gas})}$ is the absolute but indeterminate value of the chemical potential of the gas at 1 atm pressure and $p_{B(\text{over solution})}$ is the vapour pressure of B over the solution under consideration at T K. The chemical potential of pure solid B can be referred to the same standard state, if the vapour pressure of pure B, $P_{B(\text{pure solid})}$, is known thus:

$$\mu_B^{\circ(\text{solid})} = \mu_B^{\circ(\text{gas})} + RT \ln p_{B(\text{pure solid})}.$$

Now if the chemical potential of B in the solution is referred to that of the pure solid as the new standard state,

$$\mu_B = \mu_B^{\circ(\text{solid})} - RT \ln p_{B(\text{pure solid})} + RT\, p_{B(\text{over solution})},$$

i.e. $\quad \mu_B = \mu_B^{\circ(\text{solid})} + RT \ln {_r}a_B. \qquad (6.50)$

Since the absolute values of the chemical potentials of B cannot be evaluated in either of the standard states, the terms $RT \ln p_B$ or $RT \ln {}_r a_B$ become the measure of the chemical potential of B relative to the selected standard state, i.e. $RT \ln a$ is the difference between the chemical potential in the dissolved state and that in the selected standard state. Since $RT \ln p_B$ and $RT \ln {}_r a_B$ above have different numerical values μ_B referred to gaseous B at 1 atm and μ_B referred to pure solid B have different values but only because the reference point has altered.

Referring again to Fig. 38, equation (6.49) can be rewritten

$$\mu_B^Q = \mu_B^{\circ(Q)} + RT \ln {}_r a_B. \qquad (6.50\text{a})$$

Consider now a change in standard state from Q in Fig. 38 to the non-physical state represented by point R. The absolute potential is not, of course, altered but its value, relative to that of the new standard state, is reduced by the difference between the chemical potentials of the new standard state and the old one, i.e. by

$$\mu_B^{\circ(R)} - \mu_B^{\circ(Q)} = RT \ln \gamma_B^{\circ} \qquad (6.50\text{b})$$

so that $\qquad \mu_B^Q \quad$ or $\quad \mu_B = \mu_B^{\circ(R)} - RT \ln \gamma_B^{\circ} + RT \ln {}_r a_B. \qquad (6.50\text{c})$

Hence $\qquad \mu_B \quad$ or $\quad \mu_B^R = \mu_B^{\circ(R)} + RT \ln \dfrac{{}_r a_B}{\gamma_B^{\circ}}, \qquad (6.50\text{d})$

i.e. $\qquad \qquad \mu_B^R = \mu_B^{\circ(R)} + RT \ln {}_h a_B. \qquad (6.51)$

If the standard state is again moved, to S, the value of the chemical potential relative to that at S is smaller again, this time by

$$\mu_B^{\circ(S)} - \mu_B^{\circ(R)} = RT \ln({}_h a_B \text{ when wt\% } B = 1)$$

$$\cong RT \ln(0 \cdot 01 \text{M.W.}_A / \text{M.W.}_B) \qquad (6.51\text{a})$$

where M.W. signifies molecular weight and x_B is small. It follows that—

$$\mu_B^R = \mu_B = \mu_B^{\circ(S)} - RT \ln(0 \cdot 01 \text{M.W.}_A / \text{M.W.}_B) + RT \ln {}_h a_B,$$

i.e. $\qquad \mu_B = \mu_B^S = \mu_B^{\circ(S)} + RT \ln {}_\% a_B. \qquad (6.52)$

Obviously in the general case of equation (6.48), ΔG_T must be the difference of the sums of the chemical potentials of the products and reactants each relative to its own standard state and the corresponding value of K_T is the ratio of the activities of the products and reactants referred to the same standard states. The standard states used should always be clearly indicated but this is not, unfortunately, always done in the literature so that the unwary reader is liable to find himself in occasional difficulties. The suffixes r, h and $\%$ are not in common use, the standard state usually being indicated or implied in some other manner. Henrian activity, for example, is sometimes designated h.

It will be appreciated that the chemical potential of any substance in a particular thermodynamic state has an indeterminate value which is expressed as the sum of the equally indeterminate value when the substance is in an arbitrarily selected reference state and a difference term which is the change in chemical potential which can be observed and measured when the substance is transferred from the reference state to the state under consideration; and that a change from one standard state to another results in the values of the indeterminate standard state potential and the difference term changing in opposite directions by exactly the same amount.

When two or more phases are in equilibrium with each other, the chemical potential of each species present is the same in each phase, i.e. for a species i in phases I and II, which are in equilibrium,

$$\mu_i = \mu_{i(\text{I})}^{\circ} + RT \ln a_{i(\text{I})} = \mu_{i(\text{II})}^{\circ} + RT \ln a_{i(\text{II})}$$

where $\mu_{i(\text{I})}^{\circ}$ and $\mu_{i(\text{II})}^{\circ}$ are the chemical potentials of the reference states appropriate to the species i in the two phases. Should these reference states be chosen so that they are in equilibrium with each other then $\mu_{i(\text{I})}^{\circ} = \mu_{i(\text{II})}^{\circ}$ so that under any other conditions in which the phases are in equilibrium with each other $a_{i(\text{I})} = a_{i(\text{II})}$. (See page 151.) More usually the standard states in the two phases are not chosen so that their chemical potentials are equal and consequently the activities of species in solution in phases which are in equilibrium are not necessarily numerically equal. In the general case,

since $\quad \mu_{i(\text{II})}^{\circ} - \mu_{i(\text{I})}^{\circ} = RT \ln \dfrac{\text{(activity of } i \text{ in the standard state of phase II}}{\text{referred to the standard state of phase I)}}$

(cf. equation 6.50), then

$$\frac{a_{i(I)}}{a_{i(II)}} = \begin{array}{l}\text{(activity of } i \text{ in the standard state of phase II} \\ \text{referred to the standard state of phase I)}.\end{array} \quad (6.53)$$

The ratio of the concentrations of a species i in two phases which are in equilibrium with each other is sometimes referred to as the partition coefficient L_i. It will be obvious that this is not a fundamental function but its value is related to the ratio of the activities in equation (6.53) through the appropriate activity coefficients $\gamma_{i(I)}$ and $\gamma_{i(II)}$, and depends upon the units in which the concentrations are expressed. The important point is that it is the chemical potentials of each component in the two phases that are equal at equilibrium. The rest is arithmetic.

Activities in Multi-component Solutions

The treatment of activities in multi-component systems is more difficult than in binary systems because interactions between atoms of different solutes usually alter their activities from the values they would have had in binary solutions at the same concentrations. In a few cases (see page 226) activities have been determined by experiment over large parts of ternary systems but in general this approach is inadequate to meet the degree of complexity likely to be encountered in industrial applications. Two methods of estimating activity coefficients in dilute solutions (e.g. molten iron or steel) are indicated below.

If the solution contains mole fractions $x_P, x_Q, x_R \ldots$ of species P, Q, $R \ldots$ respectively, dissolved in a solvent A the activity coefficient of any one of them, say $_h\gamma_P = {}_h a_P/x_P$, can be determined by Wagner's method.[14] Let the activity coefficient of P in a binary solution with A containing mole fraction x_P be $_h\gamma_P^P$ and let the addition of Q up to x_Q alter this activity coefficient to $_h\gamma_P^{PQ}$ so that—

$$_h\gamma_P^{PQ} = {}_h\gamma_P^P \times {}_h\gamma_P^Q \quad (6.54)$$

where $_h\gamma_P^Q$ is a factor which can be found by experiment. There will be corresponding formulae for $_h\gamma_P^{PR}$, etc. It has been observed that the factors like $_h\gamma_P^Q$ are usually substantially independent of the concentrations of the solutes other than, in this case, Q but that $\log {}_h\gamma_P^Q$ is usually proportional to x_Q as long as this is small so that

$$\frac{\partial \ln {}_h\gamma_P^Q}{\partial x_Q}$$

is a constant within the Henry's law range and at constant temperature. Wagner showed that these constants can be incorporated in a Taylor series expansion of $\ln {}_h\gamma_P^*$, the activity coefficient of P in the multi-component solution—

$$\ln {}_h\gamma_P^* = \ln {}_h\gamma_P^\circ + x_P \frac{\partial \ln {}_h\gamma_P^P}{\partial x_P} + x_Q \frac{\partial \ln {}_h\gamma_P^Q}{\partial x_Q} + x_R \frac{\partial \ln {}_h\gamma_P^R}{\partial x_R} + \ldots$$

$$+ \tfrac{1}{2}x_P^2 \frac{\partial^2 \ln {}_h\gamma_P}{\partial x_P^2} \text{ and other second-order terms.} \tag{6.55}$$

At infinite dilution the second order terms may be neglected and the term $\ln {}_h\gamma_P^\circ$ can be made zero by putting ${}_h\gamma_P \to 1$ at that condition. Then putting the terms

$$\frac{\partial \ln {}_h\gamma_P^Q}{\partial x_Q} = \epsilon_P^Q, \text{ etc.,}$$

$$\ln {}_h\gamma_P^* = x_P\epsilon_P^P + x_Q\epsilon_P^Q + x_R\epsilon_P^R + \ldots \tag{6.56}$$

and only the values of the "interaction coefficients" ϵ have to be obtained by experiment. Since $x_Q\epsilon_P^Q = \ln {}_h\gamma_P^Q$, equation (6.55) can be extended in logarithmic form to give the Henrian activity of P in the complex system

$$\ln {}_h a_P^* = \ln({}_h\gamma_P^* x_P) = \ln {}_h\gamma_P^P + \ln {}_h\gamma_P^Q + \ln {}_h\gamma_P^R + \ldots + \ln x_P$$

$$= \ln {}_h\gamma_P^P + x_Q\epsilon_P^Q + x_R\epsilon_P^R + \ldots + \ln x_P. \tag{6.57}$$

The corresponding formulae in terms of ${}_\%\gamma_P = {}_\%a_P/\text{wt\%} P$ are

$$\log {}_\%\gamma_P^* = e_P^P \cdot \text{wt\%} P + e_P^Q \cdot \text{wt\%} Q + e_P^R \cdot \text{wt\%} R + \ldots \tag{6.58}$$

where $e_P^Q = \dfrac{\partial \log {}_\%\gamma_P^Q}{\partial \text{wt\%} Q}$, etc., so that $e_P^Q \cdot \text{wt\%} Q = \log {}_\%\gamma_P^Q$ and

$$\log {}_\%a_P^* = \log {}_\%\gamma_P^P + e_P^Q \text{ wt\% } Q + e_P^R \text{ wt\% } R + \ldots + \log \text{wt\% } P. \qquad (6.59)$$

Obviously these may be repeated to find ${}_na_Q^*$, ${}_na_R^*$, and ${}_\%a_Q^*$, ${}_\%a_R^*$, etc., but the calculation of these may be simplified by use of the relationships—

$$\epsilon_P^Q = \epsilon_Q^P, \qquad (6.60)$$

$$e_P^Q = e_Q^P \cdot \text{M.W.}_P/\text{M.W.}_Q \qquad (6.61)$$

and $$e_P^Q = \epsilon_P^Q \frac{\text{M.W.}_P}{\text{M.W.}_Q} \cdot \frac{1}{230.3}. \qquad (6.62)$$

These formulae are strictly applicable only at infinite dilution but can be used in the region of the validity of Henry's law to give a fair estimate of activities required. Where practicable an occasional experimental check would justify their use. Note that the validity of Henry's law is likely to be more restricted as the number of solutes and the total amount of solute increase but no check on this is usually possible.

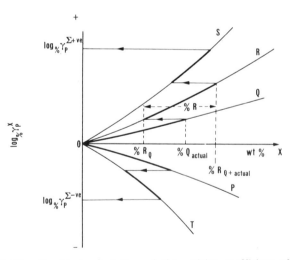

FIG. 39. Graphical calculation of the activity coefficient of a solute P in the presence of other solutes Q, R, S and T (see text).

A graphical method suggested by Morris and Buehl[21] gave closer agreement with experimental values than the analytical method when applied to pig iron[22] in which total solutes may reach 10%. Graphs of $\log_\% \gamma_P^X$ versus wt% X are drawn as in Fig. 39 for $X = Q, R, S$, etc., giving lines of increasing positive slope, and for $X = P, T, \ldots$, etc., giving lines of increasing negative slope. To find $\log_\% \gamma_P^*$ first find the value of wt% R_Q whose $\log_\% \gamma_P^X$ value is the same as that of wt% Q_{actual}. To this add wt% R_{actual} and find the value of wt% S_{Q+R} which has the same $\log_\% \gamma_P^X$ value as that of wt% $R_{(Q+actual)}$. Add wt% S_{actual} and repeat in order through all the lines with positive slope to find $\log_\% \gamma_P^{\Sigma+ve}$. Repeat for the lines of negative slope to find $\log_\% \gamma_P^{\Sigma-ve}$ and finally obtain $\log\% \gamma_P^*$ as the algebraic sum of these two. The construction lines on Fig. 39 show that this can all be done graphically much more easily than it can be explained in words.

Solutions and Thermodynamic Functions

Referring back to Fig. 38 it has been noted that solutions may be ideal, or, more usually, display either positive or negative deviations from Raoult's law. In an ideal solution the attractive forces between like and unlike atoms or molecules are identical, and the entropy of mixing ΔS_m can be calculated by statistical mechanics for complete randomness as—

$$\Delta S_m = \Delta \bar{S}_A + \Delta \bar{S}_B = -R(x_A \ln x_A + x_B \ln x_B) \qquad (6.63)$$

where the symbols under the $^-$ indicate partial molar quantities and x, A and B have the usual meanings for a binary solution. ΔS_m is numerically positive and is a maximum when $x_A = 1 - x_B = 0.5$ (see Fig. 40).

In analogy with equation (6.49) the free energy of dilution of each species as it enters the solution is $\Delta G_i = RT \ln x_i$ per mole so that—

$$\Delta G_m = \Delta \bar{G}_A + \Delta \bar{G}_B = RT(x_A \ln x_A + x_B \ln x_B). \qquad (6.64)$$

Consequently, by (6.14)

$$\Delta H_m = \Delta \bar{H}_A + \Delta \bar{H}_B = 0 \qquad (6.65)$$

and it can also be shown that there is no volume change when an ideal solution is formed. The vapour phase over such a solution contains molecules of A and B in numbers proportional to x_A and x_B so that

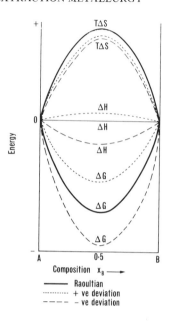

FIG. 40. The thermodynamic functions of mixing for ideal solutions (full lines) and for solutions showing positive and negative deviations from Raoult's law. (Not to scale.)

partial pressures are proportional to mole fractions. Raoultian activities are equal to mole fractions.

Normally the attractive forces between like and unlike atoms or molecules are not the same and there is some tendency toward either ordered lattice formation or compound formation on the one hand or immiscibility on the other.

If the forces between the unlike atoms are higher than those between like, atoms of one kind are preferentially surrounded by the other kind (as far as these are available) and in this way the minimum free energy is achieved in the system. ΔG_m is in this case more negative than if the solution were ideal. At the same time the spatial assembly is less random than the ideal one so that ΔS_m is not so large as if the solution were ideal. Consequently ΔH_m must be negative. In fact because ΔH_m is negative the number of available energy levels is smaller and ΔS_m is even smaller than would be expected from spatial considerations alone. Atoms, by their association in a low energy state with the other kind, become less available for evaporation or reactions than those at the same concentration in ideal solution, i.e. there is a negative deviation from Raoult's law.

In a similar manner, when the attraction between like atoms is stronger than between unlike, they are surrounded preferentially by their own kind. ΔG_m is less negative than the value calculated for ideal solutions. The spatial arrangement is again less random than in the ideal case but the effect on $T\Delta S_m$ is less than that on ΔG_m so that ΔH_m becomes positive and the total entropy change may be little different from that of the ideal case because of an increase in the number of energy levels in the system. Vapour pressures and activities are higher than in the ideal case, because atoms are more like undissolved atoms, and the deviation from Raoult's law is positive.

In either of these non-ideal cases the appropriate relationships are—

$$\Delta G_m = \Delta \overline{G}_A + \Delta \overline{G}_B = RT(x_A \ln a_A + x_B \ln a_B) \qquad (6.66)$$

(cf. equation (6.64)),

$$\Delta H_m = \Delta \overline{H}_A + \Delta \overline{H}_B = RT^2 \left[x_A \left(\frac{\partial \ln a_A}{\partial T} \right)_{x_B} + x_B \left(\frac{\partial \ln a_B}{\partial T} \right)_{x_A} \right] \qquad (6.67)$$

(from the Gibbs–Helmholtz relationship, equation (6.43))

and $\qquad \Delta S_m = \dfrac{\Delta H_m - \Delta G_m}{T}$

$$= \frac{x_A \Delta \overline{H}_A}{T} + \frac{x_B \Delta \overline{H}_B}{T} - R(x_A \ln a_A + x_B \ln a_B). \qquad (6.68)$$

Regular Solutions

In non-ideal solutions, in the range of compositions in which neither Raoult's law nor Henry's law is obeyed, it is difficult to find useful relationships between activities or activity coefficients and mole fractions of the components. In principle one might expect there to be some general relationship of which the cases so far discussed were but special cases but any such relationship has so far eluded detection. There is, however, a class of solutions which is called "regular". In these the deviation from ideality is moderate. The heat of mixing is not zero but the entropy of mixing is indistinguishable from the value that would be calculated for an ideal solution. The simple test of regularity is that

$$RT \ln \gamma_A = bx_B^2 \qquad (6.69)$$

where b is a constant which is independent of composition and is also assumed to be independent of temperature. This equation is derived from statistical mechanics. It is found to be satisfied by a wide range of binary systems particularly among organic compounds but including some metallic as well as oxide systems (slags). If a solution can be shown to conform to this equation it is possible to estimate activity coefficient values at temperatures other than those at which measurements were actually made. If regularity is assumed, integration of the Gibbs–Duhem equation can be completed and the composition/activity curve of the second component in a binary solution can be calculated after that of the first component has been determined, but this can usually be carried out more easily by alternative methods. Evidence of regularity provides some evidence about structure. Where equation (6.49) does not apply over the whole range of composition it may be found that it is valid over a limited range of composition near $x_A = 1$ or $x_B = 1$ or separately in both regions. The solution will then be called semi-regular and within the limited composition range a constant b can be used for extrapolations and for entering into other thermodynamic relationships. Other solutions have a lower degree of regularity in which b is not a constant but a function of composition. Clearly the more complex the relationship the less useful it is likely to be.

The Gibbs–Duhem equation mentioned above can readily be derived from first principles and in one of its more useful forms is expressed as

$$x_A \, d \ln a_A + x_B \, d \ln a_B = 0. \qquad (6.70)$$

If the relationship between x and a for one component is known, that for the other component can be calculated if the equation can be successfully integrated and valued. Integration gives–

$$\ln a_A = - \int_0^{x_B} \frac{x_B}{x_A} \, d \ln a_B \qquad (6.70a)$$

or $$\ln \gamma_A = - \int_0^{x_B} \frac{x_B}{x_A} \frac{d \ln \gamma_B}{dx_B} dx_B. \qquad (6.70b)$$

Graphical evaluation of (6.70a) by plotting x_B/x_A against $\ln a_B$ is difficult because $x_B/x_A \to \infty$ as $x_A \to 1$. It is at this point that the assumption of regularity can sometimes be useful.[23] Another method[24] requires equation (6.70b) to be integrated by parts to give–

$$\ln \gamma_A = - \int_0^{x_B} \frac{\ln \gamma_B}{x_A^2} dx_B - \frac{x_B}{x_A} \ln \gamma_B \qquad (6.70c)$$

which can be graphically evaluated with only a small amount of extrapolation. The Gibbs–Duhem equation can also be applied to ternary systems using a modified form. [24]

Thermodynamics and Electrochemical Reactions

It is well known that electric currents can be produced by electrolytic cells comprising two electrodes in an ionic solution. One of the electrodes is usually consumed by solution yielding electrons which pass through a conducting wire to the other electrode where they discharge positive ions. A net chemical reaction occurs which is the sum of the two electrode reactions. In the Daniell cell the anode* is zinc immersed in dilute sulphuric acid and the cathode is copper in dilute cupric sulphate solution, the two solutions being in electrical contact through a wall of porous pot.

At the anode, $Zn^0 \to Zn^{++} + 2e^-$.

At the cathode, $Cu^{++} + 2e^- \to Cu^0$

and the net reaction is

$$CuSO_4 + Zn = ZnSO_4 + Cu. \qquad (6.71)$$

If the electrodes are connected by a conductor of zero resistance there is no e.m.f. between them and the current passing can do no work. If, however, a potentiometer is interposed it will measure an e.m.f. under conditions of zero current. This e.m.f. will have a maximum value charac-

*The anode is the electrode toward which the anions migrate and the cathode that toward which the cations migrate.

teristic of the net reaction under reversible conditions. The potentiometer makes its measurement by opposing the cell e.m.f. with a known variable e.m.f. until no current flows. The cell e.m.f. is then equal to the opposing e.m.f. If the measured value is E and the opposing e.m.f. is lowered to $(E - \Delta E)$ a small current will flow. If this is allowed to pass until 1 g atom of zinc has dissolved, $2F^*$ coulombs of electricity will have passed and $2F(E - \Delta E)$ watt-seconds of electrical energy will have been generated. The corresponding free energy loss from the system will be at least $2F(E - \Delta E)$. If, however, the applied e.m.f. is raised to $(E + \Delta E)$ and 1 gramme-atom of zinc deposited with the passage of 2 faradays of electricity, the amount of electrical energy supplied is $2F(E + \Delta E)$ watt-seconds and the free energy gain cannot exceed this amount of energy. Now if ΔE is made to approach zero on each side it is obvious that for the reaction (6.71) carried out reversibly

$$\Delta G = -nFE \qquad (6.72)$$

where n is the number of electrons per atom (here 2) and ΔG is in joules.
In general

$$\Delta G = \Delta G^\circ + RT\Sigma \ln a_i. \qquad (6.45)$$

If the cell is organized so that all components are in their standard states the e.m.f. measured would then be E°, related to the standard free energy according to—

$$\Delta G^\circ = -nFE^\circ \qquad (6.72a)$$

and hence

$$E = E^\circ - \frac{RT}{nF} \Sigma \ln a_i \qquad (6.73)$$

where

$$E^\circ = \frac{RT}{nF} \ln K_T.$$

The standard states are usually the pure elements for the electrodes and 1 atmosphere pressure for any gas involved in the reactions. The standard state for acids, bases or salts in aqueous solutions is the ideal state of unit molality (1 mole per 1000 g solvent). This choice has the advantage that the molal strength is independent of temperature but it is inconvenient

*The faraday (F) = 96,489 coulombs.

because the standard state is unreal. This is also the case for the Henrian standard state and others. The inconvenience arises in this case because the real molal solution is within the range of concentrations met in practice, but there is no way of making the activity and concentration numerically equal as in other systems. That correspondence occurs only at very low concentrations below the usual range of interest. It is not possible to estimate the probable value of a salt or acid in aqueous solution. A suitable reference book must be consulted.[68] Activities referred to this particular standard state are called mean molal activities. They are presented in this book as $_{aq}a_i$ and the coefficient as $_{aq}\gamma_i$ to distinguish them from activities with respect to other standard states. When the electrolyte is a molten salt or a solution of oxides the Raoultian standard state is usually suitable. E.m.f. can be measured more accurately than any quantity of heat so that this theory offers the possibility of making very accurate determinations of free energy changes – and hence of changes in the values of other thermodynamic functions and of activities. Unfortunately it is not easy to design cells in which no side reactions occur so that although some very accurate measurements have been made in this way the method is rather restricted in its applications. Values of E can, of course, be derived from thermal data if they are not available by direct measurement.

In Europe, Standard Electrode Potentials are conventionally quoted as the *negative* of the value calculated from ΔG_T° above. The sign of the correction term must obviously also change, so that equation (6.73) becomes

$$E = E^\circ + \frac{RT}{nF} \Sigma \ln a_i. \qquad (6.73a)$$

In American literature the sign derived from thermodynamics is retained in electrochemistry but is incompatible with the conventional polarity in the electrical industries. Unfortunately these differences lead to much confusion especially when different textbooks are being used together. The student who is aware of the difficulties is in a strong position to avoid them particularly if he is willing to go back to first principles occasionally to confirm that his calculations are leading to a reasonable conclusion.

Any cell reaction can be split into two partial reactions (or groups of partial reactions) which can be assigned to the anode and the cathode respectively (6.71). Each of these has its free energy change and its corresponding electrode potential on a scale of volts in which the potential of the "standard hydrogen electrode" is zero (see page 309). The e.m.f. of a cell is the difference between the electrode potentials on that scale — the cathodic potential minus the anodic potential. Thus, in the Daniell cell discussed above, the potential at the cathode is the standard electrode potential for copper, +0.34 V, and that at the anode is the standard electrode potential for zinc, −0.76 V, so that the cell voltage is 1.10 V.

The standard hydrogen electrode is in the form of a piece of platinum coated with colloidal "platinum black" which catalyzes the reaction $H_2 = 2H^+$ between a stream of gaseous hydrogen at one atmosphere pressure and hydrogen ions in aqueous solution at normal concentration. Other more practical standard electrodes such as the calomel electrode are available as reference standards — "half-cells" of known "half-cell potential" which can be coupled with half-cells of unknown potential for the purpose of determining first experimentally the e.m.f. of the combination and then, by difference, the unknown value. The e.m.f. of any half-cell does not depend on what other half-cell it is coupled with but is dependent on the temperature and on the concentrations of its components.

The standard electrode potential E°_{Cu} for pure copper in a standard solution of cupric ions is +0.34 V corresponding to the partial reaction $Cu^\circ = Cu^{2+} + 2e^-$ for which $\Delta G^\circ_{298} = 65,000$ J. This value must be changed, however, either if the copper is impure or if the solution is different from the standard concentration, the adjustment being made by equation (6.73a), i.e. in this case

$$E_{Cu} = E^\circ_{Cu} + \frac{RT}{nF} \ln \frac{a_{Cu^{2+}}}{a_{Cu^\circ}}$$

so that E_{Cu} becomes less positive on dilution. Since the appropriate value of R is 8.314 in volt-coulombs/°C mole, the value of F is 96,500 coulombs, the usual temperature for reference is 298 K, and for Cu, $n = 2$, further simplification is possible to

$$E_{Cu} = E^\circ_{Cu} - 0.0295 \log \frac{a_{Cu^{2+}}}{a_{Cu^\circ}}$$

and in general the correction term for activities is

$$-\frac{0.059}{n} \Sigma \log a_i$$

so that a change in the activity of the ions in solution by a factor of 10 will alter the cell e.m.f. by only 0.059 V if n = 1, 0.0295 if n = 2 and so on.

Care must be taken when applying equation (6.73a) because the standard solution at which the electrode potential is E^o is not the standard state to which published values of $_{aq}a_i$ are referred. E_i^o is measured in a solution of strength 1 g-ion l^{-1} in which $_{aq}a_i$ would never be unity, the value for the activity of i which is implied in equation (6.73a). If the activity–concentration relationship for i is known and if $_{aq}a_i$ has the value $_{aq}a_i^*$ at the standard concentration of 1 g-ion l^{-1}, equation (6.73a) can be modified by putting

$$a_i = \frac{_{aq}a_i}{_{aq}a_i^*}$$

when i is an ionic species. In effect this moves the standard state for i to the standard solution of strength 1 g-ion l^{-1}. If reliable activity data are not available, a_i must be replaced by the concentration c_i in g-ion l^{-1}. This has a similar effect in that the standard state is moved to the standard solution strength but is less satisfactory because it involves an implicit but unjustified assumption that there is a linear relationship between concentration and activity.

Thermodynamics and Phase Diagrams

Phase diagrams, being graphical representations of equilibrium conditions, are quantitatively related to the thermodynamic functions of the systems they describe. The Clausius–Clapeyron equation is well known for its use for calculating the elevation of the boiling point of a liquid, on the addition of a solute. It is in fact the van't Hoff equation (6.47) applied to evaporation–

$$\frac{d \ln p_1}{dT} = \frac{\Delta H_e}{RT_1^2} \qquad (6.74)$$

where p_1 is the vapour pressure of the solvent at T_1 K and ΔH_e is its molar heat of vaporization. A small addition of a solute up to the mole fraction x_s will alter the vapour pressure by an amount, calculable by Raoult's law, from p_1 to p_2 so that $(1 - x_s) = p_2/p_1$. ΔH_e can be shown to be unchanged so that x_s can be related to the two temperatures T_1 and T_2

at which the vapour pressures over the solvent are p_1 and p_2, i.e.

$$\frac{d \ln(1 - x_s)}{dT} = \frac{\Delta H_e}{R} \left(\frac{1}{T_2^2} - \frac{1}{T_1^2} \right).$$

Integrating and putting $\ln(1 - x_s) \simeq -x_s$ (when small)

$$x_s = \frac{\Delta H_e}{H} \left(\frac{1}{T_1} - \frac{1}{T_2} \right) \simeq \frac{\Delta H_e}{RT_1^2} (T_2 - T_1)$$

since $T_2 \simeq T_1$. Putting $(T_2 - T_1) = \Delta T$,

$$\frac{\Delta T}{x_s} \cong \frac{RT^2}{\Delta H_e}. \tag{6.75}$$

This equation has been derived on the assumption that Raoult's law is obeyed, that ΔH_e is constant, and that the two approximations made in the later stages are justifiable. It shows that the elevation in the boiling point of a liquid is the same for equal *molar* additions of all solutes and it

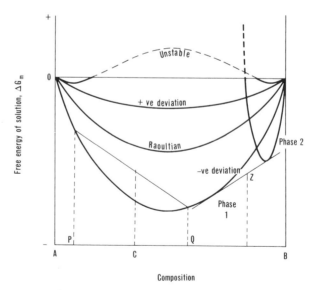

FIG. 41. Free energy of solutions showing how the most stable state can be determined when either one or two phases may be present.

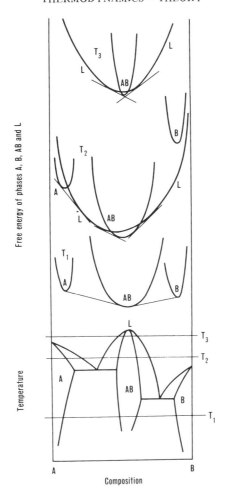

FIG. 42. Relationships between free energies and phase diagrams.

affords a means of determining molecular weights of species as they are actually present in solution.

Very similar arguments can be used to apply the Clausius–Clapeyron equation to sublimation, melting and other phase changes including the precipitation of phases at solubility limits, p being replaced where appropriate by activity a and ΔH_e by the molar heat of the appropriate transformation. Relationships analogous with equation (6.75) can be used to calculate at least the initial slopes of solidus, liquidus and other phase boundary lines. If a phase diagram has been determined by an independent method the same relationship can occasionally be used to calculate the activity of a solute, at least within a short range of concentration.

The free energy–composition curve at constant temperature for any solution has one of the forms shown in Fig. 41. Curves which are concave downwards indicate immiscibility or separation into two phases, whereas curves which are concave upwards indicate stability in the solution particularly when the concavity exceeds that for an ideal solution.

Considering the lowest curve at any composition such as C, the *single phase* whose composition is C has obviously a lower free energy and is therefore more stable than any mixture of solutions such as P and Q which might have the same net composition. (This applies only when the curve is concave up.) It also follows that if a second phase is present in the system it must be possible to find the free energy of the mixture of the two phases on a common tangent to the separate free energy curves as indicated at Z in Fig. 41. The points at which this tangent touches the free energy curves indicate the composition of each of the phases when they are in equilibrium with each other and the proportions present will be in the inverse ratio of the segments of the common tangent.

Following this preamble the relationship between the free energy–composition curves for the phases present and the phase diagram shown overleaf in Fig. 42 should be clear. It is not being suggested that this is a normal way of determining phase diagrams but it is possible to use free energy data to check or improve phase diagrams or to use reliable phase diagrams from which to deduce elusive thermodynamic quantities.

CHAPTER 7

THERMODYNAMICS – APPLICATIONS

APART from the application of thermochemical data to energy balances the most important use of thermodynamics in extraction metallurgy is the prediction of equilibrium conditions from free energy data through equations (6.46) and (6.48). A minimum possible temperature may be sought for a reaction or a temperature at which one mineral will decompose, but not another, to permit differential roasting, for example.

Alternatively a critical composition for the atmosphere in a furnace or reaction vessel may be required for the same purpose. The equilibrium concentrations of impurities may be required corresponding to different conditions in a refining operation; or it may be necessary to calculate the amount of a reagent which must be supplied to achieve a particular result, this being the sum of the stoichiometric requirement and the excess necessary to impose the chemical potential needed to maintain the equilibrium at the desired level. (See examples 14 and 15 in Appendix 1.)

Before the equations can be used figures must be found to replace the symbols. A large amount of data has been produced during the past 30 years and is continually being augmented and improved. The principal difficulty arising in the application of these equations is the selection of standard states and the evaluation of the appropriate values of activities, especially in complex solutions. Another difficulty encountered by "beginners" in the field is knowing when it is justifiable to make estimates and simplifying assumptions such as seem to be made so confidently by the established experts. This is largely a matter of following their example (with understanding) and some personal trial-and-error experimentation. An intelligent guess at an activity value, for example, may lead to a reasonable estimate of a reaction temperature whereas no guess would lead to no answer at all. It is, of course, always necessary to be clear – and to make clear – what is based on guesswork and what on hard fact and if

147

possible to estimate the probable accuracy of quantities calculated from the thermodynamic data. (See also page 228 and ref. 64.)

Experimental Methods for Evaluating Thermodynamic Functions

The basic information has been obtained by calorimetry — specific heats or heat capacities of elements and simple compounds from near 0 K upwards; heats of transformation; heats of reaction. The other important class of information concerns solutions — the chemical potentials and activities of components in solution in slags, mattes and alloys. These data are obtained in many ways from calorimetry (heats of solution), vapour-pressure measurements, equilibrium determinations and e.m.f. measurements.

Vapour-pressure measurements are made easily only when the values are high as over amalgams and zinc alloys when simple or differential manometers may be used. More often the dew-point method or methods involving rate of evaporation or rate of effusion through a tiny aperture (as in the Knudsen cell) are used for their greater sensitivity. It is essential that the whole of the pressure measured can be attributed to one species, however, so that the effusion cell could be used for measuring the activity of silica in a $CaO-SiO_2$ melt over which the vapour pressure of CaO is negligible but not in Na_2O-SiO_2 melts because Na_2O has a high vapour pressure at quite low temperatures.

Equilibrium studies often involve equilibration with a gas at controlled pressure or with a gaseous mixture, the activities of whose components can be calculated from composition if ideality can be assumed — as it usually can be at high temperatures. In the case of a mixture of CO and CO_2 an equilibrium is reached (under suitable conditions) according to—

$$2CO + O_2 = 2CO_2 \; ; \Delta G_T^\circ = -565,528 + 174T \text{ J} \qquad (7.1)$$

the equilibrium constant for which is

$$K_T = \frac{p_{CO_2}^2}{p_{CO}^2} \cdot \frac{1}{p_{O_2}} . \qquad (7.2)$$

At $1500°C = 1773$ K,

$$\Delta G_{1773}^\circ = -257,030 \text{ J} .$$

By equation (6.46)

$$K_T = \exp -(\Delta G_T/RT)$$

$$= \exp -(-257{,}030/8.315 \times 1773)$$

$$= 3.7 \times 10^7.$$

At $1500°C$ when $CO/CO_2 = 1, p_{O_2} = 2\cdot7 \times 10^{-8}$ atm. Clearly a simple gas analysis provides a means of measuring an extremely low p_{O_2} value and hence the "oxygen potential", with respect to O_2 in its standard state at 1 atm pressure, as $RT \ln p_{O_2}$ in accordance with equation (6.44c). Carbon monoxide circulating over a metallic oxide MO at an elevated temperature will react with it until equilibrium is reached in the reaction

$$MO + CO = M + CO_2.$$

At equilibrium and provided that both the metal and oxide phases are present at the end of the experiment, if the standard state of each of these phases is taken as being the pure substance and that of each gas as the gas at one standard atmosphere pressure, then the equilibrium constant is

$$K = \frac{p_{CO_2}}{p_{CO}}.$$

If this ratio is determined experimentally, it can be combined with equation (7.2) above to yield the corresponding value of p_{O_2} for the temperature at which the experiment was carried out. For success it is necessary that the CO_2/CO ratio is neither so high nor so low that an accurate analysis cannot be performed. It is also assumed that the metal and oxide phases are pure at the end of the experiment, and in particular that the metal has not taken into solution either oxygen or carbon. It will be appreciated that this method of determining the value of p_{O_2} characteristic of a metal/oxide pair is not universally applicable – but all methods have their limitations. A similar method for making the same measurement is to equilibrate with H_2/H_2O mixtures. The advantage of using gas mixtures is, of course, that ideal behaviour can fairly be assumed for them over a wide range of conditions.

When p_{O_2} has been measured, the oxygen potential at equilibrium, $RT \ln p_{O_2}$, can be calculated. This is, or 'course, the oxygen potential for the CO_2-CO equilibrium mixture and also for the metal–oxide pair coexisting in equilibrium in any proportions. It is also the free energy of formation of the oxide from the metal—

$$2M + O_2 = 2MO$$

for which $\qquad K_T = 1/p_{O_2}$

and $\qquad \Delta G_T^{\circ} = -RT \ln K_T = RT \ln p_{O_2}$

(see page 124).

In a similar way an H_2/H_2S mixture can be passed over molten iron until equilibrium is reached for the reaction

$$H_2 + S \text{ (in Fe)} = H_2S \qquad (7.3)$$

for which $\qquad K_T = \dfrac{p_{H_2S}}{p_{H_2}} \cdot \dfrac{1}{a_s} \qquad (7.4)$

where a_s is the activity of the sulphur in solution in the metal. Now

$$K' = \frac{p_{H_2S}}{p_{H_2}} \cdot \frac{1}{\text{wt} \% \text{ S}}$$

is not constant but, assuming Henry's law to hold, as wt % S $\to 0$, $_\%a_s \to$ wt % S and $K' \to K$. If log K' is plotted against wt % S and the graph extrapolated to find log K as its value when wt % S = 0 the true value of K can be found and hence also the value of $_\%a_s$ corresponding to any value of the H_2/H_2S ratio. This experimental determination of the activity of sulphur was extended[17,18] to complex ferrous alloys and led to the determination of interaction coefficients for carbon, silicon, manganese, phosphorus and other elements affecting the activity of sulphur in iron.

It will be noted that in this case, although the chemical potential of sulphur is the same in the gas and metal phases at equilibrium because of the choice of standard states, the activities are not numerically equal and there is no convenient way of expressing the potentials so that they appear to be equal.

The partition of an element between two immiscible liquid phases can be used to determine its activity in one phase if that in the other is known. The activity of iron oxide in a slag, relative to pure FeO as the standard state, can be determined by equilibrating the slag with pure iron and then analysing the iron for oxygen. Pure FeO comes into equilibrium with iron containing 0.23% oxygen at 1600°C. Assuming Henry's law to be obeyed up to saturation in the Fe–O solution (which is very nearly the case) and adopting the saturated solution in pure Fe as the standard state for oxygen in the metallic phase,

$$\text{sat}a_O = \frac{\text{wt \% O in Fe}}{0.23}$$

(see page 131).

Like the Raoultian activity this must obviously have a range of values from zero to one, but it is clearly not exactly the same as the Raoultian activity so that the prefix "sat" (for saturation) is adopted here to acknowledge the difference. If, in the oxide or slag phase, the standard state for oxygen is in combination with Fe as pure FeO, it is clear that the two standard states are at the same chemical potential and therefore (see page 131) when the slag and metal phases are in equilibrium at any other oxygen potential,

$$\text{sat}a_O = {}_ra_{FeO} = \frac{\text{wt \% O in Fe}}{0.23} \; .$$

In carrying out this work its authors had to devise means of containing both iron and slag in one crucible without contamination. The solution was to use a rotating magnesia crucible in which vortex action brought the surface of the iron into a paraboloidal form in which the slag could be held out of contact with the crucible wall.[25]

This device of using the saturated solution as the standard state for the solute is quite common, particularly among iron alloys. Considering pig iron with about 4% carbon, the saturated solution with about 5% carbon is a convenient standard state but when dealing with liquid steels with about 0.1–1.0% carbon it is more convenient to use the 1% solution by weight as the standard state – or even 0.1%.

This particular case has been examined by bringing molten iron into equilibrium with carburizing mixtures of CO and CO_2. From the reaction

$$CO_2 + C \text{ (in Fe)} = 2CO, \tag{7.5}$$

$$K_c = \frac{p_{CO}^2}{p_{CO_2}} \cdot \frac{1}{a_c} \; . \tag{7.6}$$

If p^2_{CO}/p_{CO_2} can be determined when $_{sat}a_c = 1$, i.e. in equilibrium with a saturated solution of carbon in iron, the value of K_c can be determined from which to calculate $_{sat}a_c$ corresponding to other values of p^2_{CO}/p_{CO_2}. Similarly if p^2_{CO}/p_{CO_2} can be found when wt % C has a known value in the range of validity of Henry's law another value of K'_c can be obtained from which $_{\%}a_c$ can be calculated from the same data. This case is examined numerically as a worked example on page 409 where note is taken of the limited range of validity of the equality between wt % C and $_{\%}a_C$.

Partition between immiscible metal phases provides another possible method of determining activity and the coupling of molten iron and silver has been used with some success[26] for investigating interactions between carbon in iron and copper and manganese as second solutes. The technique works best when the activities of the solutes being examined are already known in the second solvent. As more data become available this technique may become very valuable for extending them further.

A useful source of equilibrium data lies in equilibrium diagrams. In another study[27] of the Fe—S system the free energy of melting of pure iron was calculated as a function of temperature from the latent heat of fusion (15,360 J/g-atom), at the melting point (1535°C) and the change in heat capacity on melting (1.26 J K^{-1}).

Using equation (6.19) to obtain an expression for ΔH_T° and evaluating ΔS_{1808}° from 15,360/1808 and hence ΔS_T° by equation (6.23) ΔG_T° for the reaction Fe$_{(\delta)}$ = Fe$_{(liq)}$ is obtained by equation (6.14) as—

$$\Delta G_T^\circ = \mu_{Fe(liq)} - \mu_{Fe(\delta)} = 13,100 - 1.26T \ln T + 2.2T \text{ J} \qquad (7.7)$$

where the standard state is the pure liquid iron. It follows that at 1400°C (1673 K), $\Delta G_{1673}^\circ = RT \ln a_{Fe(\delta)} = -1134$ J so that log $a_{Fe(\delta)} = -0.035$ and $a_{Fe(\delta)} = 0.9225$ relative to the supercooled liquid at the same temperature.

Referring now to the liquidus on the Fe—S diagram (see Fig. 43) it will be seen that this must also be the Raoultian activity of the liquid iron solution containing 6.9% S the liquidus temperature of which is 1400°C. At this composition the mole fraction x_{Fe} is 0.87 so that the activity coefficient r^γ_{Fe} has the value 1.06 and log $_r a_{Fe(\delta)} = 0.025$. This calculation can be repeated for a range of temperatures and compositions but these are necessarily tied to the liquidus curve. To obtain activity coefficients and hence activities at temperatures above the liquidus at any

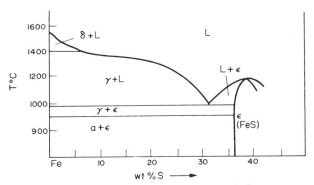

FIG. 43. Part of the Fe–S equilibrium diagram.[28] (From *Metals Handbook*, 7th ed., 1948. Amer. Soc. Metals.)

composition, the solution was assumed to be regular, so that by equation (6.69) log $_r\gamma_{Fe}$ is inversely proportional to T K. When activity values of Fe are established in the molten solution the Gibbs–Duhem equation can be used to calculate the activities of sulphur. In this way Chipman demonstrated in 1948[28] how quite a small number of measured values could be manipulated to yield a large amount of information. The derived values can be no more reliable than the experimental results on which they depend, however, and in a case like this where the liquidus curve is of very long standing its revision might show these derived values to be in serious error without the curve itself having been radically altered.

The design of electrolytic cells in which to make e.m.f. measurements for conversion into free energy values is difficult. Side reactions, polarization, contamination, and thermo-electric effects have all to be avoided so that good results are obtainable only under rather favourable circumstances. One system which can be studied fairly easily by this method is the $PbO–SiO_2$ system, at least within the narrow range of temperature, available due to difficulties in finding a crucible material to withstand the attack of PbO.[29] An anode of "Nichrome" wire dips into the pure lead and a cathode of platinum touches the surface of an oxide layer covering the lead and is flooded with a stream of oxygen. The partial reactions at the electrodes are $2Pb \rightarrow 2Pb^{2+} + 4e$ and $O_2 + 4e \rightarrow 2O^{2-}$ when the electrodes are connected by a conductor. When a voltmeter of infinite resistance is interposed the reactions take place reversibly so that the measured e.m.f. is the E of equation (6.72). If however the three components Pb, PbO and O_2 are all in their standard states pure Pb, pure PbO, and O_2 at 1 atmosphere pressure, the value of the e.m.f. measured is E° from which ΔG° can be calculated by equation (6.72a). This is ΔG_T° for the net reaction

$$2Pb + O_2 = 2PbO$$

at the temperature at which the measurement was made. If the measurement is repeated using a $PbO-SiO_2$ slag in place of the molten PbO, the measured e.m.f. will be rather lower, being—

$$E = E^\circ + \frac{RT}{nF} \cdot 2 \ln {_r}a_{PbO} \text{ (cf. (6.73a)).}$$

Hence it is possible to determine both the free energy of formation of the oxide and its activity over a range of compositions in solution in silica using a fairly simple technique (see also page 228). Recent developments in the field of solid electrolytes may enable a wider range of systems to be studied by this technique with success. Values of E° can, of course, be derived from thermal data if available.

Whatever methods are used to determine the values of thermodynamic functions it is always desirable that cross-checks should be made by independent methods. It can hardly be over-emphasized that even small errors can be greatly amplified in calculations so that the sooner an empirical confirmation can be obtained for a value initially calculated from remote measurements, the more satisfactory the situation becomes.

Sources of Thermodynamic Data

Thermodynamic data have been collected in a number of publications, a few of the more accessible of which are listed in the reference list.[30-37] They are usually presented in tabular form. Heat capacities are given in the form of equation (6.20), the coefficients a, b, c, d and e being tabulated along with a note on the valid temperature range, an assessment of accuracy and references to sources. Heats of transformations, fusion and evaporation are also listed for elements and simple compounds with notes on temperature, accuracy and sources. Some heats of reactions are also listed, usually for simple reactions of the type $A + B = AB$, i.e. heats of formations of compounds from their elements. Entropy changes are also sometimes listed for simple reactions. Both of these are calculated for a standard temperature 25°C or 298·16 K, using equations (6.19) and (6.3) and they are designated ΔH_{298} and ΔS_{298}, whether the reaction can occur or the relevant phases exist at that temperature or not. The enthalpy of all elements is conventionally zero at 25°C to enable this calculation to be made. Standard free energy changes for selected reactions are also tabulated in the form $\Delta G_T^\circ = A + BT \log T + CT$, values of A, B and C being tabulated with temperature range, accuracy and source. From these the value of ΔG° can readily be calculated at any temperature. The units adopted are J/mole for heat of formation but J/g-formula weight for ΔG so that the value quoted depends on how the equation is expressed.

There is no systematic presentation of activity coefficients available but in one textbook[38] interaction coefficients of interest in ferrous metallurgy are listed in an appendix. Henrian activities in dilute binary and more complex alloys of iron can be extracted from this compilation, but in general information about activities must usually be sought in original papers. (See also page 228.)

Estimation and Calculation of the Values of Thermodynamic Functions

If values sought are not to be found in reference books they must be estimated, calculated or measured (by methods already discussed).

Estimates can be made on the basis of well-established empirical rules, a few of which are given here.

1. For solids, Dulong and Petit's law gives heat capacity per g atom as about 26 J K^{-1} at room temperature (or rather less in substances of high melting point) rising to about 30 at the first transition temperature.

2. The heat capacities of liquids are only slightly higher than those of solids near the melting point — by, say, 0.8 J atom.

3. For monatomic gases $C_v = 12.5$ J K^{-1} mol^{-1} and $C_p = 21$ J K^{-1} mol^{-1}.

4. For diatomic gases C_p ranges from 29 at room temperature to 33 for light gases and 38 for heavy gases at about 2000°C.

5. The entropy of vaporization of all substances is, by Trouton's rule, approximately $92 \text{ J K}^{-1} \text{ mol}^{-1}$ so that if the boiling point T_e is known the latent heat of vaporization can be calculated as $L_e = 92 \, T_e \text{ J mol}^{-1}$.

6. For metals the entropy of fusion is about $9.2 \text{ J K}^{-1} \text{ g-atom}^{-1}$, so that the heat of fusion can be estimated as in (5) if the melting temperature is known.

These and other approximations are discussed critically in reference 30. A typical calculation would be that of the standard free energy of the reaction

$$CaCO_3 = CaO + CO_2 \qquad (7.8)$$

from the heat capacities and heats of formation of each of the products and reactants. It will be seen that one other piece of information is needed to complete the evaluation, which will now be carried out.

The heat capacities[20] in $\text{J mol}^{-1} \text{K}^{-1}$ are—

for CaO $\qquad C_p = 41.85 + 20.1 \times 10^{-3} T - 4.5 \times 10^5 T^{-2}$,

for CO_2 $\qquad C_p = 26.0 + 43.5 \times 10^{-3} T - 14.8 \times 10^{-6} T^2$,

and for $CaCO_3$ $\quad C_p = 82.9 + 49.8 \times 10^{-3} T - 12.9 \times 10^5 T^{-2}$,

whence $\qquad \Delta C_p = C_{p \, CaO} + C_{p \, CO_2} - C_{p \, CaCO_3}$

$$= -14.5 + 13.9 \times 10^{-3} T$$

$$+ 8.4 \times 10^5 T^{-2} - 14.9 \times 10^{-6} T^2. \qquad (7.9)$$

The heats of formation[30] in J mol^{-1} are:

for CaO $\qquad \Delta H_{298} = -635,700$,

for CO_2 $\qquad \Delta H_{298} = -393,400$

and for $CaCO_3$ $\quad \Delta H_{298} = -1,206,950$ from Ca, C and $1\frac{1}{2}O_2$,

whence for the reaction, by Hess's law

$$\Delta H_{298} = \Delta H_{CaO} + \Delta H_{CO_2} - \Delta H_{CaCO_3}$$

$$= 177,653 \text{ J mol}^{-1}. \qquad (7.10)$$

This is a simple application of Hess's law which states that if a reaction can be written as occurring in several stages the heat of the reaction is the sum of the heats of the several stages (all being referred to the same temperature). This law can be applied also to entropy, free energy or any other thermodynamic function. (It is likely that ΔH_{298} for this reaction would be determined by experiment rather than by calculation in this way and that it would be the heat of formation of $CaCO_3$ that would be deduced by Hess's law.)

Now, by Kirchhoff's law (equation 6.19)

$$\Delta H_{298} = \Delta H_0 + \int_0^{298} \Delta C_p \, dT,$$

therefore $\qquad \Delta H_0 = 177,653 - (-6450) = 183,103.$ $\qquad$ (7.11)

This is a fictitious figure obtained by assuming that extrapolation back to 0 K could be effected without any phase changes or other disturbances in the $C_p - T$ relationships used.

It can now be written that—

$$\Delta H_T = 183,103 + \int_0^T \Delta C_p \, dT$$

$$= 183,103 - 14.5T + 6.95 \times 10^{-3} T^2$$

$$- 8.4 \times 10^5 T^{-1} - 7.35 \times 10^{-6} T^3. \qquad (7.12)$$

The Gibbs–Helmholtz equation may now be applied—

$$\Delta G_T = -T \int \frac{\Delta H_T}{T^2} \, dT \qquad (6.43a)$$

$$= 183,103 + 14.5T \ln T - 6.95 \times 10^{-3} T^2 - 4.2 \times 10^5 T^{-1}$$

$$+ 3.7 \times 10^{-6} T^3 + IT. \qquad (7.13)$$

This is a general expression for ΔG_T but it involves an integration constant which can be evaluated only if ΔG is known at some particular temperature. It is known that pure $CaCO_3$ decomposes spontaneously at about 850°C (1123 K) under 1 atm pressure of CO_2. Since the carbonate, the oxide and the CO_2 are then all in equilibrium in their standard states,

$\Delta G_T^\circ = 0$ at that temperature and therefore the integration constant $I = -262$. Therefore,

$$\Delta G_T^\circ = 183,103 + 14.5T \ln T - 6.95 \times 10^{-3} T^2 - 4.2 \times 10^5 T^{-1}$$
$$+ 3.7 \times 10^{-6} T^3 - 262T. \qquad (7.13a)$$

If this is plotted against the absolute temperature the line obtained is very nearly straight giving an approximate equation

$$\Delta G_T^\circ = 177,100 - 158T \text{ J}. \qquad (7.13b)$$

This gives values for ΔG_T° as accurate as can be justified by the underlying experimental data. The actual form of such calculations obviously depends upon the selection of data available. For example, if the free energies of formation of products and reactants were available Hess's law would give a similar result directly.

The choice of basic data will depend on the quality of the data available. The most accurate – or that believed to be most accurate – should be used. The heats of formation of CaO, CO_2 and $CaCO_3$ might be obtained either by experiment or by calculation from C_p equations for C, Ca and O_2. If there is any serious discrepancy when alternative data are available there is need for the basic measurements to be repeated. This is not usually practicable for the solution of a single problem and as an interim measure the choice of data should be critically examined and the value chosen which seems to have been most carefully determined.

In the course of such a calculation experimental errors accumulate. The published accuracies of ΔH_{298} of CaO, CO_2 and $CaCO_3$ are ± 400, ± 2000 and ± 3000 J mol^{-1} respectively. The error in ΔH_{298} for the reaction is therefore ± 3400 J mol^{-1} or ± 3%. This is not bad. In extreme cases an application of Hess's law can give an error greater than the absolute value calculated – usually when that is rather small, of course, Experimentally determined values of ΔH_{298} will usually have been observed at quite different temperatures and reduced to the standard temperature using Kirchhoff's law (equation 6.19). This operation alone can introduce errors because of uncertainties in the values of C_p used. Any calculations should employ the miniumum number of steps in order to minimize error build-up during computation.

In equation (7.13b) the first term on the right is very nearly ΔH_T° and the coefficient of T can be considered as approximately ΔS_T° both of these being apparently independent of T in the range of validity of *all* the C_p equations used – i.e. where there are no phase changes among any of the reactants and products. In fact there are variations in ΔH_T° and ΔS_T°, but they almost cancel each other out in nearly all cases. When they do not, the term in $\alpha T \ln T$ must be retained in the $\Delta G_T - T$ K relationship (equation (6.24)).

Free Energy of Formation

The free energy change in a reaction of the type $A + B = AB$ is called the free energy of formation of the compound AB from its elements and is important as a measure of the stability of the compound. In the case of metallic oxides the reactions are of the general type–

$$(2x/y)M + O_2 = (2/y)M_xO_y \qquad (7.14)$$

for which

$$K_T = \frac{a_{M_xO_y}^{(2/y)}}{a_M^{(2x/y)}} \cdot \frac{1}{p_{O_2}} .$$

If the metal and oxide are both pure solid or liquid, these states can be used as standard states along with pure oxygen at 1 atm pressure. Then the easiest way to determine K_T is by finding the value of p_{O_2} at which the pure metal and oxide are in equilibrium with each other (page 149) so that

$$K_T = \frac{1}{p_{O_2}} ,$$

the activities of the condensed phases being unity. By equation (6.46), $\Delta G_T^{\circ} = RT \ln p_{O_2}$ (p being in atmospheres). This is general for any metal–oxide pair irrespective of the valency of the metal. ΔG_T° will change, of course, if the standard state changes, if the solid metal becomes liquid, for example. ΔG_T° calculated in this way is usually negative – i.e. the stable compound has a lower free energy than the elements uncombined. For this reason it is convenient and conventional to refer to $(-\Delta G_T)$ as the standard free energy of formation so that a positive quantity is being discussed and a "high" free energy of formation means a large negative value.

$\Delta G_T^\circ = 0$ only at that temperature at which the pure metal and the pure oxide are in equilibrium with oxygen at 1 atm pressure, i.e. ln $p_{O_2} = 0$. The value of the pressure at which the oxygen is in equilibrium with the pure metal and its oxide at any other temperature can be calculated if ΔG_T° is known at that temperature.

If the standard free energies of formation of two oxides are known at a particular temperature the pressures at which O_2 is in equilibrium with each metal–oxide pair can be calculated. If these pressures, the dissociation pressures of the oxides, are different, it is obvious that both metal–oxide pairs cannot be in equilibrium with the same atmosphere and that, given suitable conditions, one of the metals will reduce the oxide of the other metal.

Referring to Fig. 44 let $p'_{O_2} > p''_{O_2}$ when equilibrium is reached on *both* sides with the tap closed and the temperature the same throughout the system. If the tap is then opened oxygen will flow from chamber 1 to chamber 2. The pressure will fall in chamber 1 and rise in chamber 2. In 1 the oxide will dissociate to maintain the pressure locally at the equilibrium value, p'_{O_2}, while in chamber 2 the metal will oxidize as long as the pressure locally exceeds p''_{O_2}. As long as the tap remains open both of these reactions will continue until *either* all the oxide in chamber 1 is used up *or* all the metal in chamber 2 is oxidized. The metal whose oxide has the numerically greater standard free energy of formation ($-\Delta G_T^\circ$ greater) has the lower equilibrium p_{O_2} value, has the higher "affinity" for oxygen, has

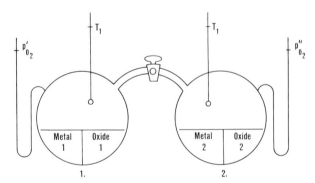

FIG. 44.   The reduction of a metal oxide by another metal through the medium of a gaseous phase in contact with both.

the more stable oxide and can reduce the other oxide to metal under kinetically favourable conditions.

Free Energy Diagrams

Information concerning oxides has been assembled[33, 39] in the "Free Energy Diagram" for oxide formation (see Fig. 45) or "Ellingham Diagram" in which ΔG_T° is graphed versus temperature for a large number of metal–oxide pairs. Each pair is represented by a line which is straight between the temperatures at which either the metal or its oxide undergoes a phase change, at which points the slope of the line usually changes. As in the case of the decomposition of $CaCo_3$ (see page 156) the equations of these lines can usually be written in linear form without loss of accuracy—

$$\Delta G_T^\circ = \Delta H^\circ - \Delta S^\circ T, \qquad (6.25)$$

the intercepts on the temperature axis through 0 K giving the approximate ΔH° values and the slope in any section being approximately the value of $(-\Delta S^\circ)$ for the reaction – each of these functions being apparently constant. Since $\Delta G_T^\circ = \Sigma \mu_i n_i$ (at equilibrium) $- \Sigma \mu_i^\circ n_i = RT \ln p_{O_2}$, when the metal and its oxide remain in their standard states at equilibrium, $RT \ln p_{O_2} = \mu_{O_2}$ (at equilibrium) $- \mu_{O_2}^\circ$ and is called the oxygen potential of the system.

It will be noticed that the slopes of almost all the lines in Fig. 45 are similar. This is because the entropy change in all these cases is similar being almost entirely due to the condensation of the gas oxygen. When the metal and its oxide are both condensed phases the loss of volume in the reaction (equation (7.14)) is always the same – 1 g molecular volume or 22.4 l./mole ($0.0224 \text{ m}^3 \text{ mol}^{-1}$) of oxygen reacting. The loss of entropy is almost entirely due to the loss of disorder inherent in this volume.

The noticeable exceptions are the lines for the oxidation of carbon. Carbon dioxide forms according to $C + O_2 = CO_2$ with only a very small volume change so that $(-\Delta S^\circ) \simeq 0$ (actually 0.84 J K^{-1}) and the line is almost horizontal. Carbon monoxide forms by the reaction $2C + O_2 = 2CO$ involving an increase in volume and therefore in entropy so that the slope $(-\Delta S^\circ)$ is negative. Carbon is unique in having this oxide whose free energy of formation becomes increasingly negative as the temperature increases – which is therefore more stable at high temperatures than at low ones. This means that carbon can reduce any other oxide provided a high enough temperature can be reached.

The discontinuities at transition temperatures are due to changes in $(-\Delta S^\circ)$ consequent upon the difference in the absolute entropy of that component of the system whose phase changed, in its two states. If the metal melts before the oxide its entropy increases slightly and the loss of entropy on its reacting with oxygen to form the solid oxide is slightly increased so that $(-\Delta S^\circ)$ becomes greater and the slope of the line on the diagram becomes steeper above the melting point of the metal. When the oxide melts, at a rather higher temperature, the entropy of the product of reaction is increased so that the entropy change as the molten metal reacts with oxygen to form molten oxide probably reverts to a value similar to that when both metal and oxide were solid. The line on the diagram becomes less steep and approximately parallel to the section below the melting temperature of the metal. If the metal boils its entropy increases dramatically. The loss of entropy upon oxidation is now much greater as the volume loss on reaction is that of metal and oxygen together. $(-\Delta S^\circ)$ is increased and the line on the diagram (shown broken on Fig. 45) becomes very steep. This is seen to be the case for the alkali metals, Cd, Zn, Mg and Ca.

In the case of Mg, which boils at $1100\,^\circ$C (1373 K), the entropy change can be calculated from the following values of entropies at 1373 K. For liquid Mg, $S_{1373} = 86.5$, for gaseous Mg, $S_{1373} = 180$, for O_2, $S_{1373} = 255$ and for solid MgO the value of S_{1373} is 97, all in J mol^{-1} K^{-1}.

Hence for　　2Mg(liquid) + O_2 = 2MgO,　$-\Delta S_{1373} = 234$ J mol^{-1} K^{-1}

and for　　2Mg(gas) + O_2 = 2MgO,　$-\Delta S_{1373} = 421$ J mol^{-1} K^{-1}.

It will be noticed that the entropy change is not simply proportional to the volume change.

The case of H_2/H_2O should be mentioned. Here the element is gaseous but so is the oxide and the slope of the line is similar to that of many of the solid metals. Except at low temperatures the S/SO_2 line is similar, but the entropy of gaseous sulphur, like that of gaseous Mg, is lower than that of the permanent gases, so that the slope of the sulphur line is lower than that for hydrogen. This probably reflects a tendency of molecules in gaseous sulphur to associate in larger than diatomic groups.

An element can reduce the oxide of any other element appearing above it in the free energy diagram for oxide formation (Fig. 45) at any particular temperature. Metals like silver whose oxides are known to be readily reduced are found at the top and Ag_2O can be seen to come into equilibrium ($\Delta G_T^\circ = 0$) with oxygen at 1 atm pressure at about 470 K (200°C) so that it must dissociate at higher temperatures. Silver oxide can be reduced by thermal decomposition alone.

Other noble metals behave in a like manner and even ferric oxide dissociates at about 1500°C and MnO at 2100°C. Most other base metal oxides are stable to much higher temperatures. The upper third of the diagram is occupied by the noble and common base metals with the line for H_2-H_2O lying at the lower edge of this group indicating that hydrogen can reduce the oxides of Sn, Cu, Pb, Fe and many other metals under favourable conditions. Lower in the diagram are lines of metals whose oxides are more difficult to reduce – Si, Mn, Cr, Ti, V, which are important for their deoxidizing powers in steelmaking. The oxides of Al, Zr, Ca, Mg and others appearing at the bottom of the diagram are extremely stable and refractory and difficult to reduce.

The order from top to bottom of this diagram has already been referred to as helping to explain the order in which the metals were discovered and brought into use by man (page 1). It can also be used to provide part of the explanation of the distribution of elements during magmatic segregation (page 11) and during the original separation of the crust and lithosphere from the chalcosphere and core (page 8), but the very high overall pressures at which most magmatic reactions occur alter the chemistry considerably and Fig. 42 and other free energy diagrams and data must be applied with caution. In particular, boiling points are greatly reduced and ideality in gases cannot be so readily assumed. Reactions involving volume changes will be affected differently from those with no volume change. Minimum temperatures for reactions, solubilities, solubility products and partition coefficients are likely to be altered by different amounts depending on how volume changes are affected by the pressure.

At some points lines in the diagram cross, indicating that the relative stabilities of the two oxides change with rise in temperature. This may apply to oxides of different metals (e.g. CoO and Fe_3O_4) or to different oxides of one metal (e.g. FeO and Fe_3O_4). The case of the line for CO which cuts across many of the others has already been mentioned.

The stabilities of any pair of oxides may be compared by writing down the equations for their formation with the free energy-temperature relationships thus—

$$4/3Al + O_2 = 2/3Al_2O_3 \; ; \quad \Delta G_T^\circ = -1,077,600 + 185T, \quad (7.15)$$

$$4/3Cr + O_2 = 2/3Cr_2O_3 \; ; \quad \Delta G_T^\circ = -747,000 + 173T. \quad (7.16)$$

By inspection it is clear that the free energy of formation of Al_2O_3 is always much greater numerically than that of Cr_2O_3, the lines being almost parallel, and that aluminium will reduce Cr_2O_3 under favourable conditions. It is used to do so in the thermit process in which powdered oxide and aluminium are ignited and react with an evolution of heat almost sufficient to fuse the chromium metal and a slag of fluxed alumina.

Alternatively data may be evaluated quantitatively thus—

For CO $2C + O_2 = 2CO; \quad \Delta G_T^\circ = -223,500 - 175T \quad (7.17)$

and for MnO, $2Mn + O_2 = 2MnO; \quad \Delta G_T^\circ = -798,500 + 164T. \quad (7.18)$

$(7.17) - (7.18)$ gives—

$$2MnO + 2C = 2Mn + 2CO; \quad \Delta G_T^\circ = 575,000 - 340T \quad (7.19)$$

for which $\Delta G_T^\circ = 0$ when $T = 1690$ K or $1417°C$ so that the carbon can reduce pure MnO only at $1417°C$ or above. (It is obvious that it is only above 1690 K that ΔG_T° is negative, which is the condition necessary that the reaction shall proceed to the right, as the equation is written.) Examination of the diagram shows that the CO and MnO lines do cross at this temperature. The diagram cannot show more than can be calculated, but does summarize a large amount of information in a convenient manner.

Grids on Free Energy Diagrams

The usefulness of a free energy diagram can be extended by superimposing on it one or more "grids".[33] In the case of the diagram for oxides appropriate grids might show (a) the values of p_{O_2} at all points on the diagram, (b) the values of the ratio CO/CO_2 which would be in equilibrium with oxygen at these pressures or (c) the corresponding values of the ratio H_2/H_2O. For any point on the diagram there is a unique set of values of T, p_{O_2}, CO/CO_2 and H_2/H_2O.

For case (a) consider the expansion of 1 mole of O_2 from the standard pressure of 1 atm to a pressure p_{O_2} atm. The free energy of expansion of 1 mole of any ideal gas is $\Delta G_T = RT \ln p$ (eqn. 6.44c). If p_{O_2} is put equal to 10^0, 10^{-1}, $10^{-2}, \ldots, 10^{-n}$ atm, ΔG_T becomes 0, $-2.303RT$, $-4.606RT, \ldots, -2.303nRT$ and can therefore be represented by a series of straight lines radiating from the origin of the diagram and having gradients 0, $-2.303R$, $-4.606R, \ldots, -2.303nR$. If each line is annotated with the appropriate value of p_{O_2} the value corresponding to any point in the diagram can be read off (with interpolation if necessary). This is quicker than calculating p_{O_2}, though not necessarily so accurate.

For case (b) consider the reaction

$$2CO + O_2 = 2CO_2 \tag{7.1}$$

for which $\quad \Delta G_T^\circ = -RT \ln \dfrac{p_{CO_2}^2}{p_{CO}^2} \dfrac{1}{p_{O_2}} = -565{,}500 + 174T. \tag{7.20}$

Rearranging,

$$RT \ln p_{O_2} = -565{,}500 + \left(174 - 2R \ln \frac{p_{CO}}{p_{CO_2}}\right) T. \tag{7.20a}$$

When $p_{CO} = p_{CO_2} = 1$ atm, $RT \ln p_{O_2}$ is the standard free energy of the reaction (7.1). When the CO and CO_2 are at other pressures, $RT \ln p_{O_2}$ is the free energy of the reaction with the gases in non-standard states but with the ratio of their pressures as indicated. For any CO/CO_2 ratio, therefore, ΔG_T is a linear function of T and a family of straight lines can readily be constructed radiating from ($\Delta G_T = 565{,}500$ J, $T = 0$ K) and with gradients equal to ($174 - 2R \ln CO/CO_2$). Normally these are drawn for CO/CO_2 ratios ranging in geometric progression from 10^{-7} to 10^{+15}. This grid indicates at a glance the CO/CO_2 ratios of the gas in equilibrium with any metal–oxide pair at any temperature, or conversely whether any gaseous mixture will be oxidizing or reducing toward the metal–oxide pair.

The grid for the H_2/H_2O ratios can be derived in a similar way and again a family of straight lines is obtained radiating from a point on the temperature axis at $\Delta G_T = -494{,}000$ J and this grid would show whether a gas of known composition with respect to hydrogen and water vapour would oxidize or reduce any metal–oxide system at a given temperature. The standard free energy of the reaction $2H_2 + O_2 = 2H_2O$ is $\Delta G_T^\circ = -494,000 + 112T$ J mol^{-1} so that the general equation of the grid line is

$$RT \ln p_{O_2} = -494,000 + \left(112 - 2R \ln \frac{H_2}{H_2O}\right) T. \qquad (7.21)$$

Combining the CO/CO_2 and H_2/H_2O grids provides a means of calculating the equilibrium composition of combustion gases in a furnace from the fuel analysis and a value for excess air or oxygen at any temperature.

Grids may be printed over the free energy diagram or may be prepared on transparent paper to be laid on top of the diagram when required. The other possibility is to provide a focal point and a radial scale round the edge of the diagram. A ruler or a taut thread can be used to indicate temporarily the position of any grid line. This means has been used in Fig. 45. Other free energy diagrams can, of course, be fitted with appropriate grids. An H_2/H_2S grid could be made for Fig. 46, for example, or an $H_2/2HCl$ grid for Fig. 80.

Free energy diagrams can be drawn for any class of reaction provided that there is a common reactant present, always in the same number of moles in the equation – for example for the dissociation of carbonates and sulphates, and for the formation (or dissociation) of sulphides. The free energies of formation of a selection of sulphides are presented graphically in Fig. 46 which shows that the order of the stabilities of the sulphides is rather similar to that of the oxides. A notable exception is, however, that CS_2 is one of the least stable sulphides so that carbon cannot be used as a reducing agent for sulphides as it can for oxides. On the other hand, SO_2 is one of the most stable sulphides. A visual comparison of Figs. 45 and 46 suggests that the free energies of formation of most oxides are greater than those of the corresponding sulphides, from which it can be deduced that most sulphides are probably easily oxidized by heating in air to the corresponding oxide and SO_2 (see also Fig. 70). Diagrams are not usually drawn for making comparisons of the stabilities of, say, an oxide and a sulphide but in principle one could be drawn for the free energy of formation of "ferrides", for example, to compare the stabilities of FeS, Fe_3C, FeO, Fe_4N, $FeCl_3$, etc. Normally such comparisons are made by

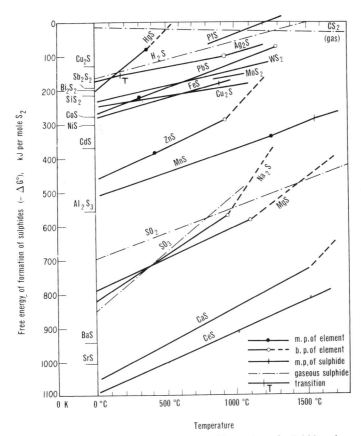

FIG. 46. The standard free energies of formation of sulphides, the standard states being the pure condensed phases and gases at one atmosphere pressure. Where the value is not known exactly or its inclusion would lead to confusion the sulphide only is indicated by its formula at the approximate value of $\Delta G°$ at 0°C. (Based on diagrams by Ellingham[39] and Richardson and Jeffes.[33])

calculation. Other examples are given of free energy diagrams in Fig. 70 for the oxidation of sulphides and in Fig. 80 for the formation of chlorides.

Limitations of the Free Energy Diagrams – and the Phosphorus Reaction in Steelmaking

Free energy diagrams and the kind of information they present are limited in their application in two ways.

First, they show the direction in which equilibrium lies but they cannot show the conditions under which it will be reached. According to Fig. 45 many metals might oxidize in air at room temperature but only in a few cases like that of sodium is the danger of spontaneous combustion great. Under certain conditions zirconium may ignite and even carbon (as coal) sometimes catches fire when stored carelessly, but most metals, including aluminium, magnesium, etc., are oxidized only on the surface at room temperature and can be heated and even melted in air with impunity. Under other conditions, in the presence of a catalyst or simply when the specific surface is extremely high, oxidation may occur with violence but thermodynamics and free energy diagrams cannot predict these conditions. Finely divided iron can be pyrophoric but massive iron is very stable in dry air at room temperature and oxidizes only very slowly at moderately elevated temperatures. At room temperature and in moist air, however, a new reaction mechanism is available by which oxidation to a hydrated oxide proceeds readily. This corrosion reaction is not fast but is difficult to bring under control and accounts for a massive reversion of iron which has been brought out of its low free energy state as an oxide by extraction metallurgists, back almost to the same low free energy state.

The second limitation is that in many processes the reacting species are not in their standard states but in solution at concentrations which may be continuously varying and at activities or pressures which may be very different from unity. In Fig. 47 the case of the oxidation of phosphorus in steelmaking is considered. Line A for the oxidation of Fe to FeO is as in Fig. 45. Line B is that for the oxidation of P_2 to P_2O_5, the standard states being the gaseous element at 1 atm pressure, the oxide as the pure liquid rather than as the gas as in Fig. 45, and 1 atm pressure of oxygen. The pure oxide would be a gas at steelmaking temperatures but the data used here are preferred to those available on the basis of gaseous P_2O_5 because the oxide enters the slag in liquid solution. It has been calculated, from data given by Turkdogan and Pearson[40]

$$2/5P_2(\text{gas}) + O_2 = 2/5P_2O_5(\text{liquid}); \quad \Delta G_T^\circ = -613{,}940 + 202T. \quad (7.22)$$

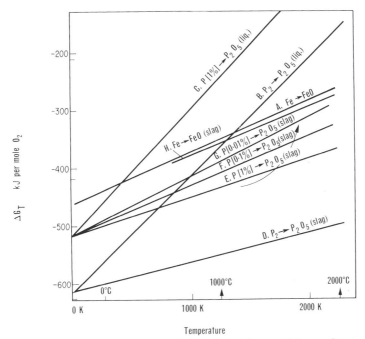

FIG. 47. Free energies of oxidation of phosphorus and iron under various conditions of interest in steelmaking. The arrow indicates the progress of the conditions in the bath during the processing of a heat of steel (see text).

These lines A and B indicate that iron would be oxidized in preference to phosphorus at, say, $1600°C$. If the phosphorus is in a 1% solution in iron another line C can be calculated from[41] –

$$2/5P_2 \text{(gas)} = 4/5P(1\% \text{ in Fe)}; \quad \Delta G_T = -97,900 - 11.8T. \quad (7.23)$$

Subtracting (7.22) from (7.23)

$$4/5P(1\% \text{ in Fe)} + O_2 = 2/5P_2O_5 \text{(liquid)};$$

$$\Delta G_T = -510,000 + 218T \quad (7.24)$$

which gives the equation for line C which is seen to lie above both A and B showing that the oxidation of phosphorus from a solution in iron to form pure liquid P_2O_5 is not thermodynamically possible. In practice P_2O_5 would form a slag with FeO and some removal might be possible depending on the activity of the P_2O_5 in this slag. In steelmaking, "acid" slags containing SiO_2, FeO and MnO but little CaO do not in fact dissolve P_2O_5 but where its removal is necessary a "basic" slag is prepared very high in CaO in which P_2O_5 has an activity which has been estimated[42] to be of the order of 10^{-20} (standard state pure liquid P_2O_5). The free energy of dilution of $\frac{2}{5}P_2O_5$ in this slag is—

$$\Delta G_T = RT \ln 10^{-20} = -153T \tag{7.25}$$

whence combining (7.22) and (7.25)

$$2/5P_2(\text{gas}) + O_2 = 2/5P_2O_5 (\text{in slag at } _r a_{P_2O_5} = 10^{-20});$$
$$\Delta G_T = -613,980 + 49.4T \tag{7.26}$$

which gives line D well below the Fe/FeO lines showing that such a slag is a good "sink" for P_2O_5.

Combining (7.24) with (7.25),

$$4/5P(1\% \text{ in Fe}) + O_2 = 2/5P_2O_5 (\text{in slag});$$
$$\Delta G_T = -516,000 + 64.9T \tag{7.27}$$

and adjusting for further dilution of phosphorus in the metal to 0.1 and 0.01%,

$$4/5P(0.1\% \text{ in Fe}) + O_2 = 2/5P_2O_5 (\text{in slag});$$
$$\Delta G_T = -516,000 + 84.1T \tag{7.28}$$

$$4/5P(0.01\% \text{ in Fe}) + O_2 = 2/5P_2O_5 (\text{in slag});$$
$$\Delta G_T = -516,000 + 103T. \tag{7.29}$$

These give lines E, F and G which represent the oxidation of phosphorus at three stages of its elimination from iron under a good basic slag. All three lines lie below the Fe/FeO line A showing that the elimination is thermodynamically possible if such a slag can be prepared. These results are only approximate. The activity of P_2O_5 in the slag is not constant at 10^{-20} but decreases in the early stages as lime comes into solution in the slag and later increases as the P_2O_5 concentration rises up to about 15%.

The interaction of other elements in the iron has been ignored but carbon, silicon, and oxygen will increase the activities of phosphorus above the values implicitly used, the effects of carbon and sulphur diminishing and that of oxygen increasing as the process continues. Line H represents better than A the oxidation of iron into the basic slag where $a_{FeO} \simeq 0.2$. The progress of conditions in the bath during a heat may be represented by the arrow in Fig. 47.

It will be observed that the Fe/FeO and P/P_2O_5 lines converge and approach rather closely as the temperature rises and as the phosphorus content of the iron falls below about 0.1%. Intersection of these lines would mark a limit to the degree of dephosphorization possible with a particular slag at the temperature at the intersection. Even a close approach to the equilibrium position would lead to a slowing up of the reaction rate so that it is to the advantage of the steelmaker not to allow the temperature to rise too high at this stage nor to attempt to raise the activity of FeO beyond an optimum value. On the other hand, high basicity in the slag will reduce the activity of the P_2O_5 with the effect of widening the spacing of the appropriate iron and phosphorus lines.

The Reduction of Iron in the Blast Furnace

A free energy diagram can be adapted to represent graphically the sequence of chemical changes occurring in a process, with respect to the chemical potential of one element. Figure 48 represents the variations in oxygen potential which are possible as a small package of air enters an iron blast furnace at its tuyère level and traverses the bosh and stack filled with ore, coke and limestone. The diagram is basically the same as Fig. 45 but restricted to the lines for the oxides of Fe, Mn, Si, Al and C with the focus and marginal scale of the CO/CO_2 grid. Superimposed on this there is a branched line following a number of paths each of which represents the changes in oxygen potential of typical small packages of gas which, starting as air, are presumed to remain coherent throughout their passage of the furnace, and which are also presumed to react "completely" on making contact with a reactive surface. These packages are also presumed to lose their temperature at an "average" rate to the burden as they pass through it. A more detailed treatment might allow for the effects of the

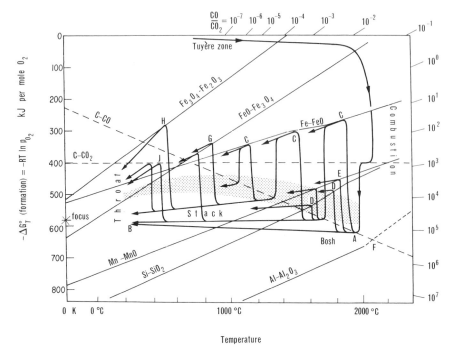

Temperature

FIG. 48. The reduction reactions in the iron blast furnace represented on a free energy diagram for the formation of oxides (compare with Fig. 45 and see text). The CO/CO_2 grid radiates from the focus on the temperature axis. The shaded area indicates the normal range of gas analyses encountered in the process.

separate reactions being endothermic or exothermic. Nitrogen, being inert, has been ignored throughout.

Initially, the gas reacts with coke, probably in two stages, with an increase in temperature and a reduction in oxygen potential to that in equilibrium with carbon at hearth temperature (point A). The condition of the gas will then follow the CO/CO_2 grid line, as its temperature falls, without change in composition, until the package collides with another reactive surface. This may not occur at all and the package may proceed through the furnace unchanged (B). If the gas reacts with FeO, however, its oxygen potential will rise suddenly to that in equilibrium with FeO (C). It may, however, react with SiO_2 or MnO at high temperatures or with

P_2O_5 (line not shown), in which case the increase in oxygen potential will not be so great (D and E). It is obviously impossible for reaction between the gas and Al_2O_3 to occur at the temperatures reached in this kind of furnace (F). After each reaction the condition of the gas proceeds along the CO/CO_2 grid line towards the focus until another impact occurs. This may be with either oxide or carbon and the direction of the ensuing reaction is in accordance with thermodynamic principles. The higher oxides of iron will not be available for reaction except in the upper part of the furnace (G and H). Coke will become less effective for reducing the oxygen potential of the gas packages at some temperature below about $800°C$ and at still lower temperatures, below $700°C$, the reverse reaction $2CO = CO_2 + C$ can occur (J). The gas actually reaching the top of the furnace will have a composition which is the weighted mean of all the little packages of which it is supposed to be composed, and the shaded area indicates the progress of the average gas condition through the furnace.

It will be appreciated that Fig. 48 is not a complete description of the reactions in a blast furnace. The reduction of phosphorus has been omitted and the reduction of SiO_2 and MnO has been treated on the basis of the standard lines corresponding to pure element and oxide. In all three cases adjustments like those made in Fig. 47 for the case of phosphorus in steelmaking would give a more accurate representation of the reactions, each line, however, being replaced by a group of lines covering all stages of the process. The presence of H_2 and H_2O in the blast furnace has been ignored. These would really need another diagram. Sulphur removal has no place in this diagram. The effects of local movements of temperatures at the sites of exothermic or endothermic reactions have not been represented. Much more could be included in such diagrams at the expense of much greater complexity but the objective at this stage is to show what can be done to summarize graphically a fairly complex thermodynamic system. It should also be pointed out that the diagram shows (like all thermodynamic data) what can happen rather than what does happen. The conditions under which reactions will happen and the rates at which they will proceed are considered in a later chapter.

CHAPTER 8

REACTION KINETICS

THERMODYNAMICS tells nothing about the conditions under which a system will proceed toward equilibrium; nothing about the mechanism of the reactions. Since, in extraction processes, production rates often depend on chemical reaction rates, there are good economic reasons for studying reaction mechanisms and factors controlling reaction rates.

The velocity of a reaction depends on several factors – the pressure, concentration or activity of each reactant; the temperature; the nature of the solvent if any; the presence of catalysts; and the mechanism of the reaction. It depends on the degree of difficulty in severing existing chemical bonds and creating conditions of energy and geometry necessary for forming new ones. It may be controlled by an over-riding difficulty of transporting chemical species to or from the site of the reaction.

Order of Reaction

Reaction mechanisms are classified as first order, second order, third order, etc., where the ordinal number indicates the number of chemical species whose concentrations determine the velocity of the reaction. These are often synonymous with unimolecular, bimolecular, termolecular, etc. reactions indicating the number of molecules involved in every unit reaction. The number of molecules involved can be obtained by examining the left-hand side of the equation for the reaction. When it is three or more it is most likely that the reaction will proceed by two or more stages, one of which will act like a "valve" controlling the overall reaction rate and determining its order. Most reactions are of first or second order. A few can be contrived to proceed at a rate independent of the concentration of any reactant and these are called zero-order reactions. They are nearly all

174

reactions between gases and surfaces. One example is where the rate depends on the proportion of surface sites which are occupied by adsorbed gas molecules. If the gas pressure is high enough all suitable sites will be occupied so that at higher gas pressures the reaction rate cannot be increased and the reaction is reduced from first to zero order. Third-order reactions are occasionally reported and higher-order reactions even less frequently. These claims to have found high-order reactions are seldom confirmed. Fractional order reactions are also reported (e.g. 1.5), these probably being compromise values where two mechanisms are operating simultaneously. Sometimes the order of a reaction changes as the concentration changes.

In a first-order reaction in which the initial concentration of the determining species is a and the decrease in its concentration at time t is x the reaction velocity is given by—

$$dx/dt = k(a - x) \qquad (8.1)$$

where k is the *specific reaction rate constant*, i.e. the reaction velocity is proportional to the instantaneous concentration.

Integrating up to time t,

$$kt = \ln [a/(a - x)] . \qquad (8.1a)$$

In this form the equation is useful for determining the value of k for a reaction from observed values of concentration changes with time.

In a second-order reaction in which the initial concentrations of the determining species are both a (for simplicity) and supposing the diminution of concentration at time t to be x in each case—

$$dx/dt = k(a - x)^2 \qquad (8.2)$$

whence, integrating up to time t,

$$kt = x/[a(a - x)] . \qquad (8.2a)$$

This time the velocity is proportional to the square of the instantaneous concentration and in general the velocity of an nth order reaction is proportional to the nth power of the concentration at time t. This provides a means of determining the order of a reaction if its rate can be measured at two or more concentrations. Alternatively the time required to transform say half of the reactants may be determined from several initial concentrations a. In first-order reactions this time is independent of a ($t_{0.5} = (\ln 2)/k$). In second-order reactions $t_{0.5} = 1/(ka)$ and in third-order reactions $t_{0.5} = 1/(2ka^2)$. The purpose of such an exercise is usually to determine, along with other evidence, something about the mechanism of the reaction or at least about its rate-controlling stage. For example, if the order is apparently first, it is possible that a diffusion process controls its rate.

In these discussions it has been assumed that the reaction products are cleared from the site of the reaction and can have no effect on its progress. In practice this is often not the case but products remain in whole or in part at the reaction site where their concentration may increase as the reaction proceeds. This is particularly so when a reaction occurs within a lump of ore during roasting when the escape of products via intergranular porosity can be quite slow. For a reaction like

$$A + B = C + D$$

in an enclosed space, equilibrium is reached when the rates of the forward and reverse reactions are equal, i.e. when

$$k_f x_A \cdot x_B = k_r x_C \cdot x_D,$$

k_f and k_r being the specific reaction rate constants for the forward and reverse reactions and x_i in this case being the equilibrium concentration of i. At equilibrium,

$$K = \frac{k_f}{k_r} = \frac{x_C \cdot x_D}{x_A \cdot x_B} \tag{8.3}$$

which is obviously the equilibrium constant for the reaction. This was essentially the argument used by Guldberg and Waage in their derivation of the Law of Mass Action.

If we look at a simpler, first-order reaction,

$$A \rightarrow B$$

where there is initially x_0 of A and no B present, and x_A is the change in the concentration of A at time t, the rate of the forward reaction will be

$$r_f = k_f (x_0 - x_A) \tag{8.4}$$

and that of the reverse reaction

$$r_r = k_r x_A \tag{8.4a}$$

so that the net reaction rate will be

$$r = r_f - r_r = k_f(x_0 - x_A) - k_r x_A. \tag{8.4b}$$

At equilibrium this will be zero so that

$$k_f (x_0 - x_{A(e)}) = k_r x_{A(e)}, \tag{8.4c}$$

where $x_{A(e)}$ is the change in the concentration at equilibrium. Combining (8.4b) and (8.4c),

$$\frac{dx}{dt} = r = k_f(x_0 - x_A) - k_f \frac{x_A}{x_{A(e)}} \cdot (x_0 - x_{A(e)})$$

$$= k_f \frac{x_0}{x_{A(e)}} \cdot (x_{A(e)} - x_A). \tag{8.5}$$

The integrated form of this equation corresponding to (8.1a) is

$$k_f \frac{x_0}{x_{A(e)}} \; t = \ln \frac{x_{A(e)}}{x_{A(e)} - x_A} \tag{8.5a}$$

or, since from (8.4c), $k_f(x_0/x_{A(e)}) = k_f + k_r$,

$$(k_f + k_r)t = \ln \frac{x_{A(e)}}{x_{A(e)} - x_A} \tag{8.5b}$$

which has a similar form to equation (8.1a). The rate of reaction is, by (8.5), proportional to the difference between the instantaneous concentration and the equilibrium concentration and therefore falls rapidly to zero as the equilibrium position is approached and by (8.5b) the time required to reach equilibrium appears to be infinite. This particular example is oversimplified but it is sufficient to make the principle clear. In the general case there must be more than two species involved and they are likely to be present in varying concentrations at the beginning of the reaction. The numbers of each kind of molecule taking part are unlikely always to be the same. If we consider the reaction $A + B = C + D$ with initial concentrations

a, b, c and d and with the concentrations of A and B falling by x at time t,

$$\frac{dx}{dt} = k_f(a - x)(b - x) - k_r(c + x)(d + x).$$

Although still a simplified case this is difficult to treat mathematically and the matter will not be pursued further here.

The Arrhenius Equation

Reaction rates increase with temperature. In 1889 Arrhenius gave the empirical relationship—

$$k = A \exp(-Q/RT) \quad \text{or} \quad \log k = \log A - Q/2.303\,RT. \quad (8.6)$$

This is generally accepted as a good approximation which fits most experimental data well. Reactions which do not obey it have very complex mechanisms. It has been improved for some purposes by the addition of a third term ($B \ln T$) to the logarithmic expression, and the Arrhenius form has been derived as an approximation from consideration of the activated complex theory (see below) conditional only on its not being extended over too wide a temperature range.

A stable molecule is a configuration of minimum free energy — that is, small changes of state can only increase its free energy. If one or several such molecules undergo spontaneous rearrangement to form another stable configuration at, usually, a lower minimum free energy, it is obvious that the system must pass through a state in which the free energy is higher than that of either of the stable states. This higher energy state is called an "activated complex". It is presumed to be very unstable and to decompose or react almost instantaneously toward the formation of the product. The life of such a complex is presumed to be not greater than the period of its thermal vibrations. The specific reaction rate constant is given by statistical mechanics as $RT/N_0 h$ or kT/h s^{-1} where N_0 is Avogadro's number, $k = R/N_0$ is Boltzmann's constant and h is Planck's constant. This means that the life of a complex is h/kT seconds. As $h = 6.62 \times 10^{-34}$ J s and $k = 1.38 \times 10^{-23}$ J K^{-1}, this time is $(4.77 \times 10^{-11})/T$ s which is, at room temperature, 1.6×10^{-13} s, at $1000°$C, 3.7×10^{-14} s, at $1600°$C, 2.5×10^{-14} s and at $2000°$C, 2.1×10^{-14} s. The total range of possible values over the industrially used range of temperatures is remarkably narrow, the largest being less than 10 times the smallest. Some of the complex may revert to reactants. Spectroscopic examination of flames reveals the presence of some very unlikely molecular groups and this is taken as confirming the presence of transient complexes. The complex need not have the character of a chemical compound, however. It may be a crystallographic arrangement of atoms of critical size or of peculiar geometry; it may be a tiny bubble of gas or droplet of liquid called a nucleus (see page 188); it may be a molecule in an absorbed state. In diffusion the activated state is reached when the diffusing atom is at what may be thought of as the narrowest point in its traverse between one lattice site and a neighbouring vacant site.

The situation is expressed graphically by line A in Fig. 49(a). The "activation energy" Q by which the energy of the complex exceeds that of the reactants is indicated. It is clear that the activation energy of the reverse reaction is higher still by ΔG for the net reaction – so that the reverse reaction is less likely to occur.

This activation energy is in fact the Q of the Arrhenius equation (8.4). At any temperature T there is an average vibrational (rotational, translational, etc.) energy per molecule around which the energies of the individual molecules are distributed in a statistical manner expressed quantitatively in the Maxwell–Boltzmann law. According to this law, the fraction of all the molecules having energies which exceed the average by at least Q J mol^{-1} is always proportional to $\exp(-Q/RT)$ – the Boltzmann factor. In certain simple cases, with respect to the kind of energy involved the proportionality factor is 1 so that the fraction of molecules with energies exceeding Q is equal to $\exp(-Q/RT)$. This would be the case when only the combined transitional energies of two gaseous molecules was involved, or only the combined vibrational energies of adjacent atoms in a liquid solution. If more than two energies contribute to the situation (e.g. mixtures of vibrational and rotational as readily happens in all but the simplest molecular arrangements) a proportionality factor becomes necessary which depends on both Q and T and can be very large. For many purposes, however, the fraction can be accepted as being equal to $\exp(-Q/RT)$. Only this fraction can participate in the formation of the activated complex and hence in the reaction as a whole. The energy level of each molecule is continuously changing as it interacts with its neigh-

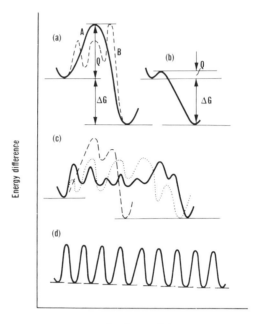

FIG. 49. Graphical representations of activation energy. (a) A single
value of Q may be the energy required to form the single necessary
complex (A) or the net energy to form the complex which has the
highest energy of several in a sequence (B). (b) When Q is low the
reaction is easily triggered off. Catalysts reduce Q to a low value. (c)
There may be several mechanisms for a reaction, of similar probability.
(d) In diffusion a similar reaction is repeated many times.

bours but the statistical distribution of energy remains constant at a given
temperature. Whether a particular molecule will react or not while temporarily energized to a sufficient level depends on (a) whether or not it
encounters another suitable molecule sufficiently energized, and (b)
whether in such an encounter the mutual orientation of the molecules is
such as to permit the necessary reaction to take place. The coefficient A in
the Arrhenius equation (8.6) is sometimes written as a product ZP where Z
is called the collision number and P the steric (shape) factor for the
reaction. P is near unity and seldom less than 0.1 for simple molecules but
can be very small ($\sim 10^{-10}$) where complex organic molecules are

involved. It might be low in some slag reactions too although there is no direct evidence of this. The Maxwell–Boltzmann law proportionality factor mentioned earlier, if not unity, would also be incorporated within A.

Most reactions probably proceed in several steps as indicated by line B in Fig. 49(a). In such a case it is usually presumed that the kinetics are controlled by the complex which has the highest net heat of formation from the reactants, but the number of steps, the values of the individual energies required in these steps and the prevalence of alternative "blind alley" reaction sequences must all contribute to the observed rate of a reaction. There may be several possible mechanisms for a reaction (Fig. 49(c)). It may be that one of these has so much lower an activation energy than the others that most of the reaction goes that way but in general alternative mechanisms are analogous with electrical resistors in parallel – the easy route carries most of the load but all routes carry some in inverse proportion to the resistance offered. If in a particular mechanism there are two or more steps whose individual activation energies are similar and not much smaller than that for the reaction as a whole, they must behave as resistances in series reducing the probability that a group of molecules will complete the course in a given time, for they will linger in the low energy intermediate states, and the probability that reverse reactions occur is relatively high. An extreme example of this is the case of diffusion (Fig. 49(d)) in which a similar energy barrier has to be surmounted for every advance of the diffusing atom by one inter-atomic distance. The resistance to mass transfer by diffusion is obviously the sum of a very large number of individual resistances offered at each step. One way of increasing the rate of a diffusion process is to reduce the distance across which it operates and hence the number of these individual steps (see page 198).

The activation energies of reactions which occur explosively like the decomposition of fulminates in detonators is very low (Fig. 49(b)) so that the molecules are rather easily brought to a condition in which almost all of them are sufficiently energized for reaction simultaneously. Some reactions are rather easily energized by radiation of a particular frequency which may be in the visible range or near it. Some reactions can be accelerated or even made possible by the use of a catalyst which is a reagent necessary to produce an activated complex of low energy of formation that will enable the reaction to proceed at a reasonable rate.

Since the complex always decomposes, the catalyst appears unchanged at the end of the reaction.

The Activated Complex Theory

An expression for the rate of a reaction can be derived from the rate of formation of the activated complex which equals its rate of decomposition and is equal, according to statistical mechanics, to $(RT/N_0 h) c_x$ where c_x is the concentration of the complex X during the reaction.

If the reaction starts

$$A + B = X = \dots \tag{8.7}$$

the equilibrium constant is

$$K = \frac{a_X}{a_A a_B} = \frac{c_X}{c_A c_B} \cdot \frac{\gamma_X}{\gamma_A \gamma_B}$$

where a is activity, c is concentration and y is activity coefficient.

Therefore the reaction rate,

$$(RT/N_0 h)c_X = (RT/N_0 h) \cdot K \cdot \frac{\gamma_A \gamma_B}{\gamma_X} \cdot c_A c_B \tag{8.8}$$

and since $\Delta G^\circ = -RT \ln K = \Delta H^\circ - T \Delta S^\circ$ the reaction rate

$$= \left[(RT/N_0 h) \frac{\gamma_A \gamma_B}{\gamma_X} \exp(\Delta S/R) \right] \exp(-\Delta H/RT) c_A c_B$$

$$= \alpha T \exp(-\Delta H/RT) c_A c_B \tag{8.9}$$

(where the constant αT replaces the contents of the square bracket), so that

$$k = \alpha T \exp(-\Delta H/RT). \tag{8.9a}$$

This is not the Arrhenius equation but the preferable form mentioned on page 178. It should be pointed out that if the reaction is of the first order the number of species which must appear on the left side of the equation (8.7) can be only one, while if of the second order there must be two species. One reason for determining the order of a reaction is to enable this theory to be applied successfully to a particular case. If the range of T is short and includes T_m we can put

$$T = T_m(1 + (T - T_m)/T_m)$$
$$\simeq T_m \exp((T - T_m)/T_m)$$
$$\simeq T_m \exp((T - T_m)/T)$$
$$\simeq T_m \cdot e \cdot \exp - (T_m/T)$$

and substituting for (the first) T in equation (8.9a),

$$k \simeq [\alpha T_m e] \exp [-(\Delta H + RT_m)/RT] = A \exp(-Q/RT) \qquad (8.9b)(8.6)$$

which is the Arrhenius form, valid only over a moderate temperature range round T_m. In fact the equation is found to be applicable over greater temperature ranges than seem to be justified by this somewhat contrived derivation. (See examples 19 and 20 in Appendix 1.)

These equations are not convenient from which to calculate absolute reaction rates. They demonstrate, however, that A, containing the entropy term, is in part a probability factor reflecting spatial and energy distributions and also that A contains an inconstant factor in the activity coefficients which may alter during the course of a reaction. It is also clear that Q is not exactly the heat of formation of the complex except at T_m.

The value of A can be obtained experimentally by determining the specific reaction rate at several temperatures and plotting log k against $1/T$ K. The slope of this line, $Q/2.303R$, gives the activation energy of the reaction while the extrapolated intercept in $1/T = 0$ (i.e. at infinite temperature) gives A as the rate of reaction when all the molecules are energized. Dimensional analysis shows that the units of A are the same as for k and depend upon the order of the reaction—collisions per second for first order, collisions per unit of concentration per second for second order, etc. (see equations (8.1), (8.2)).

The Arrhenius equation can, of course, be used for comparing rates of reaction, particularly at different temperatures. (See example 20 in Appendix 1.)

The Collision Theory

Moelwyn-Hughes[43] has shown how collision factors can be calculated from the equations of the kinetic theory of gases (applied alike to gaseous and liquid systems). If there are n molecules per cm^3 of mass m their average velocity at T K is $u = (kT/2\pi m)^{1/2}$ where $k = R/N_0$ is Boltzmann's constant. The number of molecules reaching unit area of a plane boundary surface is the same as the number in a cylinder of unit cross-sectional area and length u, i.e. nu, so that the number of collisions per unit area is

$$Z_1 = n(kT/2\pi m)^{1/2} \quad \text{or} \quad n(RT/2\pi M)^{1/2}, \qquad (8.10)$$

M being the molecular weight.

If the system contains two solute species A and B in a solvent S the frequency of collision between A and B depends on their numbers n_A and n_B per cm^3, their molecular radii r_A and r_B, and their masses m_A and m_B. Collisions occur when molecules approach to $(r_A + r_B)$.

The number of times an A molecule strikes a B molecule per second is given by

$$Z_B = n_B(r_A + r_B)^2 \sqrt{[8\pi kT(1/m_A + 1/m_B)]} \qquad (8.11)$$

so that the total number of collisions between A and B molecules per second is—

$$_AZ_B = n_A \cdot n_B(r_A + r_B)^2 \sqrt{[8\pi kT(1/m_A + 1/m_B)]}. \qquad (8.12)$$

The number of solvent molecules encountered per second depends on the diffusivity D or on the viscosity η of the solution and has been given as—

$$_AZ_S = \frac{8RT}{3\pi M_A D} \quad \text{or} \quad \frac{3\pi \eta r_A}{m_A}, \qquad (8.13)$$

M_A being the molecular weight of the solute. Note that the dimensions of Z_1 and $_AZ_S$ imply first-order reactions while those of $_AZ_B$ imply a reaction of the second order.

The Calculation of Reaction Rates

These equations enable collision frequencies to be calculated if proper estimates can be made of molecular radii. These can be calculated from the molecular volumes of condensed phases at low temperatures. Some typical results, in which the frequencies may seem to be surprisingly high, are given below.[44]

1. At atmospheric pressure 10^{24} molecules per second of any gas collide with every cm^2 of bounding surface.
2. At the surface of a liquid about 10^{15} molecules each vibrate about 10^{14} times every second, so making about 10^{29} attempts per cm^2 per second to escape or react.
3. Within a liquid the frequency of collision is also high at about 10^{30} per cm^3 per second in pure liquids. If collisions between solute molecules are considered, this will depend on their concentrations. Molecules of two solutes each at a concentration of mole fraction = 0.01 would collide 10^{26} times per second in every cm^3.

Since these frequencies are proportional to $T^{1/2}$ from (8.10) their magnitudes are only moderately affected by increases in temperature.

The fraction $\exp(-Q/RT)$ of all possible collisions which have enough energy to effect reaction depends on T and Q according to the values given in Fig. 50 which covers the ranges of both factors which are of interest in extraction metallurgy.

If a reaction occurs at an interface between two liquids and the reactants are in solution at a concentration of 1 mol. % on each side, the number of "tries" per cm^2 per second on each side is 10^{27} but as only 1% of these find the right type of molecule the number of collisions between suitable pairs is 10^{25} per cm^2 per second. If the steric factor $P = 1$ and $Q = 25$ kcal (105 kJ) at room temperature, the fraction of collisions in which the energy is high enough to cause a reaction is 10^{-18} from Fig. 50, so that only 10^7 molecules per cm^2 per second are effective in causing a reaction. This is a very small amount of matter and would probably not be detectable. At $1500°C$, however, $\exp(-Q/RT) = 10^{-3}$ and the number of effective collisions would be 10^{22} per cm^2 per second, corresponding to 0.1 g-molecules of each

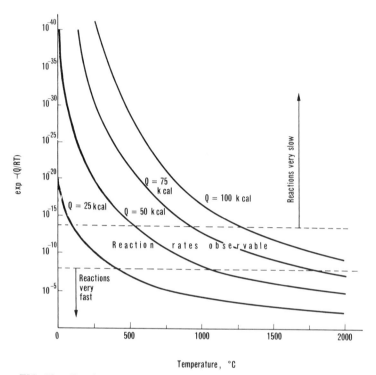

Temperature, °C

FIG. 50. Graphical representation of the relationship between Q and T and the value of $\exp(-Q/RT)$ showing the ranges in which reaction rates are likely to be very high, very low, and of intermediate value. (After Goodeve[44].)

reacting species. From solutions with a molar concentration of 1%, this rate is unreasonably high and would undoubtedly be prevented by failure to transport the reacting species to the surface. Such a reaction, where the activation energy is 25 kcal/mole (105 kJ mol^{-1}), would proceed with adequate speed at about 800°C where the effective collisions would be at a rate of about 10^{20} per cm^2 per second. Even this rate means about 2 metric tons per m^2 per hour of, say, iron, a rate which might often be difficult to reach because of mass transfer difficulties. High temperature operation may be dictated by melting point but this need not lead to *very* high reaction rates, for the reason given.

In all of this discussion it would be more accurate if concentrations were replaced by activities. In this last example where the concentrations were 1 mol.% if the activity coefficients were, say, 0.1 in each case, the effective concentrations would be down to 0.01 and the reaction rates calculated would be reduced by a factor of 10^2. Sometimes activity coefficients are not known but where they are available or where they can be reasonably estimated they should always be used.

Heterogeneous Reactions

Most reactions in extraction metallurgy are heterogeneous — that is they occur at the interface between two phases like that just discussed. The most important of these are the gas—solid reactions as in roasting and reduction processes, the gas—liquid reactions of the converter processes and the liquid—liquid reactions of the slag-refining processes. Liquid—solid reactions occur in hydrometallurgical extraction processes and in electrometallurgy and incidentally elsewhere as in the erosion of refractories by slag. Combustion reactions may be gas—solid, gas—liquid or may appear to be homogeneous in the vapour phase but even in this case much of the reaction may occur at an invisible and diffuse interface between large bodies of different gases.

Homogeneous reactions are seldom met in extraction metallurgy, largely because of the nature of the reactions which are concerned with separations and transfers of species from one phase to another across interfaces. When a soluble reagent is added to a solution in order to precipitate a solute as in the case of adding manganese as ferromanganese to steel to precipitate oxygen as MnO, the reaction will occur very rapidly, and reaction fronts will form around zones rich in reagent, near the point at which it was introduced. In these zones the solubility product of the precipitating phase is likely to be grossly exceeded, in which case nucleation of the new phase will be spontaneous. The localized reaction proceeds rapidly but there is occasionally some difficulty in dispersing the reagent to ensure uniform treatment of large batches of materials. It can occur that two solutes are present in concentrations which do not exceed the solubility product of a compound formed between them but that a small change of temperature can cause that product to be exceeded so that precipitation of the compound would be expected. This precipitation can be delayed because of difficulty in the production of nuclei of the new phase. Nuclei cannot possibly form at the point of equilibrium but only when some degree of supersaturation has been reached. The formation of stable nuclei is in effect a high-order reaction of high activation energy. A large number of atoms or molecules must be assembled before a nucleus is stable but once that condition has been reached it will grow rapidly, the reaction now being conventionally heterogeneous. In molten metals and slags reactions are rapid but mixing can be slow and atoms cannot react until they are brought into contact. As extraction processes develop toward greater use being made of low-temperature leaching techniques, more use of organic reagents and more use of halides, including gaseous phase reactions, we may find that, at the lower temperatures employed, reaction rates will be generally lower and more homogeneous reactions will come to be of interest.

Nucleation of a new phase in a solution involves an accidental convergence of a number of molecules of the precipitating species from atoms or molecules of its components in the solution. In a highly supersaturated solution such an accident becomes very probable and some solutions may contain a high population of points around which precipitating phase molecules are conveniently clustered at any moment. As the new phase is formed there will be a free energy change ΔG made up of two components – ΔG_B and ΔG_S where ΔG_B is the difference between G in the bulk of the solution and G in the bulk of the new phase while ΔG_S is the difference between the free energy in the solution and that of molecules at the surface of the precipitate. ΔG_B will always be negative if precipitation is possible but G_S will be higher than G_B so that ΔG_S will be less negative than ΔG_B and when ΔG_B is very small, near to the equilibrium temperature, ΔG_S may well be slightly positive. In a small nucleus, therefore, where nearly all the molecules are on the surface, the free energy of its formation might be positive but a larger accumulation of the same molecules would include many more in interior positions, so that the average free energy of formation would be smaller. At some critical size, the addition of one more molecule will actually lower the chemical potential of the group of molecules. In a supersaturated solution it will begin to grow. If the solution is sufficiently supersaturated the value of ΔG_S will be negative, in which case nuclei will form spontaneously throughout the solution.

If nucleation is limiting a reaction rate, it can be accelerated by providing "seed" crystals of the precipitating phase (or one with similar crystallographic characteristics) or gas bubbles, if it is a gas which is to be precipitated. Open-hearth steelmaking would seem to depend on the fortuitous presence of growth sites for carbon monoxide bubbles in crevices in the hearth bottom. In the iron blast furnace it may be the absence of such sites that restricts the rate of transfer of sulphur to the slag from the metal in which reaction the removal of an equivalent amount of oxygen from the system as carbon monoxide is necessary.

The velocity of a heterogeneous reaction is nearly always determined by the rate at which one or other of the reactants can be transported to the reaction sites at the interface (or one of the products transported away). In solids, transportation can be by diffusion only. This is very slow and solid–solid reactions are avoided. In fluids, transport is by convection over long distance except when the fluid is very viscous or quite still, in which cases diffusion is the much slower but only alternative mechanism. Convective transfer rate is measured as the product of the concentration of the species and the velocity of the flow of the fluid in the required direction. Adequate convective transfer can usually be maintained easily by mechanical, electromagnetic, pneumatic or hydraulic stirring or agitation. Flow parallel to the interface should be fast enough to be turbulent so that a strong component of flow *toward* that surface is maintained and so that the "stagnant film" associated with the surface should be as thin as possible. Within this film, fluid flow is laminar, parallel to the surface only, so that mass transfer across it, i.e. normal to the surface, can only be by diffusion. Its thickness depends on the Reynolds number for the flow and decreases with increasing velocity. It is also least when viscosity is least and is therefore lower in metals (0.001–0.005 cm) than in slags (0.01–0.02 cm) and would be expected least of all in gases. The thickness and stability of the film also depend on the "roughness" of the surface with which it is associated, and upon whether attachment by some sorption mechanism is operating.

Diffusion across this thin film is governed by Fick's laws, particularly the first. The "flux" J_i (in g/cm² s) of a species i diffusing past any point in the x direction is given by—

$$J_i = -D_i(dc_i/dx) \qquad (8.14)$$

where D_i is the diffusion coefficient for i in the phase through which it is diffusing, and dc_i/dx is its concentration gradient in the x direction. The value of a diffusion coefficient depends not only on the diffusing species and the host medium but also on the concentration level and on the temperature. Data so far available are inadequate to justify complicating the formulae to account for the variation with concentration. D_i varies with temperature according to—

$$D_i = D_0\exp(-Q/RT) \qquad (8.15)$$

where D_0 is a constant and Q is the activation energy of diffusion. This is the energy required to bring the diffusing atom or molecule into the intermediate position between two lattice sites from which it is as likely to fall into an empty site adjacent to that being vacated as it is to return to its original position. Activation energies of diffusion in metals are rather low — up to about 20 kcal/mole — and seem to depend mainly on the nature of the solvent. These values are rather lower than the activation energies for chemical reactions. Values for diffusions in slags on the other hand are considerably higher, perhaps approaching 100 kcal/mole, probably reflecting the greater range of viscosity in slags.

In the present context Fick's law becomes—

$$J_i = -D_i(a_b - a_s)/\delta \qquad (8.16)$$

where a_b and a_s are the activities of the diffusing species i in the bulk of the fluid and at the reaction front respectively, δ being the effective thickness of the film across which the diffusion takes place. The use of activities for concentrations is strictly more correct and it does occasionally happen that diffusion down an activity gradient is actually *up* a concentration gradient. Where activities are known in this field they should be used. The dimensions of D are $L^2 T^{-1}$ and the units of J are determined by those of a (or c).

In recent years some progress has been made in the development of a theory of mass transfer under turbulent-flow conditions and the above treatment is considered by some workers to be inadequate. One reason is that the boundary film thickness δ seems to be of doubtful physical significance. At its best this film lacks a sharp boundary and its thickness cannot be measured. Estimates are made from the probable concentration profile toward the surface of an "effective film thickness" but this quantity depends on the value of the diffusivity D_i of the solute being considered and so it can be different for different solutes in the same solution and different again in a value calculated for use with an analogous heat-transfer equation. It is considered better to combine D_i/δ_i as a single mass transfer coefficient k_M so that equation (8.10) becomes simply

$$J_i = k_M \ (a_s - a_b).$$

Studies of the surfaces of liquids in turbulent motion suggest that small elements of liquid are continually and randomly being thrown up to the surface bringing new material into brief contact with the contiguous phase. Particularly at high temperatures, the newly exposed material comes quickly into equilibrium with the opposing phase and mass transfer occurs in accordance with the laws of diffusion until the element is displaced by another being forced up from below. If the liquid is in a moderately turbulent state the element thrust into the surface may be considered to be essentially the "stagnant film" material, in which case mass transfer to or from it will be diffusion controlled. Liquid in extremely turbulent motion may, however, thrust into its surface elements which are themselves fully turbulent in which case the whole element will quickly come into equilibrium with the opposing phase. This gives a faster surface reaction rate.

In the first case the rate at which any species, i, traverses the interface is given by

$$dx_i/dt = A_e(a_s - a_b) \ 2 \ (D_i/\pi t')^{1/2}$$

where A_e is the area of the element of liquid being considered and t' is the time during which it remains on the surface. On the other hand, in the second case the reaction rate is controlled by the component of the velocity of the fluid V_n normal to the surface and is given by

$$dx_i/dt = A_e \ (a_s - a_b) \ V_n.$$

The first equation would apply when the velocity of the liquid parallel to the surface was fairly low and turbulence moderate. The surface would appear to be smooth or only slightly disturbed. The second equation would apply where the flow rate was fast and the surface obviously swirling and eddying. Many real cases must lie between these limiting cases and can be described by a joint equation apportioning the effects according to the relative areas of surface affected. Other cases must arise where there is no turbulence at all, when Fick's law is clearly valid, or when there is so much disturbance that elements of liquid are ejected into the opposing phase. In extreme cases emulsions may be formed, giving a high rate of reaction because of the large interfacial area created.

In general we can write

$$J_i = k_M \, (a_s - a_b)$$

where k_M is a mass transfer coefficient which must be determined empirically either directly or indirectly on some analogous system, the results from which can be interpreted using the theory of dimensionless groups. This tends to restrict the treatment to a few fairly simple geometric arrangements.

There are, of course, two sides to every interface and the overall rate of transfer may be controlled entirely by conditions on one side or the other or jointly or by chemical control (high activation energy). Control on one side can be due to a very small concentration difference between the surface and the interior, for example in a refining operation when the bulk concentration is already low and the surface value cannot fall below zero. Alternatively a k_M value may be low because of a low level of turbulence on one side. If one phase is solid, transport within it can be only by solid-state diffusion which is inherently slow and must sometimes be effective over quite long distances. Chemical control is occasionally effective. When it is, there will be no boundary layer because all species can be replenished at the interface or cleared from it faster than necessary for the reaction to proceed unhindered.

In practice there are many complexities which make quantitative predictions of reaction rates very difficult. Of these, the production of foams and emulsions is probably the most important. Temperature differences across the system and sometimes between the phases on either side of an interface make even the estimation of activities almost impossible. Despite the quantitative work which has been done it is still important to have a good qualitative understanding of factors affecting reaction rates and how the effects are produced.

Gas–Solid Reactions – Calcination

The progress of gas–solid reactions like calcination, roasting and reduction of ores is complicated by the fact that chemical reactions are proceeding while the ore is being heated. A reaction cannot occur at all until the temperature is high enough at the reaction interface.

Most of these processes are carried out in shaft furnaces or rotary kilns in which the counterflow principle operates. The burden to be heated and the combustion gases travel through and over each other in opposite directions so that at any point the gas is only a little hotter than the surface of the burden – by, say, $50°C$. In this way a high thermodynamic efficiency is attained in the heat transfer process because it is a fairly good practical approach to a thermodynamically reversible process (see page 111). The degradation of heat energy, or its loss of "virtue", is minimized but the ore lumps heat up rather slowly as a consequence and the temperature at the centre of the lump may be hundreds of degrees lower than that at the surface depending on its size, shape and conductivity.

The dissociation of a carbonate cannot occur below its decomposition temperature, which can be calculated from free energy data (page 158) and varies with the prevailing partial pressure of CO_2. For $FeCO_3$ this temperature is about $400°C$ when $p_{CO_2} \simeq 1$ atm. A lump of carbonate ore in a calcining kiln will in time come to be in two zones – an outer zone which has been transformed to oxide and an inner one unchanged. The boundary between them will be an isothermal surface corresponding to the decomposition temperature or a rather higher temperature at which the decomposition is able to proceed rapidly. Except for the possible formation of nuclei of oxide on which to build crystals of the solid product, the reaction is first order. Since the solid product is less dense than the original ore, it is probably permeable to CO_2 and offers little resistance to its escape from the site of the reaction. There is therefore no build-up of back pressure and no diffusion problem. The CO_2 pressure at the reaction front is near 1 atm pressure. The overall reaction rate is tied to the rate of advance of the reaction temperature isothermal surface, that is, it depends on the transfer of heat to raise the ore temperature and supply energy for the endothermic reaction. This heat must pass through the reaction product. If this is dense and massive the thermal diffusivity will be high; if it is porous and permeable it will be low. If the product breaks away, the heating rate of the residue will obviously be increased.

The heat transfer rate depends on a number of factors — the gas-flow characteristics as they affect the thickness of the gas film at the solid surface through which all the heat must be transferred by conduction (in a manner analogous to the transfer of matter by diffusion discussed on page 190); the gas–solid temperature difference; the temperature gradient in the solid; the depth of the reaction temperature isotherm in the ore lump; and the thermal conductivity of the ore particularly after its transformation to oxide. Conditions change as the reactions proceed, and ultimately conduction of heat through the transformed ore will usually be the rate-controlling factor. The most effective means of hastening the process as a whole appear to be by reduction of the size of the ore lump. A small size reduction, however, would increase the resistance offered by the burden to the passage of the gases and might retard production rate. A reduction in size also increases the relative importance of the resistance to heat transfer of the boundary film in the gaseous phase. A large reduction in particle size would alter the whole process — possibly taking it out of the traditional types of kiln and putting it into a fluidized-bed type of reactor vessel (see page 279). Obviously, other things being equal, the greater the temperature difference between the gas phase and the centre of the lump the faster it will heat up. The classical dilemma arises, whether to use more energy or take more time. This can be resolved only by a careful analysis of all the costs involved. This discussion is applicable to drying operations and calcinations to remove combined water and effect other decompositions.

Gas–Solid Reactions – Roasting

The roasting of ore lumps to convert sulphide to oxide or sulphate differs significantly from calcination. Roasting of this kind often involves exothermic reactions and is not likely to be rate controlled by thermal diffusion. It can be presumed that the temperature is everywhere high enough to allow the desired chemical reactions to occur at a fairly early stage in the process or at least that the reaction-temperature isotherm is well ahead of the reaction front. Each particle is oxidized from the outside so that once more an inner unchanged zone and an outer shell of solid product are formed. At the interface between these zones the appropriate reaction occurs provided the ratio p_{O_2}/p_{SO_2} is locally higher than the equilibrium ratio for the reaction at the prevailing temperature. For example, if the reaction is—

$$2ZnS + 3O_2 = 2ZnO + 2SO_2, \qquad (8.17)$$

$$K_T = p_{SO_2}^2/p_{O_2}^3. \qquad (8.18)$$

In this case it is clearly necessary that at least three molecules of oxygen come to the interface for every two of sulphur dioxide which permeate away — i.e. $J_{O_2} \ll 3/2 J_{SO_2}$. These values depend on the respective diffusion coefficients and the partial pressure gradients down which the mass transfer takes place. The diffusion coefficient for O_2 will be rather greater than that for the larger molecule SO_2 but there is always a minimum value of the ratio p_{O_2}/p_{SO_2} in the atmosphere *outside* the particle which must be maintained in order that the value of the ratio at the reaction front shall always exceed the equilibrium value. The "stagnant" film on the surface of the particle is of no special importance in this case, unless the particle size is very small.

At the interface the reaction mechanism probably occurs in several stages. First, oxygen is adsorbed on to the surface of the sulphide. Secondly, it yields electrons and becomes incorporated in the lattice of the mineral. The electrons neutralize a sulphide ion on the nearby surface. This becomes an adsorbed sulphur atom which can combine with oxygen atoms adsorbed near it on the surface. The SO_2 molecule so formed desorbs and migrates away leaving a vacant site on the mineral surface. Another sulphur ion may move up to occupy this site and continue the reaction but in general the interface advances into the mineral to reach more sulphur ions. The role of adsorption is not very clear but it seems to reduce activation energies and hasten reactions. The velocity of reaction is presumably governed by the number of oxygen molecules adsorbed per unit area of interface and in accordance with the laws of adsorption this is proportional to the local value of p_{O_2} and hence depends upon the value of p_{O_2} in the atmosphere around the particle and in diffusivity. In the final stages, the larger the particle the higher is the p_{O_2} in the atmosphere outside which is necessary to maintain a particular flux of oxygen at the reaction interface. Steps should be taken to increase the oxygen potential in the gases round the ore in the last stages if total oxidation is required. A counter-flow process is obviously appropriate.

Even this treatment has been simplified in several respects. First, unless the temperature is everywhere the same, diffusion would better be considered as being down a "chemical potential gradient"

$$\frac{d(RT \ln p)}{dx}$$

in the x direction rather than a partial pressure gradient. Secondly, ore particles are not homogeneous lumps of sulphide. Inert gangue may hinder or even prevent reactions occurring mainly for geometric reasons. Thirdly, the heat of reaction may cause fusion or agglomeration of particles which could alter the diffusion coefficients and diffusion paths and ultimately slow up the process. Fourthly, side reactions may occur by design or accident. Prominent among these is sulphate formation, most probably in a reaction zone in the outer (cooler) part of the oxide shell, sulphates not being stable except at temperatures below about 800°C. By providing a "sink" for SO_2 this should permit a lower p_{O_2}/p_{SO_2} ratio to be operated without detriment to the main reaction rate. Deliberate sulphating of fine concentrate is usually restricted to a separate part of the kiln where temperatures are uniformly lower and the SO_2 content of the atmosphere is distinctly higher than in the oxidizing section.

A mass of fine concentrate must behave like a single large lump toward the reacting gases unless steps are taken to improve the gas–solid contact over that possible diffusion alone – e.g. by raking or fluidization. It will be appreciated that gases will usually penetrate the solid by way of intergranular routes rather than through crystals. At the reaction front individual crystals become converted to product "topochemically", that is the radial rate of conversion is the same all the way round. What is called a front may be several crystals wide with each of these at progressively less advanced stages of conversion toward the centre of the ore particle. As the outermost grains in this frontal zone become completely converted, so the zone moves inwards to involve more grains deeper into the particle. The rate of attack on these grains within the mineral particle may be determined by quite a different mechanism from the particle as a whole and is very likely to be a solid-state diffusion mechanism within the grain. If the grain size is coarse this mechanism can take over control of the reaction rate for the particle as a whole. In the extreme case of the ore lump consisting of a crystalline mass of igneous origin like some iron ores which are almost pure crystalline magnetite, the roasting reaction is controlled completely by solid-state diffusion.

Gas–Solid Reactions – Reduction

Reduction processes as conducted in blast furnaces are rather similar to roasting processes. Once more the "stagnant film" can offer only negligible resistance to the transfer of matter compared with that offered by the outer zone of the reduced ore. Once more the reduction of the innermost core must be rather slow unless a very high reduction potential gradient is available to maintain a high diffusion rate along a very long path. The most important industrial reduction process is the reduction of iron oxides in the stack of the iron blast furnace. The principal reducing agent is carbon monoxide in the furnace atmosphere which has a CO/CO_2 ratio of about 10^4 at the bottom of the stack falling to about 1/1 at the top. There is also some hydrogen present and in general carbon, as CO or CH_4, and hydrogen are the reducing agents in all similar reduction processes. The reactions involved are always endothermic and might be controlled by heating rate alone like calcination reactions but a comparison of the reducibilities of a wide range of iron ores has shown[45] that, for these at least, the permeability of the particle to gases is the most important factor determining rate of reduction. In some minerals there can be two or even three stages of reduction (as in the case of iron oxides) and a like number of interfaces operating within each lump of ore. Each interface acts like a valve controlling the reduction potential of the gas which can diffuse beyond it. This is of little importance, however, because the higher oxides within are usually the most easily reduced. In the case of iron ores it is the lowest oxide, FeO, that is the most refractory and in the last stage of its reduction a piece of ore consists of a core of FeO surrounded by "sponge" iron. The most difficult iron ore to reduce is the crystalline magnetite mentioned above. Its reduction is by diffusion of oxygen from the interior toward the reaction front. The reduction of Fe_3O_4 produces an FeO which has a slightly greater volume than the magnetite and forms a very compact mass around it. Its reduction is so slow that it is sometimes considered worth while to give the ore an oxidizing roast to haematite (Fe_2O_3) which on reduction gives a permeable magnetite the reduction of which through permeable FeO to sponge iron is relatively very easy.

In both roasting and reduction processes it is obvious that higher overall reaction rates can be obtained by increasing the reaction interface areas. This can readily be done by reducing the size of the ore particles which has the additional merit that the longest diffusion path is also reduced. In traditional equipment this will usually lead to the development of higher resistances to gas flow between the particles and this alone can slow up production rates (which are what matter). If the particle size is small the ore must be stirred, turned over, or rabbled to ensure the adequate exposure of the surfaces of the particles to the reactive atmosphere. Obviously there is a limit to which size reduction can be taken if traditional equipment is to be used but modern suspended solids fluidized bed techniques may be able to exploit the advantages of very small particle size without suffering from the worst of its disadvantages (see page 277). The use of small uniform pellets permits the use of moderately small particles without unduly increasing the resistance to gases passing through the ore bed.

The use of counterflow in reduction processes is virtually obligatory with the gas with the highest reduction potential (highest CO/CO_2 ratio or lowest p_{O_2} value) meeting the ore in its final stages of reduction where it can impose the highest possible chemical potential gradient across the mineral particles. At the same time gas which has been oxidized in its traverse of the system may well still have sufficient reduction potential to take the ore through its first stage of reduction. In the iron blast furnace the CO/CO_2 ratio at the bottom of the stack is about $10^4/1$ where the last of the FeO is being reduced. The equilibrium CO/CO_2 ratio is only about 10/1 at that stage but such a gas mixture would effect reduction very slowly indeed. At the top of the stack the gases have a CO/CO_2 ratio of about $1/1$ where a mixture with $CO/CO_2 = 1/1000$ could in theory reduce haematite. These large excesses of reduction potential are needed to ensure adequate reaction rates and also because the process needs an adequate capacity for reduction. A gas mixture which is near to the equilibrium composition not only takes a long time to effect reduction but must be supplied in vast quantities.

Liquid–Solid Reactions – Leaching

Leaching of ores is rather similar to gas–solid reactions but is conducted at a uniform temperature. The ore lumps contain tiny particles of valuable mineral embedded in gangue and the exact mechanism of leaching by solution and diffusion depends on the spatial distribution of the two phases and on the permeability of the gangue. If the gangue is quite impermeable the mineral particles must be exposed and solution will proceed on a narrow front, solvent ions diffusing inwards and dissolved ions diffusing out to the liquor outside by the narrow path etched out by the solution reaction. If the gangue is fissured or permeable, solvent may reach the surfaces of totally enclosed mineral particles. Solution may then be easy but the escape of the dissolved ions can only be by diffusion and must be slow. As a result the concentration of a metallic ion within the particle must be high and near to saturation during most of the process. If the process involves a reaction such as the oxidation of the mineral prior to its solution the oxidizing reagent must also reach the reaction site by diffusion. The process is almost certain to be rate controlled by the diffusion of one or other of the ions involved – whichever is the slowest having regard to both their size and the concentration gradients determining their motion.

Obviously leaching proceeds faster when the ore particles are small and if the liquor outside can be kept dilute with respect to the metallic ions being extracted, but strong with respect to reagents. The adoption of counter-flow operation would help to meet these requirements. The accumulation of high concentrations of dissolved salts in the "stagnant" film around the ore particles should be minimized by agitation to reduce the thickness of those films which are much more important in their effects in liquids than in gases. (See also page 205.)

Liquid—Liquid Reactions at Slag—Metal Interfaces

Heterogeneous reactions at liquid—liquid interfaces include the important slag—metal reactions of steelmaking which have been under close examination in recent years. Steelmaking temperatures are so high that reactions are usually presumed to be extremely fast ($\exp(-Q/RT)$ high) so that thermodynamic equilibrium is achieved everywhere (in short range) and especially at the interface where it is envisaged that the top layer of atoms in the metal and the bottom layer in the slag are at the same chemical potential with respect to all species present. The rate at which species pass through the interface depends principally on the rate of their transport under the chemical potential gradients existing between the bulk of the slag and the bulk of the metal but each chemical reaction at the interface acts like a valve restricting the flow of each species involved to chemical equivalents of the slowest to pass through.

When convective transfer is well organized (as it is in steelmaking) the composition on either side of the interface is constant up to a distance δ from the interface which is the thickness of a boundary film or "stagnant layer" across which mass transfer can be only by diffusion as already discussed on page 190. In the steelmaking case δ on the metal side is thought to have a value about 0.003 cm and on the slag side 0.015 cm, figures which seem to be justified by their success in calculating very reasonable values for the flux of oxygen and manganese during the conventional steelmaking process.[46] Under other conditions of temperature and degree of agitation they might be somewhat different.

In each side of the interface Fick's law can be applied and the flux of any species must be the same on both sides at any time so that—

$$D_{i(\text{slag})} \frac{a_{i(s-b)} - a_{i(I)}}{\delta_{(\text{slag})}} = D_{i(\text{metal})} \frac{a_{i(I)} - a_{i(m-b)}}{\delta_{(\text{metal})}} \qquad (8.16a)$$

where (I) signifies interface and $(s - b)$ and $(m - b)$ signify bulk values on the slag and metal sides respectively. The activity values should be inserted where they are available and it should be noted that the activity values at the interface $a_{i(I)}$ are *not* the same numerically when referred to different standard states on the metal and slag sides. If activity values are not available concentrations can be used. It should be realized that activities may be strongly affected by the presence of other solutes within the boundary layer.

Diffusion in liquids is difficult to determine accurately. Values in slags are lower at about 10^{-5} cm^2 s^{-1} than those in metals for which 10^{-4} cm^2 s^{-1} is more typical. Values of δ already quoted still lack direct experimental confirmation.

A further difficulty encountered in the application of this attractive and apparently simple theory is that the assessment of interfacial area usually reduces to guesswork. Agitation in steelmaking is caused by the evolution of bubbles of carbon monoxide which grow rapidly and break through the interface in large numbers. The effective area of the interface is therefore greater than that of the bath by an unknown factor of the order of 5 or 10 which depends mainly upon the amount of fragmentation or splashing which occurs, throwing small pellets of metal up into the slag layer. Secondly, as the interface is disturbed it is extended so that the atoms must move forward from rearward ranks much faster than they could diffuse. If immediate equilibration occurs when they reach the surface and before the surface re-assumes its normal shape the diffusion gradients are made more steep on both sides and the mass transfer rates are increased beyond those predicted by the simple theory. In view of the complexity of the system it seems something of a miracle that the use of D_i/δ_i as mass transfer coefficient has predicted the observed reaction rates as successfully as in the work mentioned.[46] However more elegant the theory based on turbulent flow patterns discussed earlier, this direct approach, based only on diffusion, is justified by some successful applications. This does not mean that it represents the physical processes involved better than the turbulent-flow model — only that the diffusion approach can provide a usefully realistic estimate of transfer rates under a set of simply defined conditions. (See example 21 in Appendix 1.)

The amount of agitation varies greatly from one process to another. In the iron blast-furnace hearth it is rather small. The effective area is practically the plane area of the hearth less the cross-section of the coke immersed in the metal, and the diffusion zones may be several inches thick but there is some disturbance due to descending metal. Reaction rates in the blast-furnace hearth are well known to be slow. The factors determining mass transfer rates also determine heat transfer rates, and a steep temperature difference between the slag and the metal is a characteristic of the blast-furnace hearth, the slag being perhaps $100°C$ hotter than the metal, corresponding to the large difference observed between the chemical potentials of, say, sulphur in the two layers.

The presumption that equilibrium is almost instantaneously attained at the interface remains to be examined. When a reaction consists of a simple transfer of a species like oxygen from slag to metal the presumption is probably very near the truth. Simple reactions do not often occur in isolation, however, and even the transfer of oxygen from a slag to a metal must be written—

$$(O)^{2-} \rightarrow \underline{O} + 2e^-$$

where the parentheses indicate species in the slag phase and the underlining species in the metal phase. As will be discussed in the next chapter, available oxygen in slags exists as ions. Clearly the two electrons must go somewhere and indeed it is necessary to assume that the transfer of oxygen is accompanied by the transfer of a cation also—

$$(Fe)^{2+} + 2e^- \rightarrow \underline{Fe}$$

so that the net reaction is a transfer of FeO. Equally well the accompanying cation might be that of manganese—

$$(Mn)^{2+} + 2e^- \rightarrow \underline{Mn}.$$

If it is desired to oxidize the manganese out of a solution in iron by means of an oxidizing slag, i.e. one high in FeO, the sequence of reactions will then be—

$$(FeO) \rightarrow (Fe)^{2+} + (O)^{2-}$$

$$(O)^{2-} \rightarrow \underline{O} + 2e^-$$

$$(Fe)^{2+} + 2e^- \rightarrow \underline{Fe}$$

followed by $\qquad \underline{O} + \underline{Mn} \rightarrow (MnO).$

This composite reaction might appear to involve slow stages in the dissociation of FeO and in the nucleation of MnO. The theory of slag structure discussed in the next chapter shows, however, that the ionization of FeO in steelmaking slags is well advanced and the reaction would not be slowed by that step. Further, the nucleation of MnO is not necessary at the interface because it is so soluble in the slag. Indeed it could be that the oxygen never actually crosses the interface; or that MnO never actually forms but that its components enter the slag separately but in equal numbers. It is clear that there is no part of this reaction that could control its rate slower than the diffusion to the interface of oxygen or iron through the slag, or of manganese through the metal, the slag diffusion probably being the slower, during most of the process, though perhaps not near an equilibrium condition.

If MnO is to be reduced from a slag the reaction sequence will be—

$$(MnO) \rightarrow (Mn)^{2+} + (O)^{2-}$$

followed by
$$(O)^{2-} \rightarrow \underline{O} + 2e^-$$

and
$$(Mn)^{2+} + 2e^- \rightarrow \underline{Mn},$$

but if the oxygen is allowed to accumulate in the metal phase this reaction will soon cease. It must therefore be removed by a reaction such as—

$$\underline{C} + \underline{O} = CO \text{ (gas)}.$$

In this case CO must be nucleated and as that is a stage involving a high activation energy it could be slow enough to control the rate of the overall reaction—

$$(MnO) + \underline{C} = Mn + CO \text{ (gas)}.$$

In the iron blast furnace the reduction of manganese oxide seems to be rather slower than might be expected of a simple reaction at temperatures over $1400°C$. This may be only because the reaction is never very far from its equilibrium but it is also possible that there is control by oxygen diffusing to sites where the evolution of CO gas can proceed easily, without nucleation.

Another and much slower blast-furnace hearth reaction is the reduction of SiO_2 which must proceed in several steps somewhat as follows. First there is a dissociation of silica into ions (see Chapter 9)—

$$2SiO_2 = SiO_4^{4-} + Si^{4+}. \tag{8.19}$$

At the metal surface the silicate ion may disintegrate spontaneously—

$$SiO_4^{4-} = \underline{Si} + 4\underline{O} + 4e, \qquad (8.19a)$$

the Si and O atoms dissolving in the metal (signified by underlining) while the electrons released may neutralize the positive ion—

$$Si^{4+} + 4e = \underline{Si}^0. \qquad (8.19b)$$

It is necessary that the oxygen atoms must not accumulate in the metal or the reaction will not be able to proceed, so the reaction

$$\underline{C} + \underline{O} = CO$$

must also occur.

The overall reaction is of first order and might be controlled by diffusion of oxygen to sites where CO could be evolved or by nucleation of CO, but its activation energy, 250 kJ mol^{-1}, is too high for diffusion control and suggests reaction control through the rupture of the high energy Si–O bond in the second step of the chain of reactions set out above.

The transfer of sulphur from iron to slag in the same process by

$$S^0 + 2e = S^{2-} \qquad (8.20)$$

is also very slow, of first order and of the same high activation energy, with no obvious slow stage to explain the phenomena. If it is assumed, however, that the only source of electrons for (8.20) is a share of those arising by reaction (8.19a) the unusual characteristics of the sulphur reaction and its well-known dependence on silica reduction are adequately explained.

Gas—Liquid Reactions

The nucleation of a gaseous product is a high order reaction of very high activation energy. The pressure of a small bubble is given by $P = 2\sigma/r$ where σ is the surface tension of the wall material, and r the bubble radius. In molten metal this pressure is high because the surface tension is high. In steel $\sigma = 1.6$ N m^{-1} ($= J\ m^{-2}$) or 1600 dynes/cm at 1600°C. Gas in a bubble of radius 10^{-9} m in liquid steel at 1600°C would be under a pressure of

$$\frac{2 \times 1.6}{10^{-9}} = 3.2 \times 10^9 \ N\ m^{-2} \quad \text{or} \quad 32,000\ atm.$$

Employing the relationship $P \cdot V = n R \cdot T$ (equation (6.44)), where $R = 8.205 \times 10^{-5}$ m^3-atm, it can readily be calculated that for a bubble of this size at $1600°C$ (1873 K), $n = 8.72 \times 10^{-22}$ moles which corresponds to about 525 gas molecules. Such an accumulation of molecules might be supposed to be more than enough to form a stable nucleus capable of rapid growth in a supersaturated solution of gas in metal. The internal pressure in so small a bubble is too high, however, to permit gas molecules to enter the bubble from surrounding metal with gas in solution at normal industrial concentrations. In the case of hydrogen in steelmaking the normal range of concentration is 2–10 ppm. The appropriate reaction is—

$$\tfrac{1}{2}H_2 = \underline{H}_{\text{ppm in Fe}}; \quad \Delta G_T^\circ = 31{,}973 - 44.36T \text{ J.} \quad (8.21)$$

At $1600°C$, $\qquad \Delta G_{1873}^\circ = -51{,}515 \text{ J}$

so that $\qquad\qquad K_{1873} = \dfrac{\text{ppm } \underline{H}}{\sqrt{p_{H_2}}} \qquad\qquad (8.22)$

or $\qquad\qquad$ ppm $\underline{H} = 26.6 \sqrt{p_{H_2}} \quad$ (Sievert's law). (8.23)

When $p_{H_2} = 32{,}000$ atm the concentration of hydrogen in molten iron in equilibrium with such a bubble is $26.6 \times \sqrt{32{,}000} = 4758$ ppm or almost 0.5%. At 10 ppm, on the other hand, the equilibrium pressure is 0.14 atm and the corresponding bubble radius is about 0.2 mm. Hydrogen can be precipitated only into bubbles with a radius of curvature greater than 0.2 mm while from smaller bubbles the hydrogen would be absorbed into the metal.

The other important case is the evolution of carbon monoxide from molten steel during decarburization or deoxidation. The reaction is—

$$C_{\% \text{ in Fe}} + O_{\% \text{ in Fe}} = CO; \quad \Delta G_T^\circ = -22{,}390 - 39.7T \text{ J}$$

so that $\qquad \Delta G_{1873}^\circ = -96{,}750 \text{ J}$

and $\qquad K = \dfrac{p_{CO}}{\%a_C \times \%a_O} = 500.$

If the radius of a bubble is 10^{-9} m so that its internal pressure is 32,000 atm, then the product $_{\%}a_C \times {_{\%}}a_O = 64$. On page 409 it is shown that the highest possible value of $_{\%}a_C$ is 35.5 when the iron is saturated with carbon. The corresponding equilibrium value of $_{\%}a_O$ is 1.8 but iron is saturated with oxygen when $_{\%}a_O$ is only 0.23 (see page 151). Equilibrium between a bubble as small as 10^{-9} m containing CO and Fe–C–O alloy at 1600°C is clearly impossible.

Under more likely conditions, with 0.5% C ($_{\%}a_C = 0.53$) and 0.01% O, the equilibrium $p_{CO} = 2.65$ atm; or with 0.1% C and 0.05% O, $p_{CO} = 2.50$ atm.

The total pressure in a bubble of gas in molten steel is the sum of the external atmospheric pressure p_a, the pressure due to the head of liquid metal above the bubble, p_{Fe}, and the pressure due to the surface tension, $2\sigma/r$, i.e.

$$p_{total} = p_a + p_{Fe} + 2\sigma/r. \qquad (8.24)$$

Normally, $p_a = 1.0$ atm. p_{Fe} increases by 0.7 atm for each metre depth and would, in a bath of steel, typically have a value of 0.4 atm. If the total pressure is 2.5 atm the value of $2\sigma/r$ must then be about 1.1 atm or 110,000 N m^{-2}. The radius of the corresponding bubble is

$$\frac{3.2}{110,000} = 0.000029 \text{ m} \quad \text{or} \quad 0.03 \text{ mm}.$$

Only bubbles greater than 0.06 mm diameter can grow. Smaller bubbles are extinguished. Clearly some source of sufficiently large bubbles is required if CO is to be evolved freely from the molten steel. The usual source in open-hearth and electric-arc steelmaking is found in cracks and crevices which occur fortuitously in the hearth bottom or in the surface of unmelted fluxes. Provided the width of such fissures exceeds the critical value, gas pockets within the crevices can grow by absorbing CO from the metal. As the body of gas expands beyond the mouth of the crevice bubbles break off and rise through the steel growing rapidly as they do so. Each suitable fissure acts as a growth point, supplying a continuous stream of bubbles as long as the supply of oxygen and carbon is maintained. Bubbles can also be supplied by injection or by agitation which induces cavitation. The effect of reducing the external pressure p_a to nearly zero is to reduce the critical size of fissure by about half. Rising bubbles would also be larger and of lower total pressure, which would enhance the rate at which they would grow.

Bubbles of gas in metal rise through it because of their much lower density but very small bubbles rise relatively slowly (cf. page 63) and may very easily be swept back down in the streams of liquid which are displaced by the ascent of the more independent large bubbles. In this way the residence time of any bubble is much longer than the time required for it merely to rise to the surface through static liquid. Further, it requires a certain minimum kinetic energy in a rising bubble for it to break through a metal surface against surface tension. Small bubbles tend to collect under the surface or even "bounce" back into the metal.

While the bubble is moving through the molten metal (or slag) it is usually envisaged that it is accompanied by a boundary film on the liquid side across which all reacting molecules must diffuse. Some very successful estimates of reaction rates have been made on this basis.[46] The persistence and stability of such a boundary layer would seem to be subject to the size and shape variations of the bubble in a manner similar to that discussed with respect to liquid–liquid interfaces. Bubbles expand considerably as they grow by reaction or by rise in temperature. Their surfaces therefore incorporate large numbers of atoms from the ranks of atoms behind the interface, which enhances the mass transfer rate to the surface. This effect is of greatest importance when the bubbles are still very small. As they grow bigger the increase in surface area produced by each molecule added to the volume becomes smaller and smaller until the effect becomes of negligible importance. At larger sizes the disturbance of the boundary film by distortion of the bubble is probably much more important. Bubbles are not all perfect spheres but react with the liquid through which they pass so that they assume peculiar mushroom shapes which are continually altering. The effect of such mobility upon mass transfer across a boundary layer can at present only be guessed at.

All this presumes that a stable boundary layer is possible on the liquid side of the interface, attached somehow to the gas bubble.

Adsorption mechanisms for attaching liquid to solid, or gas to either liquid or solid surfaces are readily understood but the degree to which attachment is possible of liquid on to gas seems still open to examination.

Reactions probably occur in stages, as indicated below for the case of a bubble of CO dissolving more CO out of a supersaturated iron melt—

$$CO \text{ (in bubble)} + \underline{O} \rightarrow CO_2 \text{ (in bubble)}$$

$$CO_2 \text{ (in bubble)} + \underline{C} \rightarrow 2CO \text{ (in bubble)}.$$

In this case CO_2 appears in the role of an activated complex which might, however, have an appreciable life before finding another carbon atom with which to react, suggesting a potential controlling step for the process. The process is, however, generally thought to be diffusion controlled, suggesting that transport across an oxygen deficient or carbon deficient zone in the metal is the determining stage. The activated complex may of course be quite different from that suggested above, and the CO_2 may be adsorbed on the bubble surface or may diffuse rapidly across that surface to find the necessary carbon atom with which to complete the reaction and the small proportion of CO_2 actually found in the bubble may play the part only of stabilizing the necessary population of adsorbed molecules on the bubble wall. Obviously the detailed mechanisms of these reactions are not well understood. Gases are dissolved in metals as atoms but emerge into bubbles, with reasonably high reaction velocities, as molecules.

Another type of gas—metal reaction arises when air or oxygen is bubbled through or is blown on to the top of a metal, with the purpose of oxidizing out impurities. Most steel is being refined in this way today.

In the first case, a third phase, the molten oxide of the metal is first formed using up all the available oxygen in the bubble. The oxidation of the solutes which have to be removed probably proceeds thereafter via slag—metal reactions in a manner similar to that described earlier but at a much more extensive interface. Only those solutes whose free energies of oxide formation at the prevailing temperature are (numerically) greater than that of the solvent metal can be so removed. The mechanism of the primary reaction between gaseous oxygen and the metal may be diffusion controlled as soon as a film of oxide covers the surface—diffusion of the metal through the film probably being the slow step. The diffusion of solutes to the oxygen is certainly too slow to permit their preferential oxidation.

When discrete bubbles are blown into the metal the oxide or slag which forms will accompany the residue of the bubble (N_2) (if any) up through the metal. The buoyancy conferred by the gas gives the slag globule a high relative velocity with respect to the metal which must favour a high reaction rate between them by thinning the boundary layer on the metal side.

When oxygen is blown down on to the metal surface the geometry is, of course, very different and the reaction interface would seem to be smaller but the reactions are apparently just as fast — and indeed the limitation seems to be the rate at which oxygen can be supplied. This is because, as a result of the impact of the gas jet, metal spray is continuously being blown off the surface in a partially oxidized state and thrown into the slag layer. This has two effects. First, fresh surface is continuously being exposed, leading to an extremely high oxidation rate which is possibly controlled by the rate of delivery of oxygen to the surface. Secondly, a slag-metal emulsion is formed with an extremely large interfacial area, across which the oxidation and removal of carbon, silicon and phosphorus is rapidly completed. Adjustment of the jet characteristics to produce the right quality of emulsion becomes an important aspect of the process. Rapid breakdown of the emulsion at the end of the process is of almost equal importance, otherwise heavy metal losses to the slag will be sustained.

It will be clear that reaction rates are most often limited by factors like a diffusion coefficient which are far beyond the control of the operators. Within the limitations laid down by nature the most promising way in which processes can be speeded up is by increasing the surface area across which reactions take place — by crushing ore to smaller lumps, by emulsifying slag with metal, or by blowing smaller bubbles. In most cases, however, much has to be learned of the mechanisms of the reactions before it can be predicted whether the radical modifications to process design which might be entailed would lead to economic advantage. In Chapter 12 it will be seen that in some fields improvements suggested in the above discussions have in fact already been put to some use.

CHAPTER 9

SLAGS AND MATTES

Slag Structure

MOST metallurgical slags are molten silicates, with aluminates, phosphates, plumbates, antimonates, borates and fluorides as other possible major or minor acid constituents. Extraction slags are formed from the oxides among the gangue minerals which can be made fluid at a reasonable temperature by the addition of a suitable flux – usually another oxide and often in the form of a low-grade ore of the metal being extracted, an ore with a different type of gangue. The slag should ideally have a low solubility for the oxide of the metal being smelted, to ensure a high yield. For siliceous gangue, CaO and FeO are common fluxes, while limy gangue might be fluxed with silica sand.

Refining slags are specially prepared to be capable of absorbing, usually as oxides, impurities from the metal, and may incorporate chemical reagents for converting these impurities into soluble or volatile forms.

In recent years the term slag has also been used to describe molten inorganic salts – mainly halides – which are products of smelting operations involving the reduction of a pure metal halide by reaction with metal like Ca, Mg or Na. These are extraction slags. Their main function is to melt and separate from the metal, but in some cases even this is not necessary because the separation can be effected by leaching.

It has long been recognized that slags contain "acid" oxides (like SiO_2) and "basic" oxides (like CaO) and also that acids and bases may be weak or strong. The present view of slag structure – the ionic theory – explains the significance of these distinctions and accounts for most of the physical and chemical properties of slags. The ionic theory is based on X-ray diffraction studies of silica, silicates and glasses and on measurements of viscosity, surface tension and electrical conductivity.

210

It has long been appreciated that a compound like NaCl exists in the solid state not as molecules but as a regular array of Na^+ and Cl^- ions alternately positioned on a three-dimensional cubic lattice. The ions are held in place by electrostatic forces in the solid state. Above the melting point the vibrational energy per molecule exceeds the bonding energy but is insufficient to permit escape of ions to the vapour phase. Without the rigidity imposed by the lattice, the salt subsides into its container and assumes a horizontal surface under gravity. In the liquid, the ions are not packed so regularly nor so compactly as in the solid but electrical neutrality is maintained, even in short range as each ion continues to be surrounded by nearly 6 (the coordination number) neighbours of the opposite sign. The density of the liquid is less than that of the solid because there are many more vacant sites or "holes" than in the solid structure. Bonds are directed in a less regular manner and are powerless to resist even tiny mechanical forces.

The values of electrical conductivities of slags and glasses lie in a range which identifies them with molten salts of ionic structure rather than metals on the one hand, or covalent liquids (like CCl_4) on the other. The dominant ion in glasses is, however, the SiO_4^{4-} ion in which the oxygen atoms are tetrahedrally arranged round the silicon atom (Fig. 51). The Si–O bond is covalent and the $(SiO_4)^{4-}$ unit, having all its valency shells complete, is very stable, but like organic building units it is prone to polymerization.

Silica itself is a non-conducting covalent compound. Its structure when solid is shown in Fig. 52 to consist of a hexagonal lattice of silicon atoms, each sharing four tetrahedrally disposed oxygen atoms, and extending symmetrically in three dimensions. It can also be considered as an alternation of SiO_4^{4-} groups and Si^{4+} ions but this view is not consistent with its known properties.

(a)   (b)

FIG. 51. SiO_4^{4-} tetrahedra: (a) complete and (b) with the top oxygen atom removed. (From Richardson.[48])

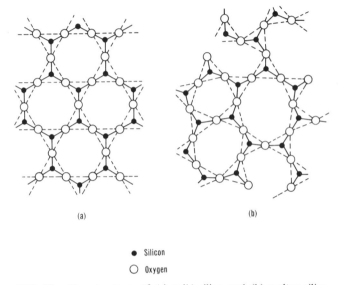

(a)   (b)

● Silicon

○ Oxygen

FIG. 52. The structures of (a) solid silica, and (b) molten silica, each built from tetrahedra as in Fig. 51 (b), the oxygen atoms having been reduced for clarity. Molten silica would contain also occasional Si^{4+} ions. (From Richardson.[48])

In the molten state the rigid lattice is lost but the SiO_4^{4-} tetrahedra persist and remain linked together to an extent which depends on temperature. A simple chain such as—

$$\left[\begin{array}{ccc} O & O & O \\ | & | & | \\ O-Si-O-Si-O-Si-O \\ | & | & | \\ O & O & O \end{array} \right]^{8-}$$

would have 4 oxygen atoms more than three "molecules" of SiO_2, each of them carrying 2 electrons to fill all the valency shells so that there is an 8— charge on this complex ion. This can be achieved only by the dissociation—

$$5SiO_2 = Si_3O_{10}^{8-} + 2Si^{4+}. \tag{9.1}$$

Ring structures require a lesser degree of dissociation. To make the ion—

$$\begin{bmatrix} \begin{array}{c} O \quad\quad O \quad\quad O \\ \diagdown \quad \diagup \diagdown \quad \diagup \\ Si \quad\quad Si \\ \diagup \, | \quad\quad | \, \diagdown \\ O \quad\quad O \\ \diagdown \quad \diagup \\ Si \\ \diagup \diagdown \\ O \quad\quad O \end{array} \end{bmatrix}^{6-}$$

the reaction is:

$$4\tfrac{1}{2} SiO_2 = Si_3 O_9^{6-} + 1\tfrac{1}{2} Si^{4+} \tag{9.2}$$

and if rings are extended to contain loops in the third dimension quite large ions can be produced which are rather like fragments of the solid lattice and which carry only small negative charges so that only a small proportion of silicon atoms need be present as positive ions. For example, the ion $[Si_9 O_{21}]^{6-}$ (Fig. 53) originally depicted by Bockris *et al.*[49] would result from the dissociation—

$$10\tfrac{1}{2} SiO_2 = [Si_9 O_{21}]^{6-} + 1\tfrac{1}{2} Si^{4+}. \tag{9.3}$$

Molten silica contains *very* large ions of this kind balanced by a small proportion of Si^{4+} ions at temperatures near to the melting point

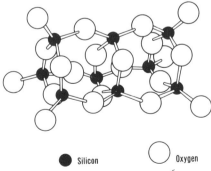

● Silicon   ○ Oxygen

FIG. 53.   A possible arrangement of a $[Si_9 O_{21}]^{6-}$ ion suggested by
Bockris *et al.*[49]

(Fig. 52). At higher temperatures anions become smaller and cations more numerous. An ultimate structure consisting of an equal number of SiO_2^{4-} and Si^{4+} ions can be envisaged but would require a very high temperature indeed. The presence of very large anionic units is reflected in the very high viscosity of silica because the anions are not easily broken up by mechanical forces.

Basic oxides dissociate when dissolved in molten silica, e.g.

$$CaO \rightarrow Ca^{2+} + O^{2-}$$

and in the presence of Si^{4+} ions,

$$Si^{4+} + 4O^{2-} \rightarrow SiO_4^{4-}.$$

This would be accompanied by some further dissociation of large anions to restore the equilibrium implicit in equations (9.2) and (9.3) so that

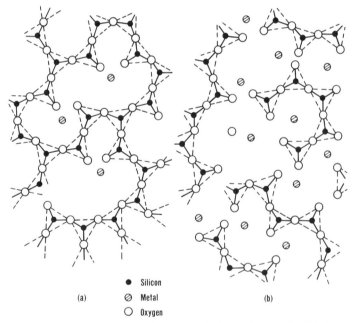

	● Silicon	
(a)	⊘ Metal	(b)
	○ Oxygen	

FIG. 54. The structures of molten solutions of a metal oxide in silica, the proportion of metal oxide being higher in (b) than in (a). (From Richardson.[48])

continued additions of CaO would ultimately reduce all the complex ions down to SiO_4^{4-} tetrahedra at the composition corresponding to the orthosilicate, i.e. $2CaO \cdot SiO_2$. Other basic oxides would behave in a similar manner (see Figs. 54 and 55). The accuracy of this picture of slag structure is, however, subject to further discussion later in this chapter.

Further additions of CaO beyond the orthosilicate proportions are presumed to dissociate into equal numbers of Ca^{2+} and O^{2-} ions. The whole arrangement must maintain short-range electrical neutrality and high entropy so that even in very short distances the composition is very close to the average for the melt. There is no suggestion of the formation of molecules.

Other slag constituents behave either like silica or lime. Al_2O_3 accepts oxygen ions to form stable AlO_3^{3-} ions; Fe_2O_3 may form FeO_3^{3-} ions; and P_2O_5 gives rise to very stable PO_4^{3-} ions. These are capable of polymerization to larger ions and Al_2O_3 and P_2O_5 are soluble in molten

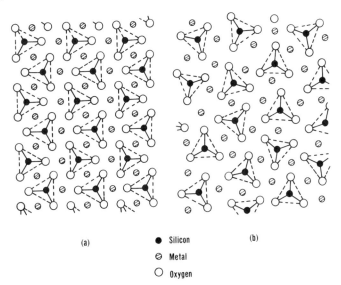

(a)   ● Silicon   (b)

                 ⊘ Metal

                 ○ Oxygen

FIG. 55. The structure of (a) solid orthosilicate (from Richardson[48]) and (b) molten orthosilicate in which dissocation of the silica network is complete. Further additions of metal oxide would produce free oxygen ions (see Figs. 52 and 54).

silicates, Al and P atoms replacing Si atoms in anions to some extent. The common basic oxides (like FeO, MnO, MgO) dissociate like CaO and donate O^{2-} ions with similar effect. Sulphides dissociate to produce S^{2-} ions which can replace free O^{2-} ions but sulphur cannot take the place of the oxygen atoms in the anionic groups. The disposition of the sulphur atoms in those acid slags such as iron blast-furnace slags which have no free O^{2-} ions is not known but they presumably associated with the Ca^{2+} ions or attached loosely to the perimeters of the silicate groups. Fluorides reduce viscosity very much. The mechanism producing this effect is not known, but it has been suggested[50] that whereas Ca^{2+} ions can link up SiO_4^{4-} groups thus—

$$
\begin{array}{ccc}
O & & O \\
| & & | \\
O-Si-O\cdots Ca\cdots O-Si-O \\
| & & | \\
O & & O
\end{array}
$$

in a weakly polymerized stucture, the formation of CaF^+ ions makes this chain building impossible and so keeps the viscosity low.

Some direct experimental support for the ionic theory of slag structure has been obtained by chromatographic separation of ions found in aqueous solutions of phosphate glasses.[47] A wide range of ion size was found from PO_4^{3-} through $P_3O_{10}^{5-}$ which was the most frequently occurring size, to $P_nO_{3n+1}^{(n+2)}$ chains with n very large – perhaps about 20 – but with rapidly diminishing frequency of occurrence. The distribution depended upon the cation present and became rather wider – favouring an increase of both the larger and smaller sizes – at high temperatures. Only linear ions were found in the phosphate glasses. Unfortunately the same technique has not been applied with success to the case of silicate ion sizes in slags or glasses because, whereas crystalline phosphates in solution gave one ion size only, pure silicates interact with water to produce ions of different sizes. Thermodynamic data suggest that ion size ranges in liquid silicates are much narrower than in phosphates and that they are strongly influenced by the cations present.

The Basicity of Slags

In terms of the ionic theory a "basic" slag is one in which there are free O^{2-} ions and an "acid" slag one in which there are no free O^{2-} ions but which have a high capacity for incorporating them into their more complex structures. Thus acid slags attack refractories made from basic oxides – and vice versa. Quantitatively the degree of basicity has been expressed in several ways. The simplest expression is the CaO/SiO_2 ratio (by weight) which has served the iron-making industry well where the range of slags is narrow. The ratio $(CaO + MgO)/(SiO_2 + Al_2O_3)$ might appear to be better but there is some doubt as to the "equality" of CaO and MgO on the one hand and that of SiO_2 and Al_2O_3 on the other. It is sometimes reduced to $(CaO + \frac{2}{3}MgO)/SiO_2$ on the grounds that MgO is a weaker base than CaO and Al_2O_3 plays little part as an acid (or receiver of O^{2-} ions). In steelmaking slags P_2O_5 must be added to SiO_2 so that the so-called "V-ratio" for the basicity becomes $(CaO + MgO)/(SiO_2 + P_2O_5)$. The exact role of each constituent may depend upon the kind of slag and it is often stated that Al_2O_3 being an amphoteric oxide behaves as an acid only in very basic slags. Fe_2O_3 is similar, being mentioned as an acid oxide only in connection with very basic oxidizing slags. The ratios mentioned above are all in weight per cent. This is quite illogical and ratios in terms of moles or mole fractions are much more likely to yield satisfactory relationships with other operational factors. Molar ratios are generally found in scientific work but industrial usage favours weight ratios for obvious reasons.

In terms of the ionic theory, basicity is expressed as "excess base" or excess moles of O^{2-} ions per 100 g slag given by—

$$n_{O^{2-}} = n_{CaO} + n_{MgO} + n_{FeO} + n_{MnO} + \cdots$$
$$-2n_{SiO_2} - 3n_{Al_2O_3} - 3n_{P_2O_5} - 3n_{Fe_2O_3} - \cdots \qquad (9.4)$$

where n_i is the moles of each species per 100 g slag. This can be extended to include all oxides present and could be adjusted to allow for sulphide and fluoride. It does not, however, acknowledge any difference in the relative "strengths" of the basic oxides, or of the acid oxides.

The force of attraction of a metallic ion for an oxygen ion is proportional to the ratio z/a^2 where z is the charge on the metal ion and a is the sum of the radii of the two ions. The force is strongest in the cases of phosphorus and silicon which have small ions carrying large electric charges and these in fact attract oxygen so strongly that very stable anions are formed. Large alkali ions with single electron charges

TABLE 3. A COMPARISON OF THE ORDER IN WHICH OXIDES LIE WHEN ARRANGED ACCORDING TO (a) INCREASING z/a^2 AND (b) DECREASING IONIC BOND FRACTION

z/a^2[42]	Ionic bond fraction[52]		Character of oxide
0.18 Na$_2$O	Na$_2$O	0.65	
0.27 BaO	BaO	0.65	Basic
0.32 SrO	SrO	0.61	
0.35 CaO	CaO	0.61	
0.42 MnO	MgO	0.54	
0.44 FeO	MnO	0.47	
0.44 ZnO	ZnO	0.44	
0.48 MgO	BeO	0.44	
0.69 BeO	Al$_2$O$_3$	0.44	Intermediate
0.72 Cr$_2$O$_3$	Cr$_2$O$_3$	0.41	
0.75 Fe$_2$O$_3$	TiO$_2$	0.41	
0.83 Al$_2$O$_3$	FeO	0.38	
0.93 TiO$_2$	Fe$_2$O$_3$	0.36	
1.22 SiO$_2$	SiO$_2$	0.36	Acid
1.66 P$_2$O$_5$	P$_2$O$_5$	0.28	

exert the least attractive force on oxygen ions and are therefore most "basic" in their behaviour. Table 3 shows a selection of z/a^2 values for oxides of interest and indicates a gradation from strong acidity to strong basicity which corresponds fairly well with their observed behaviour.

It is generally recognized that a bond need not be wholly ionic or covalent. It may be in its character intermediate between the extreme types possibly by means of a resonance mechanism. The ionic bond fraction (= 1 − the covalent bond fraction) for a binary compound can be obtained from the difference between the electronegativities of the two elements. Electronegativity is a measure of the affinity of an atom for its valency electrons and can be measured through electrode potentials. Pauling[51] gives a value on an electronegativity scale to every element and Wells[52] provides an empirical relationship between electronegativity differences and ionic bond fractions. If both elements in a compound are non-metals then their electronegativities are similar so that the difference is small and the ionic bond fraction is, by Wells's relationship, low. If a metal and a non-metal combine the electronegativity difference is larger so that the ionic bond fraction is high. For example, the non-metallic bond P–O is only 28% ionic whereas the Ca–F bond is 80% ionic.

Ionic bond fractions are included in Table 3 where they can be seen to put the oxides in a somewhat different order from the z/a^2 ratios but still with the strong acids and bases well distinguished. Neither assessment of the order of the basicities of the intermediate oxides is distinctly better than the other, possibly because of interactions between the solutes and silica as the solvent, and between the several solutes themselves. The case has some analogy in aqueous solutions. When HCl is dissolved in pure water two covalent compounds interact in such a way as to produce a highly ionic solution. Silica itself is typically covalent in character but when it has accepted oxygen ions from metallic oxide donors which are themselves about 50% covalent, typically ionic solutions can be produced. Whether or not an oxide yields up its oxygen ion seems to depend on factors in addition to those listed in Table 3 of which ion size is likely to be important as it determines which spatial arrangements are possible.

A really satisfactory way of estimating the activity of free oxygen ions in a slag has not yet been discovered. The dissociation of basic oxides is probably never complete but it is more so in alkali oxides than in those of alkaline earths, and in these, more so than in MnO, FeO, etc. Further, a weakly acid oxide like Al_2O_3 may, if there are few donated oxygen ions available, adopt basic character and yield some oxygen ions to form SiO_4^{4-} tetrahedra. P_2O_5 alone is a stronger acid than SiO_2. As will be seen in the next section most metallurgical slags are *less* basic than the orthosilicate composition, the chief exception being basic steelmaking slags.

Phase Diagrams

Early work on slag systems was concentrated on phase diagrams. Much work done on solid phases has been more useful in the fields of geology and ceramics. For metallurgists the emphasis has moved to the liquid state, but to some extent the early assumption that the behaviour of molten slags would reflect the properties and structures of the corresponding phases below the liquidus temperatures has been justified, particularly in the case of viscosity. Liquid properties, however, depend very much more on how far the temperature exceeds the liquidus temperature and so reflect the shape of the liquid surface. They never change abruptly with composition as do the solid properties at phase boundaries so that no observed correspondence of liquid properties with the solid phases can ever be particularly clear-cut. Accurate information about the position of the solidus and particularly the liquidus surfaces in ternary and higher order systems is of course very important and much of this information was recorded by the early workers in the field.

The binary diagrams between metal oxides and silica are presented together in Fig. 56 (silica rich end only). Except for the alkali oxides and barium oxide (which is borderline) small additions of metal oxides beyond about 2% lead to the separation of a second liquid phase. The order in which the oxides fall in this diagram is again similar to that found in Table 3. It appears that the alkali oxides lose their oxygen ions almost completely to the silicate anionic groups or networks and that

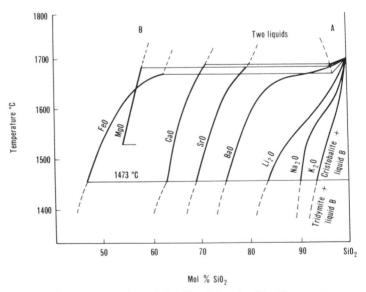

FIG. 56.   A comparison of the silica-rich ends of the binary systems between SiO_2 and several basic oxides. (After Richardson.[53])

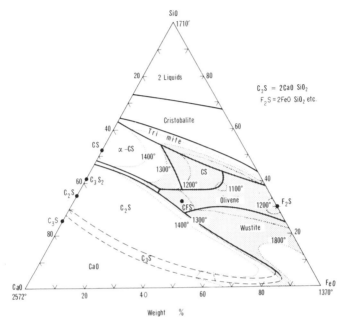

FIG. 57. The CaO–FeO–SiO$_2$ phase diagram. The shaded area shows the range of compositions which melt at 1300°C or lower. (After Bowen et al.[54] as modified by Osborn and Muan.[56])

the alkali ions themselves are accommodated in the interstices within the networks. The other metal ions, having a somewhat greater attraction for oxygen, form groups which cannot be fitted into these interstices and so must separate off with some silica in anionic groups which are smaller and which can arrange themselves along with the basic ions in such a way as will satisfy their coordination requirements.

Perhaps the most important ternary phase diagram is that for the system CaO–FeO–SiO$_2$ shown in Fig. 57 on which the isotherms on the liquidus at 1100, 1200 and 1300°C are clearly marked. The "useful" part of the system is shown shaded.

Real extraction slags will contain some Al$_2$O$_3$, some MgO and some metallic oxides derived from the ore. If these are in small proportions only it may not be unreasonable to count Al$_2$O$_3$ as SiO$_2$, MgO as CaO and some of the metallic oxides as FeO for purposes of estimating

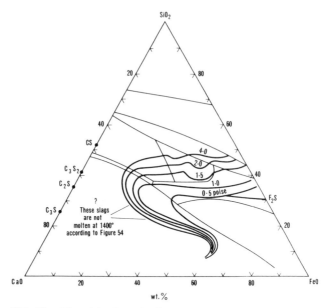

FIG. 58.   Viscosities in poise in the $CaO-FeO-SiO_2$ system at
$1400°C$. (After Kozakevitch.[55]) (1 poise = 0.1 N s m^{-2}.)

liquidus temperatures or viscosities from diagrams like Figs. 58 and 59
but there is obviously a strong case for examining the appropriate part
of such a system at least on a quaternary diagram. The amount of
information available on high-order systems is still rather limited, however,
and its interpretation is often difficult.

From Fig. 57 it is clear that the composition range available for
making slags at reasonable temperatures is rather restricted and it
becomes even more so if "free-running" temperatures are considered
rather than simply the liquidus temperature. Figure 58 shows the viscosities
of slags in this same system at $1400°C$ and, as might be expected, these
rise rapidly when the base falls below the orthosilicate proportion.
Viscosities decrease rapidly with rise in temperature, however, so that a
"free-running" condition with viscosity less than about 5 poise
(0.5 N s m^{-2}) can be achieved over quite a large range of composition if
the temperature is raised high enough.

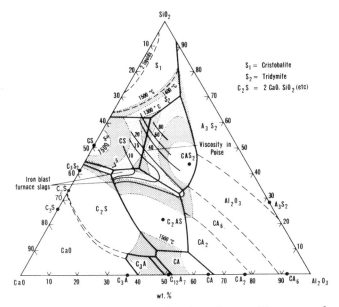

FIG. 59. The $CaO-SiO_2-Al_2O_3$ phase diagram. The ranges of composition which are molten at $1500°C$ are shown shaded. (After Greig, Rankin and Wright, modified by Osborn and Muan.[56]) The viscosities indicated are in poise at $1500°C$. (After R. Hay.[59]) (1 poise = $0.1 \ N \ s \ m^{-2}$.)

In the extraction of iron, the FeO is reduced out of the slag to a very low level and the composition is found in the $CaO-SiO_2-Al_2O_3$ system (Fig. 59), again in the metasilicate region and again restricted by melting point considerations to a small part of the diagram, shown shaded. The superimposed iso-viscosity lines indicate that much of even that small range of compositions is not available because the slags would not be sufficiently fluid. In this process, however, about 5% of MgO is usually introduced which reduces both melting points and viscosities below those in the ternary system. The quaternary system including MgO is one which has in fact been studied fairly thoroughly.

Iron oxide is reducible by carbon from any slag containing it, but the residual slag composition would normally move across Fig. 57 to viscous or even solid areas in which, for kinetic reasons, the continued reduction of the FeO would usually be very slow or halted. Iron oxide may, however, be fed into a refining slag, as in the refining of steel, for the express purpose of oxidizing out impurities – carbon, and phosphorus in the case of steel. Iron oxides are available in a fairly high degree of purity cheaply and in large quantities. Their use for this purpose in steelmaking helps to keep the "yield" of metal in the refining stage high. An oxide used in this way must, of course, be at an oxygen potential higher than the oxides which might be formed from the impurities. The oxygen potentials of the oxides of copper and lead, for example, are rather similar to those of some of the metals found in copper and lead as impurities to be removed by refining, but litharge *is* used for the oxidation of As, Sb and Sn from lead bullion.

Iron oxide should not be considered too strictly as FeO. In the solid state it is a berthollide compound called "wustite" and in the Fe–O diagram (Fig. 60) it appears as a phase with a range of oxygen content

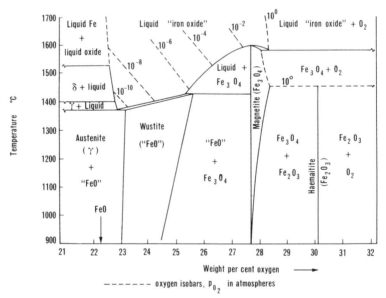

FIG. 60. Part of the Fe–O phase diagram after Darken and Gurry[57] and with oxygen isobars drawn in accordance with Osborn and Muan.[58]

rather higher than the FeO composition. Molten "FeO" as it is often designated can contain up to about 28% of oxygen at 1600°C under an oxygen pressure of 1 atm — i.e. rather more than the magnetite composition — but only 26% at 1500°C. It is customary when analysing slags to determine FeO and Fe_2O_3 and it is considered that under highly oxidizing and basic conditions the higher oxide shows itself as FeO_3^{3-} cations as already discussed. Under acid conditions, the higher oxide of iron (or a higher O/Fe ratio) brings about a higher degree of dissociation of the complex anions, per mole of iron, than does the lower oxide. The O/Fe ratio is dictated by the oxygen potential of the system — i.e. by the prevailing p_{O_2} value, as indicated by the isobars in Fig. 60.

Slags can be made reducing as well as oxidizing. Again the most important example comes from steelmaking. In the basic electric process, a slag is made up which is very high in CaO, with some CaF_2 to give it fluidity, and about 1.5% of CaC_2 is formed in it by feeding in fine anthracite or coke. This kind of slag reacts with oxygen in the melt below, reducing it to a very low concentration. It creates the best condition for removing sulphur and gives positive protection against the oxidation of alloy additions when these are made. The very high temperatures available in the electric arc furnace are needed to melt this kind of slag. The use of other reducing agents, such as silicon or aluminium, in a similar way is sometimes possible, instead of carbon.

Slags are also capable of dissolving small amounts of neutral metal atoms, presumably lodged in holes in the slag structure. They can also dissolve water in some form, probably dissociated, and can pass either oxygen or hydrogen — or both — from a furnace atmosphere through to a melt below.

The transmission of oxygen is obviously very easy in the case of a slag containing iron oxide or the oxide of any other metal which can display more than one valency. At the surface of such a slag the oxygen concentration can be raised in equilibrium with an atmosphere containing 1% of free oxygen to such a level that the O/Fe ratio is 0.37 while at the interface in equilibrium with underlying iron or steel it will be reduced to a level such that the O/Fe ratio is only 0.29 (from Fig. 60). These figures are subject to some adjustment for the effects of other components of the slag on the activities of iron and oxygen. In general, however, the difference is great enough, unless the concentration of iron is very low, to ensure a high transfer rate of oxygen through the slag layer by diffusion, with faster rates when assisted by convection. In a case when there is no iron or other bivalent metal available, the concentration differential across the slag would be much lower, depending on the physical solubility of oxygen and transfer rates would also be low.

When a slag is involved in chemical reactions, a knowledge of the activities of its constituents would always be useful. Values of these activities have been difficult to obtain because their determination involves difficult high-temperature experimentation and requires very accurate chemical analyses in an awkward field. The range of interest is often between the Raoult's law and Henry's law ranges and suitable standard states are not always easy to specify in these "strong" solutions. Probably the best work in the field is the determination of the activities of FeO in the system CaO–FeO–SiO$_2$ already discussed.

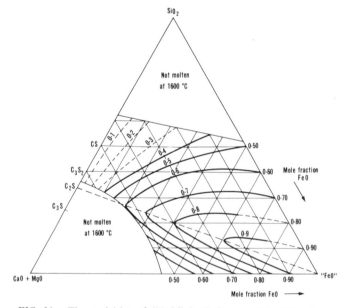

FIG. 61.   The activities of "FeO" in FeO–(CaMg)O–SiO$_2$ slags at 1600°C. (After Taylor and Chipman.[25])

The results of the work[25] are presented in Fig. 61 which should be compared with Fig. 57. The activity values are referred to $a_{FeO} = 1$ in pure "FeO" in equilibrium with iron and oxygen at 1 atm (see page 151) and determinations were made through the oxygen contents of pure iron equilibrated with many different slags. The free energy of mixing of CaO and SiO_2 is very much greater than that of FeO and SiO_2 so that it is to be expected that the Ca^{2+} ions will assume positions of lowest energy in the ternary solution leaving the Fe^{2+} ions in less stable positions but it is not very easy to see what this means in terms of a slag structure. Perhaps the Ca^{2+} ions tend to be enclosed within "cages" of SiO_4^{4-} units while the Fe^{2+} ions are excluded from these arrangements and occupy zones of greater disorder rather like the liquid equivalents of grain boundaries. The effect is to increase the activity coefficient of FeO above the value 1.0 over a large area of the diagram. This part can be seen to lie astride the tie-line between the FeO corner and the calcium orthosilicate (C_2S) composition in such a way that the effect is a maximum when the ratio $CaO/SiO_2 = 2$. Lime added to an oxidizing $FeO-SiO_2$ slag makes it even more oxidizing; or a required degree of oxidizing power can be maintained with a lower FeO content in the presence of CaO. This was used in the acid open hearth steelmaking process as a means of increasing the yield of iron.

It will be noticed in Fig. 61 that the activity coefficients of FeO are *less* than 1.0 in slags which are very low in silica. This is interpreted as showing that there is a tendency for CaO and "FeO" to form some stable configuration in very basic slag. It is not clear what this arrangement can be unless it is assumed that the slag is slightly oxidized so that FeO_3^{3-} ions have been formed. This is not, however, compatible with the conditions under which the activities were measured, that is with the slag in equilibrium with iron. At high oxygen pressure FeO_3^{3-} ions certainly form and have an important role in basic steelmaking. The affinity of Ca^{2+} ions and FeO_3^{3-} ions is sometimes referred to as "ferrite" formation but in the liquid slag it need be considered only that a low-energy arrangement of Fe^{2+}, FeO_3^{3-}, Ca^{2+} and O^{2-} ions is formed such that the availability of Fe^{2+} ions for entering into oxidation reactions is reduced. The information embodied in Fig. 58 has been extended to embrace a wider range of basic steelmaking slags presented as a pseudoternary system made up of "acids", "bases" and "FeO".[60]

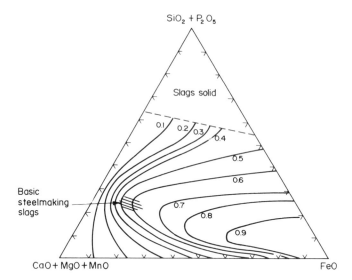

FIG. 62.   Raoultian activities of FeO in basic steelmaking slags.[60]

This is shown in Fig. 62 which is very similar to Fig. 61 in all main features. Other activity data are available for a_{MnO} in the $CaO-MnO-SiO_2$ system[61] and for a_{SiO_2}, a_{CaO} and $a_{Al_2O_3}$ in parts of the $CaO-SiO_2-Al_2O_3$ system.[62,63] The iso-activity lines for SiO_2 in iron blast furnace slags are presented in Fig. 63, superimposed on the liquidus surface of the ternary diagram shown in Fig. 59. The strong negative deviation reflects the stability of the orderly arrangement of calcium and silicate ions in the melt and clearly suggests that there will be stable silicates of lime in the solid state — which is of course the case. Much of this information had been sought in the first place to help to explain slag structures but the rate of its accumulation is increasing and the much larger volume of data needed for application to kinetic problems is rapidly becoming available.

This kind of information is difficult to establish, difficult to present and difficult to interpret and use. It is also difficult to keep up with revisions of values except in the narrow field in which one is deeply involved. The National Bureau of Standards in the U.S.A. now has its National Standard Reference Data System and the National Physical

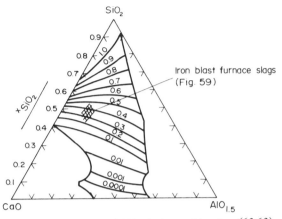

FIG. 63.   Raoultian activities of SiO_2 in ironmaking slags.[62,63] Note that concentrations are on mole fraction basis and that alumina is entered as $AlO_{1.5}$. This makes the proportions of the diagram similar to those in Fig. 59 which is drawn on a wt% basis.

Laboratory in Britain runs a critical assessment and computer storage service for thermodynamic data, the stored information being made available to organizations or individuals on request but at a fairly high price.[64]

Mattes

Mattes are solutions of metallic sulphides. They have high electrical conductivities suggesting that their structures, when molten, are ionic or possibly partly metallic. They have rather lower melting-point ranges than slags. They are much denser than slags (sp. gr. about 5 for matte and 3 for slag) and they are immiscible in both slag and metal phases, though either can dissolve sulphur. Mattes are used either for the collection of the valuable mineral in a straightforward smelting process (Cu, Ni) or for the collection of impurities in a sulphide phase from which the principal metal (e.g. Sb) has been displaced by Na, charged as soda ash (Na_2CO_3). Oxide ores can be smelted to matte with pyrites or gypsum as the source of sulphur but this is rare.

The copper mattes, which are by far the most important

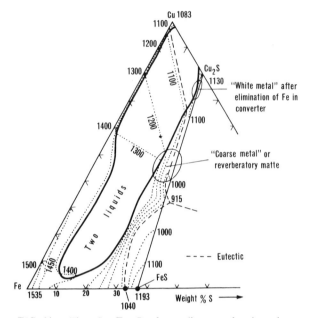

FIG. 64. The Cu–Fe–S phase diagram showing the composition ranges in which typical copper mattes are found. (After Reuleaux.[69])

commercially, have been well discussed by Ruddle.[65] Referring to the Cu–Fe–S ternary diagram (Fig. 64), the copper–iron mattes are found in a narrow band of composition running between the FeS and Cu_2S compositions. Cu_2S is immiscible in Cu but FeS is soluble in iron, the solution having, however, very marked positive deviation from Raoult's law so that it is not surprising that a wide band of immiscibility separates the Fe–Cu edge of the diagram from the FeS–Cu_2S band in which the mattes lie.

Within this band several workers have detected an eutectic structure, but the actual eutectic point lies off the FeS–Cu_2S join to the side of rather higher sulphur content and it is in fact a ternary eutectic. The activities along the pseudo-binary join between Cu_2S and FeS have been determined.[66] Both species show small deviations from Raoult's law, negative in the case of Cu_2S and positive in the case of FeS, but this latter observation has been disputed.[67] Most commercial mattes, however, contain rather less sulphur than that required to provide a mixture of the two terminal sulphides and the true eutectic would be obtained only under a sulphur pressure that is rather higher than atmospheric.

Mattes usually contain some oxygen, and experimental work on mattes should be carried out under controlled oxygen pressure as well as sulphur pressure. Analysis for oxygen in mattes is not easy as it is essentially a determination of the state of combination of iron in a sample which is usually rather difficult to dissolve without using an oxidizing agent. In low-grade mattes made under reducing conditions in a blast furnace, iron may be present as the metal with FeS and Cu_2S. High-grade commercial matte may contain FeS, Cu and Cu_2S if made under reducing conditions. Reverberatory mattes made under an oxidizing atmosphere and slag may solidify with up to 20% of Fe_3O_4 and a small amount of FeO but the state of the iron in the melt is not apparently known. The solubility of Fe_3O_4 in copper–iron mattes may be about 30% (12% oxygen) in very lean matte (i.e. one which is high in FeS and low in Cu_2S) but decreases rapidly as the copper content increases. The precipitation of magnetite in reverberatory hearths, and in converters as iron is oxidized out of the matte and slagged (so increasing the Cu content and lowering the solubility of the Fe_3O_4), is a well-known feature of these processes which gives a great deal of trouble.

The copper content ("grade") of a copper matte depends on the Cu/Fe ratio and on the O/S ratio of the concentrate of calcine being smelted. All of the copper present goes into the matte but the iron is partitioned between matte and slag in proportions which are determined by the availability of sulphur and oxygen. Sufficient iron and sulphur must be present to provide enough FeS to protect the copper in the matte from oxidation. Low-grade matte can have as little as 20% Cu (25% Cu_2S) but normally the grade is higher with about 45% Cu (55% Cu_2S) the balance being mainly FeS with iron oxide in solution.

Mattes also collect a number of other metals besides iron and copper, particularly nickel, cobalt, zinc and lead, gold, silver and platinum metals. Nickel–copper–iron mattes are produced from nickel ores and are converted into copper–nickel mattes during nickel extraction, i.e.

FeS is blown out. Cu_2S and Ni_3S_2 form an eutectic which can be produced coarse enough for the constituents to be separated by flotation after crushing.

Mattes are important only in the extraction of copper, nickel and sometimes antimony. In lead smelting copper is collected in a matte in the hearth, a controlled quantity of sulphur being included in the charge for the purpose. Sulphur is partitioned between the matte and any metallic phase present with a small amount also entering the slag. Oxygen and iron are partitioned between the matte and any slag in the system. Mattes are intermediate products and the distinct lack of knowledge about them is partly due to this fact, partly to the difficulty of investigating them, and partly to the fact that in the next stage of their processing, conversion, the problems encountered are not primarily connected with matte constitution and structure.

Dross

Dross is the product of drossing operations, the first refining stage on low melting metals and particularly on lead. Oxygen or sulphur may be stirred into bullion to refine it. Lead having a much lower melting point than oxides and sulphides, solid products form on its surface mingled with some metal. This "wet" dross is "dried" by stirring in volatile reagents like sawdust or ammonium chloride which clean the metal globule surfaces and let them coalesce to re-join the bath of refined metal below. Dross is, of course, skimmed off, squeezed to release more lead, and given further treatment to recover whatever values it may contain.

In the desilvering of lead the separating phase, which is mainly zinc, is crystalline in character and is generally referred to as a "crust".

Speiss

Speiss is basically iron arsenide but it forms a molten phase insoluble in matte, metal or slag, which may contain a variety of other elements such as nickel, copper, cobalt, lead, antimony and tin. Drosses may be melted down to produce four layers – metal, matte, speiss and slag – so effecting some degree of separation in one simple operation. The speiss layer is often important in the scavenging operations which follow the refining of base metals.

CHAPTER 10

FUEL TECHNOLOGY

ENERGY management is frequently an important aspect of the work of the extraction metallurgist because the cost of energy is usually a high proportion of the total cost of the metal produced. It is essential that energy is used as efficiently as possible but it is also necessary that the production rate shall be as high as possible so that the cost of the capital being used is apportioned among the maximum number of units of production. Where these requirements are incompatible the best possible compromise must be sought.

Energy

Energy is required in extractive processes for many purposes. These include the obvious requirements of a supply of heat to raise temperature, to supply the latent heat needed for melting and evaporation and for endothermic reactions. In particular the energy of reduction of metal oxides and sulphides to the metallic state is an inescapable energy charge on the process. Energy is also required for all handling and moving operations, for all comminution processes and to a lesser extent for separations. It is continually being dissipated as friction, as radiated and conducted heat and as sound. Any process which can recover a metal from a particular ore with a lower energy consumption than alternative processes is likely to be economically attractive but the *total* energy cost must be accounted for, including hidden costs such as the cost of energy required to produce reagents. The energy required to produce chlorine, for example, or pure hydrogen which may be used in low-temperature processes can be much more costly than the coal or oil which are used in traditional smelting. Another example is that the heat derived from the oxidation of iron in any steelmaking process is not cheaper than heat obtained from burning coke because at least an equivalent amount of coke must have been consumed to produce the iron. In the end it is the total cost that is important — other things like product quality being equal.

Energy is obtained mainly from naturally occurring "fossil" fuels — coal, oil and natural gas. These sources are supplemented by water power, nuclear energy and energy from miscellaneous other sources such as wood and wood charcoal, peat, or spent sugar cane which have occasional local significance only and are used on rather a small scale. Energy is also derived from the heats of essential exothermic reactions, of which the oxidation of sulphides is probably the most important, and pyrites is sometimes used as a fuel (with the production of sulphuric acid as a by-product). Any energy balance must take into account the heats of all reactions which occur in the process being examined.

Fossil Fuels

Energy from this source is available to industry as coal, oil or natural gas which are primary energy forms or as coke, manufactured gas or electricity which are secondary forms of energy and usually more costly.

Solid Fuels

Raw coal is used by industries to a rapidly decreasing extent. A high proportion of the coal mined is carbonized to produce coke and gas or is burned to raise steam for the production of electricity. Coke is used in large quantities in blast furnaces — and in particular in iron blast furnaces. The coal required for coke production must have rather special properties and is not plentiful. It should have between 22% and 33% of volatile matter and a high "swelling index" to be designated a "coking coal". Much research over the years has shown how weakly coking coals can be blended along with good coking coals and the carbonization rate controlled so that reasonably good coke can be produced from coals which are not all of the most suitable type. The most recent developments are now showing that weakly caking coals which could never produce conventional coke can be made into chars which can then be bonded with tars, briquetted and fired to a product called "formed coke" which behaves satisfactorily in the blast furnace. It is strong, sufficiently unreactive toward CO_2 and can be made to a specified size. This invention may well ensure the survival of the blast furnace beyond the 20 or 30 years for which coking coals are likely to be available and may widen the range of geographic areas in which it is possible to operate with iron blast furnaces economically.

Apart from the necessity that it should have a good swelling index the most important properties of a metallurgical coal are that it should have a low sulphur and ash content. Sulphur in fuel readily associates with any metals present and its subsequent removal can be costly especially when this has to be in a limy slag. Sulphur in coals should not normally exceed 1% but it is sometimes found to be 5 or 6%. In the absence of a better one, such a coal might have to be used but would not command a high price. Inorganic constituents form an inert residue after combustion, called "ash". This must be removed from the furnace either as dry ash from beneath the grate, as fly ash which must be taken out of the flue gases before they are discharged, or as slag. The slag may be the slag of the metallurgical process being conducted in which case an appropriate amount of flux must be added at some cost. Some ash has a low fusion point and melts in the furnace to form clinker which can be difficult to remove. Ash in coal should not be higher than about 10%, which rises to about 13% in coke, but available fuel may not always be as good as this and processes may have to be adapted to the use of inferior coals. The calorific value of a coal depends upon the carbon, hydrogen and ash contents. The price of a coal will depend on its calorific value, on its sulphur content and on its size as well as on its coking quality. Coking coals and high-carbon, low-volatility anthracites are the most costly. The cheapest coals are mainly used in electricity power stations which are now designed to use them in pulverized form. Very poor fuels can now be burned efficiently in fluidized beds. This development is likely to be applied mainly in the field of steam raising and electricity generation.

Coke, produced from coal by carbonization at up to $1300°C$, should be hard and strong and chemically unreactive. The strength is needed to withstand crushing in the blast furnace. The unreactive nature is necessary to inhibit its reaction with CO_2 in the interests of its efficient oxidation finally to CO_2 rather than to CO. The same qualities are required of formed coke, along with close size requirements.

Liquid Fuels

Liquid fuels embrace petroleum oils, and coal tar fuels produced during the carbonization of coal. These are put to similar uses and are used in a very similar manner but they must not be mixed either in storage or in use because they interact, forming sludges. Crude petroleum is fractionally distilled to yield many familiar products such as petroleum spirit, kerosene and diesel oil. The fraction which has not distilled off below 200°C is further distilled under reduced pressure to separate light fuel oil below 250°C and heavy fuel oil above 250°C. The residue from these stages contains greases and waxes and ultimately pitch which can be used in pulverized form as a solid fuel. The fuel oil fractions are adjusted to specified viscosity values for marketing by blending with lighter or heavier fractions as appropriate. It is important that the viscosity should be kept very close to the nominal value because the oils must be stored, pumped through supply lines and atomized in burners which are all designed to be used at particular temperatures – that is with oil of a particular viscosity. The system used is the same all over the world – to accommodate the needs of shipping, for example. The other physical property of importance is "flash-point". This is determined by a test that is designed to ensure that thinning back has not been done using highly volatile fractions which might ignite too readily. This is essentially a safety precaution. Chemically, oils should be as low as possible in sulphur but 2% is not an uncommon level in the heavier fraction. The price reflects the sulphur content. Calorific values of oils of any type do not vary much from typical values. Coal tar fuels are relatively low in sulphur. They burn with particularly good luminous flames.

Gaseous Fuels

Gaseous fuels include natural gas which is principally methane, coke-oven gas (methane, hydrogen and carbon monoxide) and blast-furnace gas (carbon monoxide, carbon dioxide and nitrogen) along with a number of other manufactured gases used in limited quantities. Of these, producer gas and water gas are perhaps the best known derivatives of coal but the trend is now toward total gasification of coal to mixtures of carbon monoxide and methane by the Lurgi or some similar process involving partial oxidation with oxygen and steam at high temperature and pressure. Refinery gases butane and propane are becoming important. They have very high volume based calorific values, are easily stored and transported as liquids and can be used to develop extremely high temperatures when burned with pure oxygen. Gaseous fuels are easily distributed and are easily controlled as to composition and calorific value. Sulphur can be removed easily if necessary. The calorific values of gases vary over a wide range depending upon how much inert nitrogen is present and which combustible species are present. If the calorific value is expressed in $J \, m^{-3}$ it is obvious that a gas which is mainly butane $(C_4 H_{10})$ will have about 4 times the calorific value of methane (CH_4), having almost 4 times as much mass per unit volume. Different gases need burners of different design. Lean gases like producer and blast-furnace gases can be burned usefully only if the air is preheated and preferably the gas also (with the outstanding exception of the Cowper stove where, however, a moderately hot flame is sufficient). Gases rich in methane must not be preheated because the hydrocarbon decomposes depositing soot. Flame propagation velocities of different gases are also very different, being highest with hydrogen. The optimum gas/air ratios (volume ratios) vary depending upon the molecular species present. Burners must be designed to accommodate these properties.

The gas used in given circumstances may depend mainly on availability. In some countries natural gas is considered a "premium fuel" and is reserved for use where it can command a high price as in the domestic market. Elsewhere, where it is plentiful and no other fuels are available, it might be used as a reducing agent. In important centres of ironmaking and steelmaking, coke-oven gas and blast-furnace gas are used very generally. These gases are at best by-products and at worst waste products but the blast-furnace gas, though very lean, is available in such vast quantities that the energy in it cannot be discarded and in any case the toxicity of the gas is such that it could not be discharged to the atmosphere without first being burned.

Sulphides as Fuels

The oxidation reactions of sulphides are almost all strongly exothermic, so that whenever sulphide concentrates are roasted, heat is evolved. Sometimes there is enough heat to sustain the process so that we have autogenous roasting or sintering with no carbonaceous fuel added. In some cases this can be an embarrassment as the temperatures tend to get out of control. In the sintering of lead sulphide, for example, part of the product is recycled in order to use up some of the heat evolved – an apparently wasteful procedure. Where a concentrate does not have quite enough sulphide in it to give autogenous roasting some iron pyrites may be added to generate heat. Pyrites may be burned to produce SO_2 as a source of sulphuric acid or of elemental sulphur, Fe_2O_3, which can be used for ironmaking and heat which can be used for steam raising. Pyrites is available in large quantities in areas rich in sulphide ores and represents a cheap source of energy, provided the SO_2 and Fe_2O_3 can be disposed of without incurring costs. The release of SO_2 into the atmosphere cannot be tolerated so that fixing it and selling it is necessarily part of the whole process. The removal from the iron oxide of metals like copper, which are undesirable in the feed to a blast furnace, is also necessary and can be profitable. The combustion of pyrites can be carried out in any roasting furnace but a fluidized bed reactor is now preferred.

Electrical Energy

Electrical energy is derived from coal, oil, water power, or nuclear energy. It is usually generated by public utility companies and purchased by domestic and industrial users alike in industrially advanced countries. Some large-scale users do, however, produce a part of their own requirements. This may be a means of putting to use some of their excess low potential heat for which there is no thermal use on the plant. This occurs in steel plant where the thermal requirements are biased toward the need for high-temperature heat. Some plants operate a stand-by generator to ensure that power will be available for certain essential purposes even if the public supply is interrupted. In a few industries the amount of electrical energy used is so great that special generating capacity must be created for its use. Aluminium smelters have traditionally been built where plentiful water power was available and many of the early hydro-electric power schemes were built by the aluminium producers for their own use. More recently aluminium smelters have been built on a site alongside a coal mine and a power station, the latter being built and operated by the public utility company but primarily to supply the smelter. A smelter and a nuclear power station have also been built together under a similar arrangement. In other circumstances where mines are being developed in areas which have not been previously industrialized it may be necessary for the mining company to generate all of its own power and probably during the development stage supply also the local population – as happened around the aluminium plants in Europe 80 years ago.

Electricity may be required for several purposes. One of the most important of these may be heating but in some processes very little heat is required but a great deal of power is needed for driving comminution mills, or for electrolytic reduction. All industrial plant needs electricity today and it is sobering to consider that the generation of electricity and its distribution as we enjoy it today have all been developed in the last 80 years with a remarkably high proportion of that development within the last 40. Apart from driving major pieces of machinery it is used in all manner of minor equipment down to hand tools and in control mechanisms, for many of which it would be impossible to find a substitute energy form.

Electricity can be supplied over a wide range of voltage. Distribution is almost always of three-phase alternating current at a frequency of 50 or 60 (U.S.A.) Hz. Long-distance transmission is at voltages up to 400 kV but this is stepped down to 400 V (phase) or 250 V (line) in many countries, 250/100 V in others, for general distribution. Industrial users can be supplied at higher voltages, however, if they wish, accepting power at, say, 3000 V at a transformer station where it will be stepped down to meet the needs of the particular plant. At an electrolysis plant part of the power supplied will be rectified to direct current but a similar plant generating its own power may install d.c. generators. In general, alternating current is preferred because it is more flexible in use – transformation of voltage is easy, transmission costs are lower and there are other big economies associated with the use of three phases. Currents and voltages are lower and motors smaller than they would be if the supply were either single phase or direct current and serving the same purposes.

Certain restrictions are put on the user of electricity supplied by public utility companies. The three phases supplied must be kept "in balance", that is, the same amount of power must be taken from each, and the power factor must not be allowed to fall below a prescribed value, often 0.9. This means that the voltage and current sinusoidal fluctuations with time must not be. allowed to get "out of phase" with each other by more than about 25° due to inductive loading causing the current to "lag" behind the voltage. Where inductive loading is unavoidable suitable capacitance must be built into the circuits to correct the lag. Similarly if the load is capacitative, inductance must be introduced to bring the power factor above the prescribed limit. These installations can be costly and the need for them should be designed out of the system as far as possible. Electricity is charged, however, on the basis of kVA while the power developed can be measured as kVA x P.F (Power Factor) so that is is in the interests of the user to maintain the power factor as close to 1.0 as is practicable. The pricing of electricity to industry is sometimes based on an agreed maximum rate of consumption rather than on the total quantity of energy consumed. Sometimes there are penalties payable whenever a prescribed maximum rate is exceeded and sometimes there is a large charge payable whether power is drawn or not on the grounds that the capability to supply must be maintained continuously. Sometimes the price will be different at different times of the day, usually to encourage the use of electricity at night, and it is obvious that with adequate forward planning in consultation with the supply company it may often be possible to reduce the cost of electrical energy for particular purposes.

The Cost of Energy

The cost of energy is probably rising steadily in the sense that the proportion of human effort devoted to its supply is steadily increasing. Gradually the easily won coal and oil reserves are being used up and we are being forced to seek further supplies in deeper strata and in off-shore oil fields, sometimes in very inhospitable waters, where the costs incurred are much higher than they have been in the past. Much of the effort and cost is now being expended on building sophisticated equipment like oil rigs rather than on direct labour costs at the oil well or mine and this trend is even more apparent when nuclear energy is considered. A very large part of the cost of producing electricity from nuclear energy is the capital cost of building the reactors and heat-exchange equipment and the associated fuel reprocessing plant. The waste products have to be disposed of at some cost also. The personnel working on these plants are not numerous but are all highly trained and include many engineers and scientists. The fuel is not as costly to mine as coal – much less so in terms of the energy it can produce – but processing the ore to fuel rods is much more costly than any preparation that may be applied to coal.

The relative cost of the various forms of energy varies widely from one country to another. The prices are often artificial because of taxes and subsidies which are applied by governments from time to time. Different classes of user may even pay different prices in one country. Apart from the price paid per ton or per joule there are secondary costs associated with storage and handling of the fuel or energy within the plant. Oil needs storage tanks and pipelines, sometimes heated to allow the oil to flow through them. Coal burns leaving an ash residue which must be removed. Electricity can be expected to be much dearer than coal or oil per joule because it takes about 4 tons of coal to produce the electricity with the energy equivalent of 1 ton of coal. Even allowing that electricity can be made from some of the cheapest grades of coal it cannot be marketed at less than about 5 times the cost of coal, per unit of energy. On the other hand, electricity can be used more efficiently than coal or oil – the inefficiency having already been incurred in the power station – so that the cost per effective joule may not be very different from that of coal. The user must consider these matters carefully when designing processes, to ensure that his energy costs are not higher than they need be.

In principle the extraction of any metal can be accomplished using any available form of energy. Traditionally iron is smelted using coke but many other ways have been suggested for reducing iron ores. Most of these are based on the use of natural gas or oil. Only a few come near to being able to compete with the traditional process using a blast furnace and these only because of special circumstances — where coke is not available, where the scale of operations is small or where it is convenient to produce a low carbon crude iron for economic reasons. The price of steel made by one such process is now so close to that of steel made by the traditional route that direct reduction plant is now being built in Britain to make pellets of reduced iron ore as feed for electric arc furnaces which are normally fed with scrap which is not always in good supply. Methods have been suggested based on the use of nuclear energy. Much of the required heat could come from a nuclear reactor direct but high temperatures could be reached only through generated electricity and reduction, in the absence of carbonaceous reductants, could be effected only by hydrogen generated by the electrolytic dissociation of water. This is not yet economically attractive. Some metals would be difficult to produce without electricity but all metals could be produced by electrolytic reduction, either in an aqueous electrolyte or a fused salt bath. The choice of energy form is made not on the basis of what is possible but on the basis of what is available. From among those available the final choice is that which when used leads to the process with the lowest overall production costs.

Furnaces

Furnaces in extraction metallurgy are used principally for calcining, roasting and melting. They vary widely in size and shape and many of them have developed into large high-temperature reaction vessels designed to facilitate chemical interactions between components of the charge or between burden and furnace gases rather than merely to raise the temperature of the charge. It is, of course, always important that the primary functions of a furnace, the combustion of fuel, the development of heat and the transfer of the heat into the burden, shall be carried out efficiently but it may be necessary to accept something less than the maximum thermal efficiency if a mass transfer process is being undertaken simultaneously.

Most furnaces used in this field are either hearth furnaces or shaft furnaces but many have developed as hybrids of these main types. Rotating kilns, Wedge-type roasters and some vertical retorts belong to that hybrid category. It is impossible to find neat categories into which to put fluo-solids reactors, sintering machines, converters, vacuum arc remelting furnaces and other specialist equipment in which reactions are carried out at high temperatures. Classifying by the kind of energy used there are three types: gas and oil, electric and coke. Electric furnaces are nearly all arc furnaces while coke-fired furnaces are nearly all blast furnaces.

Shaft Furnaces

Shaft furnaces may be used for calcining and roasting, for smelting, remelting and in the one case of zinc reduction, for the production of gaseous metal, so that it is used as a retort. The blast furnace used for smelting iron is by far the largest shaft furnace. It is about 30 m high, is refractory lined, and burns coke in a hot air blast supplied through tuyères just above the hearth. Blast furnaces for smelting other metals (lead, copper and tin) are only about 5 m high, often of rectangular section and constructed in the high-temperature zone from steel water jackets without any lining other than chilled-on slag. When in use they may be considered as operating in two zones – an upper preheating zone in which roasting may also be performed and a melting zone beneath with a crucible for the collection of molten products. When the preheating or roasting functions are not important the shaft may be shortened, so that there are "low shaft" furnaces – some of them with electric arc heating – but the distinction between these and hearth furnaces becomes somewhat blurred. The advantage of the shaft furnace over all others is that the transfer of heat from ascending combustion gases to descending solid burden can be very efficient. This advantage cannot be enjoyed in any shaft-shaped furnace heated by electric arcs in the smelting zone because there is then no large body of hot gas to carry the heat convectively up through the furnace. The efficiency of heat transfer is at its best when the heat available in the rising gases at any temperature is approximately the same as that required to bring the equivalent amount of descending burden up to the same temperature from cold. The gas will always be at a higher temperature than the solids in contact with them. That difference should be great enough to ensure that the heat-transfer rate is satisfactorily high but low enough to ensure that the transfer is taking place under conditions which are, as far as is practicable, thermodynamically "reversible". Ideally this would mean no temperature difference between gas and contiguous solids. In practice a difference of 150–200°C is a good compromise which allows a high proportion of the energy of the fuel to be transferred into the burden — perhaps up to 90% of it — at a reasonable rate. If there is a chemical reaction occurring at the same time between the ascending gases and the descending burden, the same principle holds, that the chemical potential differences between the reacting components of the gas and solids must be sufficiently large to ensure that the reactions will proceed reasonably rapidly but at the same time small enough that there will be a reasonable approach to thermodynamic reversibility in the reactions which will minimize their energy requirement. The best compromise to satisfy the thermal conditions need not be compatible with the best for the chemical conditions so that a further compromise may have to be made. In general the enthalpies of the reactions will so affect the thermal requirements that the heat demand will not be a smooth function of burden temperature. Consequently the gas/solid temperature difference will not be constant at all levels in the furnace.

In the iron blast furnace there are reduction reactions between gas and burden, slag-formation reactions between components of the burden and dissociation reactions involving one component of the burden only. Perhaps the most important, however, is the reaction

$$C + CO_2 = 2CO$$

which goes to the right at temperatures above 670°C (see Fig. 45) and to the left at lower temperatures. This determines that if iron oxide is reduced at a low temperature by CO, the CO_2 formed will be stable and will not be altered before it leaves the furnace. If the reduction of the oxide of iron is delayed until the ore has come low in the furnace where the temperature is high, any CO_2 produced will be recarbonized to CO. The enthalpy of oxidation of carbon to CO is only -113 kJ mole^{-1} while that of its oxidation to CO_2 is -393 kJ mol^{-1}. Clearly there is a considerable thermal advantage in ensuring that the reduction reactions occur as high in the furnace as is practicable. Iron ores are prepared by sizing, pelletizing, sintering and heat treatment to make them more readily reducible and if a low production rate were acceptable practically all the oxide could be reduced with the production of stable CO_2, but in practice productivity is important as well as energy economy and a compromise is adopted. The iron blast furnace can be used to illustrate another point. If the ore is rich the total energy requirement is low per unit of iron, and a high proportion of that heat is required at high temperatures so that the gases leaving the smelting zone are relatively cool and, indeed, rather close to the temperature of the contiguous descending solids. If the ore is lean much more heat is needed per unit of iron and because of the larger flux requirement a much higher proportion of that heat must be available at low temperatures. The gases leaving the smelting zone are therefore much hotter than when the burden is rich. The gas/solid temperature difference is much greater − sometimes much too great. The burden heats up very rapidly in the lower bosh with unfortunate effects on the slag-formation processes. The temperature distributions in the furnaces in the two cases are quite different. Neither is ideal, one with too small differences between gas and burden and the other with too large.

Some control can be imposed on this process through various operational factors. The particle sizes of components of the burden are important. Ore particles should be small for rapid reduction but not so small that the burden becomes impermeable to the furnace gases or that excessive amounts of ore are blown out with the top gas. Ideally the burden should flow down the stack like a fluid in "plug" flow and ideally the gases will pass up through the burden with the same velocity all the way across the furnace. Uniformly sized and well-mixed burdens behave best and pelletized ore at the right size is most satisfactory in ensuring that all the solids arriving at any level will be in very nearly the same condition. The gas and solids mass flow rates are not independent of each other but the linear velocity of the gas can be controlled within limits either by altering the overall gas pressure (high top pressure practice) or by decreasing the proportion of atmospheric nitrogen in the combustion air (oxygen enrichment). By either of these means the relative velocity of the gas past the solids and the effective contact time can be brought under some degree of control. The temperature distribution in the furnace can also be controlled by raising or lowering the flame temperature or combustion temperature at the tuyères. This can be raised using oxygen enrichment or lowered by injecting either steam or a hydrocarbon with the hot blast. In either case an endothermic reaction will reduce the flame temperature and some of the hydrogen produced will be re-oxidized in the stack, releasing some of the heat that was absorbed in the tuyère zone. The use of oxygen and steam together (effectively replacing nitrogen with steam) provided a sophisticated means of controlling flame temperature and temperature distribution. An older method, still used occasionally, was to supply some air through supplementary tuyères at the top of the bosh where some CO would be burned to CO_2, producing heat but at the same time lowering the reduction potential of the stack gases. This must slow up the reduction reactions so reducing the proportion of low temperature reduction and therefore the amount of CO_2 produced by the reduction reactions. If overdone, the use of bosh tuyères becomes counter-productive.

It will be appreciated that fine control over the operation of a large blast furnace is scarcely possible. The analytical and thermal data necessary for controlling the process are never very accurate because the problem of sampling is extremely difficult and also because of the very large time lags between the taking of action and the observation of the effects of that action. Ideally it would be desirable to control the temperature distribution of the burden to be an approximately linear function of the distance from the stock line but it is not possible to measure that parameter, far less to control it.

A good example of a shaft furnace used for re-melting is the Asarco furnace which is used for melting copper cathodes for casting to wire bars. A high production rate is required under a well-controlled, slightly reducing atmosphere. The furnace is about 7 m high and 1.5 m in diameter. It is fired with a sulphur-free hydrocarbon gas burned completely in combustion chambers outside the shaft and injected through tuyères just above hearth level. Heavy cathodes can be melted at up to 70 tons per hour and run continuously through an induction-heated holding furnace (forehearth) to casting machines. The only control on quality is through the furnace atmosphere, through which the oxygen content of the metal is adjusted to the desired value (see page 366).

Hearth Furnaces

Hearth furnaces, usually reverberatory furnaces fired with gas, oil or pulverized coal, are used widely for melting and separating slag from matte or metal. They usually work at very high temperatures. Although their principal function is melting, it is often necessary that they should hold the molten charge as a specified temperature to allow reactions and separations to take place. As the combustion gases are usually in contact with the charge, some interactions between the gaseous and liquid phases may be expected and also between gases and solids during melting. Metal may pick up sulphur from the combustion gases or the gases may be used to oxidize impurities from the metal. Hearth furnaces may operate batch-wise or continuously. Steel is made in large batches but copper is smelted continuously. In either case the thermal efficiency is rather low. Gases must leave the furnace at a temperature not lower than that of the molten charge, which in the case of steel is 1600°C. Obviously the heat in this gas must be put to use. It is most appropriately used to preheat the combustion air and about 30 per cent of the heat in the combustion gases can be circulating, through preheaters. Preheating may be done using recuperators up to about 1000°C or using regenerators at higher temperatures. Recuperators are continuously operating counterflow heat exchangers, the use of which makes single-ended firing of the hearth furnace possible. When the temperature is too high for a recuperator, two regeneration chambers are used, one of which is being heated by gases leaving the furnace while the other is being used to heat the air. These chambers are filled with a chequerwork of refractory shapes which can withstand as high a temperature as any furnace and have good heat-transfer characteristics. The hot gases pass down through the chequers in the first half of the cycle until the top runs of bricks are almost as hot as the gas. In the second half air is passed up through the chequers so achieving, in effect, a counter-current transfer of heat. The system works well, but imposes the restriction that the hearth furnace must be fired alternately from one end and then from the other. This does not allow the design to be ideal either at the burner end or at the flue end, nor does it make it practicable to arrange for the gases leaving the furnace to preheat the burden entering the furnace. Hearth furnaces have the inherent disadvantage that the area from which the burden loses its heat is very large, per ton of material melted. At high temperatures the opportunity for heat conservation by the use of insulation is very limited because the refractories are often working near to their maximum temperatures. Insulation will increase the temperature of the brickwork and reduce its life. Indeed air cooling and even water cooling are often used to preserve refractories at some cost in heat. The large surface area of the metal in the hearth furnace can offer an advantage if it affords a means of speeding up the reactions between metal and either slag or furnace atmosphere but in recent years increasing use has been made of gas jets to expedite both gas–metal and slag–metal reactions.

The continuously fired copper-smelting reverberatory furnace (see Fig. 71) can be considered to operate in two zones. The first is a melting zone and the second a settling zone. In the former a high rate of heat transfer is wanted so that the flame temperature should be as high as possible. Heat transfer is by radiation and convection. Over the settling zone the flame temperature need not exceed about 1300°C but clearly the combustion gases will again carry away a large amount of heat which must be put to use – in this case usually in waste-heat boilers. This depends on the overall heat policy of the plant but in most industrial plants steam is useful and can be produced cheaply from "waste" heat at quite low temperatures. If steam cannot be put to direct use, it can be used to make electricity.

Rotating kilns are not used for melting but can be operated up to very high temperature with solid or partly molten burden. They are operated on the counter-current principle with the burner at the lower end of a long inclined cylindrical shell while the burden is rolled gently down through it from the higher end. In this way the kiln resembles a shaft furnace but the heat transfer is not so efficient. The product is usually discharged hot but in some designs there is a short zone at the lower end in which some transfer of heat from the product to secondary combustion air is possible. Rotation of the kiln transports the burden through it and also agitates it to expose new surfaces continually to the flame. Heat transfer is by radiation and convection but would soon be limited by conduction if this stirring action were not applied. The rotation also helps to keep the refractories at a reasonable temperature as at every revolution of the shell brickwork is screened from the flame for about one-third of the time, during which it is yielding some of its heat to the stock. Heat losses from the outside of the shell are rather great and cannot easily be reduced because any insulation must be put inside the steel shell, that is between the hot face brickwork and the steel. Insulating bricks are never very strong and are not suitable for service in that situation.

A multiple-hearth furnace like the Wedge roaster (see Fig. 66) can be seen to have features of both shaft and hearth types. In so far as the roasting function is more important than thermal efficiency the main features are designed to that end. Maintaining the correct SO_2/O_2 ratio at every level is more important than saving the few units of energy lost to incoming cold air, but refractory insulation is being increasingly used. It will also be appreciated that the faster the oxidation reactions proceed the shorter the residence time of the concentrate and therefore the smaller the general heat loss per unit mass of concentrate treated so that there can be a thermal advantage in this case in running the furnace at the maximum possible production rate. Similarly sinter strands must be run primarily to produce good sinter and if this means recycling a proportion of product (in the case of blast roasting a sulphide ore), this must not be considered a misuse of energy. The general failure to recover the heat in the sinter produced is another matter but so far no satisfactory means of doing so has been developed.

Converters

The most impressive metallurgical reaction vessel is the converter which is used for converting copper and nickel matte or blast-furnace iron to steel. Air or oxygen is blown through or on to the surface of the molten charge and the net enthalpy change in the resultant reactions is the only source of energy for the process. The vessel is of compact form — not spherical but with a low surface/volume ratio which ensures a low general heat-loss rate. Modern steelmaking converters are usually top blown with oxygen. Reaction rates are high. Up to 300 tons of iron can be refined in about 40 minutes. Consequently the total heat loss to the surroundings per ton of steel is quite small. If the composition and mass of the metal being charged is known, the requirements of flux and oxygen can be calculated from simple stoichiometry and a knowledge of the physical chemistry of steelmaking and the heat released by the reactions can also be determined. The heat required to raise the temperature of the converter and its contents to the finishing temperature can be estimated along with heat losses to the surroundings and in the gaseous products. Any surplus heat must be used up by melting either scrap or iron oxide (ore or millscale) incorporated into the charge. About 10% of the energy leaves the converter with the gases. A quarter of this is as sensible heat and the remainder as the chemical energy of carbon monoxide. The gases are heavily laden with dust and fume and a complex gas-cleaning and heat-recovery system is needed to recover the heat and prevent the fume from escaping to the atmosphere. The sensible heat can be used for steam-raising and the carbon monoxide, after cleaning, is used in some plants in the same way or put to some other purpose in the plant. The intermittent nature of the converter operation makes it difficult to run the gas-treat-ment plant efficiently. Auxiliary fuel burners are used in the boilers to maintain firing when the converter is turned down. With oxygen blowing, the volume of gases produced per ton of steel made is much lower than when air was blown and hence the recoverable sensible heat is also lower but the quality of the cleaned gas is much better. The scale of the operations has been very much increased in recent years and with an increasing use of oxygen at the same time, the problem of atmospheric pollution has become severe. In most countries, regulations about the control of pollution have become very strict and the cost of the necessary gas treatment has become very high. There is no escaping these costs but it is obvious that the essentially non-productive operations involved shall be carried out with as much efficiency as the major operations if total costs are to be held down. (See also page 321 and Fig. 76.)

Combustion

The design of burners is important in all kinds of furnace. They are very different for the various fuels but in all cases they must be able to supply both fuel and air to the furnace at the necessary rates and in the proper proportions. Usually there should be a small excess of oxygen over the stoichiometric ratio to ensure that combustion is complete within the furnace space. The reactants should also be brought together sufficiently intimately to ensure that the reactions occur at the required rate. This determines the flame temperature which is always important but especially so when the furnace temperature is very high. If a furnace is running at 1600°C, and the flame is only at 1700°C, only about one-seventeenth part of its available heat above ambient temperature is available for heating the furnace. If the flame temperature were raised to 2000°C, however, one-fifth of its heat could be utilized in the furnace.

A very fast combustion reaction gives a shorter hotter flame than a slow reaction because there is less heat lost *during* the reaction, to the surroundings. Other factors are involved, including the calorific value of the fuel, the percentage of excess air and the amount of preheat applied to the air. It is also possible to obtain higher flame temperatures by enriching the air used with oxygen. This really means reducing its nitrogen content or the nitrogen supplied per unit of fuel or energy. Less energy is needed to heat nitrogen up to the flame temperature. This energy is distributed among the other constituents of the combustion gases, bringing them to a higher flame temperature.

The combustion of solid fuels is controlled mainly through particle size. Finely pulverized fuel is blown into the furnace with primary air, rather like a gas. It burns rather slowly and requires a very large excess of air to complete the combustion. Solid fuel in lump form is burned in fuel beds into which air is injected through a grate or through tuyères. The combustion of carbon is complicated by there being two oxides with very different heats of formation. CO_2 is formed first with a large evolution of heat and the development of a high temperature but in a fuel bed it may react endothermally with carbon to produce CO with immediate loss of temperature. This can be delayed by using large coke of low reactivity and, where practicable, a shallow fuel bed. Liquid fuels are "atomized" to small droplets by the action of the burner and projected into the furnace with primary air. These droplets evaporate and oxidize at a rate which depends on their size and on the distribution of secondary air in the furnace. The size, shape and temperature of the flame are determined by the characteristics of the burner. Gaseous fuels burn most efficiently when pre-mixed with air (or oxygen) as in the Bunsen burner, but such an arrangement precludes the preheating of the air. Oxy-propane lances have been developed on this principle. These give extremely high flame temperatures and could be used in steelmaking converters but the flame propagation velocity of the mixture is so high that its emergence from the mouth of the burner must be at supersonic velocity, creating an almost insuperable noise problem. More usually gases are burned in "diffusion" flames in which air and fuel gas streams diffuse through each other with the assistance of varying amounts of turbulence induced by burner design and furnace geometry. Gas flames, excepting those of propane and butane, are generally cooler than oil flames but are rather more easily controlled especially if control over the furnace atmosphere is required.

Heat transfer from combustion gases is by a combination of convection and radiation. Radiation is most effective from luminous flames (of the same temperature) but these are not the hottest flames. CO_2 and H_2O in the gases do radiate some heat but transparent flames should be passed through the furnace with sufficient turbulence to ensure efficient convective transfer to the burden and where practicable to the roof which will re-transmit the heat by radiation to the charge. This is the principle of the reverberatory furnace. Heat-transfer rates are greatest when the temperature difference between the hot gas and the burden is large and the heat-transfer process is most efficient in counter-current exchanges. When oxygen enrichment (or pure oxygen) is used to create a very hot flame, the heat transfer may be concentrated into a very narrow zone near the burner because the gas has low mass and cools very rapidly as its heat is lost. In the iron blast furnace only a very small amount of oxygen enrichment can be tolerated for this reason unless there is some compensation applied, such as steam injection, as was discussed earlier. An increase in heat-transfer rate is not always matched by a similar increase in reaction rates or productivity. In the simplest of cases an increase in the rate supply of heat to the surface of an ingot will not have an appreciable effect on the rate at which heat will pass into the body by conduction. The maximum heating rate is determined by the maximum temperature to which the surface can be raised. In general, reactions which are rate controlled by mass-transfer mechanisms cannot be accelerated merely by heating them more quickly.

Electric Furnaces

When electricity is used for heating in extraction processes it is usually an arc furnace that is employed. Arc furnaces vary in size from 1 or 2 tons up to 200 to 300 tons (in steelmaking). There is a variety of electrical arrangements possible. The three-phase a.c. arc furnace of the Heroult type is usual for steelmaking but a d.c. arc can be used either struck between two electrodes or between a single electrode and a conducting hearth. Power for such a furnace can be obtained by a d.c.

generator driven by a three-phase a.c. motor. In some copper-smelting furnaces there are three pairs of a.c. electrodes, each pair supplied by a different phase of the normal three-phase supply.

Carbon arcs operate at about 4500°C where heat transfer by radiation is dominant. Ideally, arcs should be smothered with stock so that no visible energy can escape. In steelmaking this is not possible and the heat is reflected from the roof as in a reverberatory. Electric furnaces for smelting iron (Fig. 73), copper matte or aluminium (electrolytically) all run with a black top, the electrodes being buried in the charge. Heat is generated by the current passing through plasma, slag and metal or matte in proportions depending on the design and operation of the process. The electrical circuitry for these furnaces is heavy and complex. Transformers supply energy at a range of voltages between about 50 and 500. Control gear maintains the electrode/metal gaps at a standard size but especially during melting down there is much hunting in the control. Arcs frequently break and remake. This causes the currents to fluctuate violently and it is necessary to have heavy chokes in the system to damp out these surges as well as capacitors to maintain the power factor. The current is of the order of 20,000 A in each phase in the larger furnaces so that it is necessary to reduce the transmission losses by building the transformer and furnace alongside each other, the incoming power at, say, 30,000 V being reduced to the working voltage at a point within reach of the electrode connectors through bundles of flexible conductors.

Electric furnaces produce much less gas than fuel-fired furnaces but it must still be removed and treated. The gas from the steelmaking arc furnace is largely carbon monoxide which must be burned before discharge. It is also laden with fume during the refining stage so that equipment must be available to clean it. The chemical energy in this gas is about 20% of the total energy of the process and attempts have been made to recover some of it by venting it away through a twin furnace shell filled with scrap for the next charge which can be substantially preheated. Apart from the saving in heat, this arrangement, which involves considerable investment, must improve productivity considerably.[70] The gas emitted by the copper-smelting furnace is mainly SO_2 so that too must be vented and fixed.

Electricity is not commonly used simply for heating in extractive processes but its use does seem to be increasing. For most purposes other energy forms are cheaper. In steelmaking where a high proportion of scrap is to be melted from cold the rate of melting by electric arc is so much higher than by gas or oil heating that the extra cost of the energy per joule can be fully justified. Heat losses to the surroundings are lower (per ton of steel) and better use is made of capital assets. Copper is smelted electrically but only where the cost of electricity is favourable. Iron has been smelted electrically also where hydro-electric power made it possible to do so in Scandinavia, Italy and Switzerland but as a general rule there should be some technological advantage to be gained before electrical heating is employed. This could be that a particularly high temperature is required as in the cases of certain ferro-alloy smelting processes or, as with vacuum arc remelting, that no other form of energy is suitable.

Electricity is the principal form of energy used for purposes other than heating. Electrolytic winning and refining require large amounts of electrical energy as direct current at low voltage and very high amperages. More is required for movement and transportation: to drive locomotives and cranes; to compress air for blast furnaces; to shake riffle tables or agitate the pulp in flotation cells; to open and close furnace doors or drill out tap-holes. All movement needs power and most of that power is now supplied as electricity. The total required for such purposes may seem small compared with what is consumed by a large smelting furnace but is by no means inconsiderable in any industrial plant and even small percentage savings generated by improvements in the efficiency of utilization of the equipment are usually worth a lot of money. Well-designed, efficient equipment which is well maintained and skilfully operated will usually use less power than badly designed plant which suffers neglect. A flotation cell or a ball mill which is being improperly used will take longer than it should to produce the desired result and during that extra time the driving motors are running, using up energy. If the speed of the mill can be optimized or the concentrations of reagents to the flotation cell can be properly adjusted an immediate improvement in both production rate and energy consumption per ton of product will be seen.

It is not always an advantage to rush a process. Beyond some optimum rate of production the energy cost frequently starts to rise disproportionately to the benefits in other directions. The case of the electrolytic deposition of a metal is considered on page 313 and in example 18 in Appendix 1, and that of the iron blast furnace on page 245.

Furnace Refractories

The need to use energy efficiently is well understood but it is not so well known that a great deal of effort goes into improving the performance of refractories. Furnaces are built from refractories and lined with them. They are used as pre-fired bricks, as unfired bricks and blocks to be fired *in situ* or as castable preparations to be mixed with water and used like concrete. In some applications the refractories last almost indefinitely. In others, conditions are such that the linings last only a few hours. In these cases refractories can be considered as consumables and their consumption is rated in kg per tonne of metal produced.

Refractory bricks are costly and use up much energy in their production. Their replacement can be costly in labour and also in lost production. Where possible, relining is carried out to planned schedules which must fit in rationally with the cycle of the industrial week. Thus if relines are to be carried out at week-ends after 5 days in service, a better material which will last 6 days would not be worth extra cost but another which can survive in service for 10 days could be worth more than twice the price of the original. In some applications the question is whether the lining will last one shift or two but the principle is the same — a life of one and a half shifts would not be helpful.

It is not possible here to deal with refractories technology at any great length but only to outline some more important aspects. The material for any job must be selected carefully to withstand the particular conditions of the case. The important factors are temperature and chemical environment. The refractory must be strong enough to withstand whatever load must be applied to it at the working temperature. Failure is more likely to be by creep than by any catastrophic fracture but cracks can develop either due to the effects of thermal cycling or because of thermal diffusion of minor constituents of the brick at very high temperatures. More mobile chemical species migrate from the hot fact to cooler parts where they embrittle the brick locally. Chemically, the material should be inactive toward whatever slag, oxide, dust, fume or gas may come into contact with it at the working temperature. In addition, bricks may have to withstand abrasive damage or erosion by fast-moving, dust-laden gases; they may have to undergo severe cycles of heating and cooling; and there could be a requirement about their having low or high thermal conductivity. Usually some compromise must be accepted between incompatible conditions.

Despite the wide range of refractories available the choice is usually narrow because of one or other of the operating conditions. Chemical considerations usually determine whether the brickwork will be acid or basic, that is from the silica/alumina group or the dolomite/magnesite/chromite group. Within either group the working temperature will narrow the possibilities further and the final selection is then made on the basis of special conditions to be met and on price. The wide range of bricks available was originally developed for use in the now obsolescent basic open hearth steelmaking furnaces. In recent years the production of these bricks has been reduced drastically and the silica bricks which used to go into the roofs of these same furnaces are now made only for coke ovens and a few minor applications like the domes of Cowper stoves. Firebricks remain the common bricks of furnace building but with a higher proportion of high alumina grades being built into furnaces of all types to withstand better the harder wear associated with generally increasing production rates. Alumino-silicate bricks with 72% Al_2O_3 and recently even 86% Al_2O_3 are being used over a very wide range of applications where wear resistance, spalling resistance, slag resistance and high refractoriness are all required together. These can be found where silica was formerly used or replacing chrome-magnesite elsewhere with equal success. Any type of brick can be made to a range of specifications in which the texture and porosity are varied to provide different physical properties – different thermal conductivity, permeability to gases, or different density. Steelmaking converters are now being lined with blocks of dolomite which have the great merit of being cheap both to make and to install. In recent years too, the hearths of iron blast furnaces have been made of carbon instead of firebrick. Carbon is not attacked by iron under these conditions and its high thermal conductivity enables air cooling to be applied with good effect to preserve the hearth.

Apart from their applications within furnaces, refractories are found in flues, runners, ladles and heat exchangers. They are also fashioned into small components like stopper rod sleeves, stoppers and gates on ladles, and into sheaths for protecting thermocouples and similar probes. There is also a range of insulating refractories. These are usually made of porous firebrick. They can be very refractory but lack strength at ordinary temperatures and tend to shrink if heated beyond their working limit. Their average conductivity is not as low as those of asbestos, slag wool or vermiculite none of which, however, can be used above about 800°C, whereas bricks can be used up to at least 1200°C without deterioration. Recent developments in fibrous materials made from china clay and aluminous materials have raised the maximum temperature of use to 1600°C and greatly simplified construction. The conservation of energy by means of insulation is always attractive. It is essentially a matter of designing the furnace so that the assembly of refractories used has the highest practicable resistance to the passage of heat compatible with durability. In general, insulation is most effective when applied to installations which operate at only moderately high temperatures – up to about 1300°C. External insulation alters the temperature distribution, reducing the temperature gradient behind the hot face so that a greater thickness of the hot face refractories may be working at a temperature near to or even above the safe value. Where temperatures exceed about 1400°C it is usual to find water cooling applied to the more vulnerable parts to impose steep temperature gradients on the brickwork. Heat loss to the cooling water is part of the cost of maintaining the refractories in good condition.

Energy for Transport

In the field of transportation the energy used is roughly proportional to the distance travelled and there is obviously an advantage in having the lines of communication as short and direct as possible, at least within a works area. Gravity should be used as much as possible to assist movement. Pipelines should be short, wide and straight as possible to minimize pressure drop in fluids being transmitted without being larger and costlier than necessary. Even pipelines should be designed to their purpose. When solids like ore or coal are to be transported over a distance the form of transport used should be considered very carefully. This was the subject of an excellent study by Worthington.[71] Here the first question asked was about the cost of moving t tons per day of material a distance x miles using n locomotives and w waggons, given

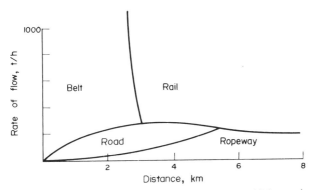

FIG. 65. Diagram showing conditions in which various transportation systems for ore were cheapest in 1962.[71]

prices for the running of locomotives and waggons and other relevant data. It was shown that the optimum size of train depended both on the distance and on the daily load. The exercise was then extended to embrace other means of transporting the same material – lorries, conveyor belts and overhead ropeways – to show how each method had the advantage of cost over the others within certain load/distance ranges. Although the costs are now of mere historical interest the summary of their findings on a predominance diagram is reproduced in Fig. 65.

Clearly this line of enquiry is aptly described by the term "optimization" and in principle any self-contained group of operations can be studied to find by mathematical computation the best way of carrying them out. This is a relatively modern tool of management now called "operational research". The methods used are probably not very different from those used traditionally by successful managers but the work has to some extent been taken over by specialists who are adept at using computers to deal with problems at a higher level of complexity than most managers would be willing to tackle. The basic approach is to develop a "model" of the process from which to establish the operational parameters which should determine the appropriate line of action. The model is then reduced to a series of equations involving these parameters. By systematically introducing series of values for each parameter within the possible range and by examining all possible combinations of these, the "most efficient" set of values can be identified – where the criteria of efficiency have been defined at the design stage of the exercise.

CHAPTER 11

INSTRUMENTATION AND CONTROL

IN ORDER to control a process it is necessary to make observations and to take appropriate action. At the most primitive level, observations are visual and action manual but between eye and hand a brain must decide what action ought to be taken in order that state of the system observed shall be altered in the right direction and by the right amount. In modern industrial plant the observations are made using instruments which accurately measure the values of a selection of parameters which collectively provide a description of the state of the system which is being controlled. These values are then fed into a computer where they are entered into a programme which has been designed to calculate what action should be taken to change the observed state of the system toward a required state. The computer may also be able to instruct machines to take the necessary action. The steps are the same but the equipment is different. The observations are more precise and more detailed and the calculations are carried out rapidly and accurately but without initiative or imagination. The calculations are more complex than a brain could cope with in a reasonable time and the computer will not become tired and inefficient but its answer is only the answer it has been told to give in any particular circumstances. The machines are more powerful than a hand and there is great convenience in having the computer set the controls, especially when several changes may have to be made simultaneously.

Ideally measurements and adjustments should be made continuously to keep a system in its optimum state or to change the state in accordance with some prearranged programme. There are two kinds of difficulty, however. One is concerned with measurements — whether they can be made with the necessary accuracy and with the necessary speed.

260

The other has to do with the speed of responses in the control system – how long does it take for the effect of a change to be measurable?

Instrumentation

The important measurements which must be made on metallurgical extraction plant are those of temperature, pressure and quantities of matter and energy. Quantities of matter are most often determined as mass or volume flow rates and the same may be used as measures of energy – where, for example, we might measure the flow of oil to a furnace in m^3/h. When measuring quantities of matter we may require some detail about the quantities of particular chemical species present – not only to know how much ore, but how much Cu, Fe, S, SiO_2, etc., is being fed into a smelting furnace, for example. Accurate chemical analyses are required of raw materials, intermediates and products, sometimes very fast and sometimes very frequently.

Some of the required measurements are made very easily or, at least, are made using equipment which is well established in use in a wide range of technologies. The important essential today is that an electrical signal can be produced the strength of which is a simple continuous function of the value of the parameter being measured. In many instruments this condition is satisfied from the beginning. The thermocouple is an obvious example. In other cases such as manometers there is no electrical output in the simple instrument so that some supplementary device must be devised if these are to be coupled up with electrically operated control equipment. A device which produces a change in an electrical circuit in response to some change in another system with which it is associated is called a "transducer". Where the electrical response is not direct as in the case of simple manometers it is usually possible to convert some small displacement into an electrical signal by moving an iron core in an induction coil so changing its impedance, or by moving a plate in a variable condenser to change its capacitance. A small change in the resistance or the reactance of a circuit can be converted into a small current or voltage using an unbalanced Wheatstone bridge in the case of d.c. signals or a suitably modified a.c. bridge for a.c. signals. By appropriate selection of the bridge ratio some amplification of the original signal can be effected. Further electronic amplification is now standard practice.

The measurement of temperature is usually easy. Techniques have been standardized for using thermocouples, resistance thermometers and thermistors within appropriate temperature ranges. Thermocouples are most easily used and can be taken to higher temperatures than the others but care must be taken to place them so that they do not give spurious indications – showing the temperature of a flame, for example, rather than that of the burden beneath it. At very high temperatures even the precious metal couples become unreliable because of diffusion across the junction. This can make their continuous use inadvisable. The support is also liable to fail at high temperatures so that we find that in steelmaking only occasional temperature determinations are made and then by dipping into the metal a specially protected probe containing a Pt/Pt–Rh couple, several inches of which are discarded after each dip. Total radiation pyrometers also produce an electrical signal but they are not very useful in a situation like a steelmaking furnace because of difficulties with emissivities, and because of fume obscuring the sighting and settling on optical surfaces. One way of obtaining a continuous measurement in difficult circumstances is to make an indirect measurement at some point at which the difference between the measured temperature and the required temperature can fairly be presumed to be constant. This technique can be quite satisfactory particularly if the interest is principally in temperature changes, provided there is not too much time lag between temperature changes at the two points.

Any of the traditional methods of measuring pressures can be adapted to provide electrical signals, but simple manometers require some kind of mechanical device to convert the levels in the tubes into a displacement suitable for activating transducers. Over a wide range of pressures, strain gauges can be used to make pressure gauges without moving parts. These are assemblies of fine wire, the electrical resistance of which is very sensitive to strain. In use, the strain gauge is attached by an adhesive to the outside of a piece of suitable tube which expands and contracts elastically as the pressure inside changes. As it does so the strain gauge wire suffers corresponding changes in its dimensions and therefore in its resistance which are easily translated into either current or potential measurements. In a similar way strain gauges could be attached to Bourdon gauge tubes. At low pressures the use of diaphragms or bellows is more common but even there strain gauges could in principle be incorporated. The piezo-electric effect can also be used for pressure measurements giving, in this case, a potential directly.

The measurement of quantities of materials can be made either by batches or continuously, whichever is appropriate. Weighbridges are used for batch measurement of quantities passing into and out of plant or from one process to another. Materials being conveyed by pipeline or on conveyor belts can have their flow rates measured continuously. These rates usually require to be averaged out and integrated over standard time intervals. The measurement of rates of flow of gases and liquids is carried out using conventional equipment of which the simplest are based on orifice meters and Venturi meters. The pressure differentials which are the basic measurements may be converted to electrical signals as discussed above. It will be appreciated that a pipe must be filled with a liquid for these measurements to be possible. Solid materials on conveyor belts can be weighed in transit. The belt passes over two sets of supporting rollers which are mounted on a platform which is in effect a balance pan. This rests on a group of load cells which are specially mounted strain gauges designed for heavy-duty operation. These measure the mass on the section of belt so supported continuously and present the information as a flow rate or as a quantity passing during any time interval. Load cells have a widening range of applications. They are now being incorporated in the trunnions of steelmaking converters to weigh continuously, during the blow, the whole converter with charge – about 600 tons. By this means the progress of the reactions can be monitored, in particular the rate at which carbon is being oxidized.

Apart from electrical energy which is measured directly with conventional equipment, the measurement of energy is essentially the measurement of quantities of fuel in solid, liquid or gaseous form. Flow rates of fuels must be converted to energy terms through the calorific values of the fuels. Volume flow rates are subject to some adjustments for ambient temperatures and pressures. Energy associated with materials, such as the sensible heat in flue gases, can be determined only by calculation from readily obtainable values of, in that case, the flue-gas volume, temperature and, through its analysis, specific heat.

Chemical Analysis

The determination of the chemical compositions of raw materials, intermediate products and final products by classical chemical methods has been practised in the extraction industries for many decades. The standard methods are rather slow, however, and for most purposes "rapid" methods were developed which were faster if not quite so accurate. These were used where the result was required before a process could be completed. In an open-hearth steelmaking process in which a heat of steel might take 10 hours to complete, a 20 minute delay for the final analysis was not considered to be unduly long. In a modern steelworks a heat of steel is made in 40 minutes and the 2 minutes required for a spectrographic analysis is still a higher proportion of the total processing time than was 20 minutes of 10 hours. There is still a need for faster and if possible continuous analytical techniques to be developed.

Continuous gas analysis is possible in certain cases. Hydrogen can be monitored easily because it has a uniquely high thermal conductivity. A hot wire loses its heat to its surroundings at a rate which increases with the concentration of hydrogen in the atmosphere around it so that hydrogen concentration can be measured indirectly through a measurement of the resistance of such a wire — which changes with temperature. Of all the industrial gases only oxygen, nitrous oxide and nitrogen peroxide are paramagnetic. This property is used to measure the oxygen concentration in flue gases on a continuous basis. Gas is passed through a strong magnetic field which diverts the free oxygen over a hot wire. The more oxygen present the more heat is taken from the wire so that once again its resistance is altered. Carbon dioxide can be measured by the absorption of infra-red radiation in a waveband in which other normal flue gases have no absorptive power and dry steam can be determined similarly on another wave band. Not all gases are possessed of such useful properties. Carbon monoxide is difficult and can be determined only by first absorbing carbon dioxide chemically, converting the carbon monoxide to the dioxide and then measuring that. Instruments set up to carry out such a sequence of operations can be made to report CO/CO_2 ratios on an almost continuous basis. Total analyses of gases usually have to be carried out in a laboratory but compact mass spectrographs are made which can be used for almost continuous analysis of gases for several constituents. Industrial gas must usually be cooled and cleaned before being analysed.

The continuous analysis of aqueous solutions is also possible, at least for most constituents and taking them one at a time. The atomic absorption spectrometer is most useful for metallic elements at low concentrations. A solution is sprayed into a hot flame and a beam of light, from a lamp, emitting radiation of a wavelength characteristic of the element being determined, is passed through the flame. The intensity of radiation of this wavelength is reduced by an amount which depends upon the concentration of the element in the sample. The equipment is rather delicate for on-line use and needs rather much adjustment and maintenance. Another possible tool is the selective ion electrode. This is used in much the same way as the more familiar pH electrode and, like it, produces a potential which is proportional to the logarithm of the activity of the ion being determined. Electrodes are available for cations and anions. They are costly and not very robust being subject to both mechanical and chemical damage.

Solid materials are most frequently analysed by chemical methods. In order to ensure accuracy, the primary sample of an ore, flux or fuel is initially very large and it must be crushed, reduced and mixed and the procedure repeated until there are only a few grammes of fine powder. This may be taken into solution for analysis by chemical methods or by physical methods such as the atomic absorption spectrometry mentioned above. Alternatively it may be analysed by emission spectrography. Modern direct-reading spectrographs are quick and accurate in skilled hands but their operation demands that great attention be paid to the details of the standard procedures. Spectrography is better suited to minor elements and trace elements than to major constituents of an ore like, say, silicon. Modern instruments can handle up to twenty or thirty elements at once. Another instrument which is perhaps easier to use and more flexible in operation is the X-ray fluorescence spectrometer. The sample must be solid, and have one flat face of a specified minimum size. It can be a compacted powder or indeed it may be compacted from a slurry or pulp and can be examined without drying. The sample is exposed to X-rays of short wavelength. Each element present fluoresces at its own energy levels and the instrument analyses the spectrum produced, measuring the intensity of the emission of each element of interest. This method is not good for trace elements but is very satisfactory at all concentrations above about 1%. Specimens must be prepared in a simple manner, after which results can be available in a few minutes. Metals can be analysed by either of these methods very quickly, a similar area of flat surface being required in each case. Where minor elements are required, the emission spectroscope is preferred, and indeed it is widely used for monitoring impurity levels in finished products. For the analysis of steels the spectroscopy is carried out *in vacuo* to allow short-wavelength ultra-violet light to be used so that carbon and phosphorus can be included in the analyses.

Liquid metals are almost always sampled, cast into moulds and analysed as solid metal but in two, as yet, exceptional cases electrodes have been developed with which to determine oxygen in solution in the melt. The two cases are steel and copper. In steel the determination is made by dipping a disposable electrode into the bath. A potential is developed over the first 10 seconds and the probe survives about as long again. The potential is a measure of the oxygen in the steel from which the necessary quantities of deoxidants can be calculated. In copper refining a similar electrode is used in the launder between the refining furnace and casting machine. It is coupled with the oxygen/fuel ratio control valve of a burner which plays a flame on to the metal a short way upstream. This flame can be so adjusted automatically that it can oxidize the metal slightly or reduce oxygen out of it, maintaining the concentration within the narrow limits demanded for wire bars of electrical quality copper. These electrodes can survive several days of service. They use solid electrolytes which are solid solutions of CaO or Y_2O_3 in either ZrO_2 or ThO_2. These solutions have CaF_2-type structures which contain anionic vacancies which permit the conduction of electricity only by the migration of oxygen ions. In use this electrolyte is sandwiched between an electrode in which oxygen is at a standard potential on one side and the solution of oxygen in iron (or copper) on the other. The potential measured between the electrodes is proportional to the logarithm of the activity of the oxygen in solution in the metal.

The elements hydrogen, oxygen and nitrogen are best analysed by a vacuum fusion method. A sample is melted under high vacuum in a small carbon crucible. The gases hydrogen, carbon monoxide (equivalent to the oxygen present) and nitrogen are pumped to an analytical section of the apparatus. Hydrogen is determined by thermal conductivity, carbon monoxide using infra-red absorptiometry and nitrogen by difference or all together by mass spectrography.

All of the analytical methods discussed are reasonably accurate in so far as they determine the composition of the sample taken but it is necessary to ensure that the sample is really representative of the bulk of material being tested. Large masses, particularly of solids, are not easily sampled and errors which look small in percentage terms can correspond to large errors on a mass basis. Costly analytical equipment must be matched by adequate sampling arrangements.

Particle Size

Methods of determining particle-size distributions in the products of comminution and classification processes have been mentioned in Chapter 4. They are all time consuming and unsuitable for controlling these processes except on a day-to-day basis. It is now possible to obtain a measure of particle-size distribution in a suspension of particles in a fluid on a continuous basis using ultrasonic sound waves. The energy of these is absorbed by the particles with an intensity which depends on the relative size of the particle and the wavelength of the vibration. Sound having a wavelength which is large compared with the diameter of a particle found in the path of the beam is affected very little but short-wave vibrations striking a large particle are either reflected or absorbed. By transmitting beams of sound energy at two or more frequencies and comparing the degree to which each is absorbed it is possible to make quantitative deductions as to the distribution of particle sizes present. These can be checked against the results of sizing tests using sieves. The important aspect of the ultrasonic test is that it provides an immediate measurement that can be used to control the mill or classifier with which it is coupled.

Process Control

Many processes are controlled simply by making measurements of a single parameter on the output side and adjusting one process variable in such a way as will bring the measured value closer to a standard desired value. For example, the temperature of a furnace or its product may be measured and the energy input rate raised or lowered to bring the temperature closer to a specified value. Even in such a simple case it is not easy to avoid "hunting". This fluctuation above and below the required temperature is an unintentional overshoot which is due to thermal inertia in the system. For example, refractories lying between the heat source and the charge where the thermocouple is situated are subjected to thermal gradients. When the control temperature is indicated and the heat is cut back, surfaces opposite the thermocouple will continue to rise in temperature by conduction of heat from within the brickwork. Not until the temperature gradients are reversed will the cooling be sensed by the instrument. The ceramic sheath on the thermocouple has its own small temperature gradient which must also be reversed. The frequency with which observations are made is important. If the interval is too long a very large deviation may build up which is difficult to correct quickly. In many cases continuous measurement is possible but the remedial action may only be made at intervals and may require several seconds or even minutes to complete. In this case the action taken might be insufficient by the time it was made. Modern control systems avoid these possibilities by employing anticipatory techniques having varying degrees of sophistication. The heat-input rate to a furnace will be reduced gradually as the observed temperature approaches the required value, minimizing overshoot and in principle reaching the ideal input rate just as the correct temperature is reached. Subsequent alterations will only be made to compensate for changes in external conditions, the opening of furnace doors and similar events. Alternatively when the temperature is expected to fluctuate over a wide range because of alterations in the furnace load the rate of change of temperature may be computed and used to determine how the energy input rate should be adjusted. The secret of success is to achieve the correct "tuning" of the control system to the characteristics of the operation being controlled, that is, the timing of the checks and the magnitude of the changes to input effected on each occasion must be carefully chosen to match the thermal capacity of the furnace and its load and the rate at which heat can be transferred within it. The same principle applies in other systems than furnaces — to grinding circuits, flotation cells or leaching vats.

Whenever possible the control of an extractive process should be total in the sense that every factor that can affect the outcome of the process should be taken into account simultaneously and given its due weight. Most processes must be controlled on the basis of an energy and materials balance and it is very important that all the information needed to strike an accurate, detailed and complete balance shall be available. It is also important that operators are clear as to which operational factors are the causes and which are the effects, particularly in complex multi-phase systems where the trains of causes and effects can be very difficult to follow.

Consider a simplified roasting process. Suppose that a concentrate is to be partially roasted to a specified degree in a rotating kiln where it is exposed to and reacts with combustion gases from a burner, the concentrate and gases flowing in opposite directions through the furnace. The reaction has to be incomplete so that the time during which concentrate is in contact with the gas must be regulated.

There are only a few operational parameters which can affect the quality of the product. These are:

1. The composition of the concentrate.
2. The physical condition of the concentrate (e.g. coarse or fine).
3. The rate of feed (= throughput) of the concentrate.
4. The rate of fuel supply.
5. The fuel/air ratio.

External observations can be made as follows:

6. The composition of the calcine.
7. The temperature of the calcine.
8. The composition of the gases leaving the kiln.
9. The temperature of the gases leaving the kiln.

Measurements of these temperatures and compositions may also be made at intermediate positions within the kiln.

In such a simple process there are not many degrees of freedom. Suppose initially that items 1 and 2 are constant, that 6 must be kept steady and that the chemistry of the process demands that 7 is also held to a particular value. It becomes necessary to decide the best chemical potential (oxygen potential, for example) to maintain in the gases leaving the kiln. There must be some excess potential over the minimum required by the thermodynamics of the reaction but too great an excess is likely to be costly in fuel. For example, if the reaction is oxidizing, too large an excess of air in the combustion gases entails heating up an unnecessarily large amount of inactive nitrogen. If, on the other hand, the reaction is reducing in nature, the use of furnace gas with too high a CO/CO_2 ratio will lower the effective calorific value of the fuel further than necessary. In either case avoidable fuel costs will be incurred. To find the best compromise between a high chemical potential in the gases, coupled with high energy consumption, high reaction rate and good productivity and a low chemical potential which is economical of fuel but leads to a poor production rate, it is necessary to carry out an investigation into the relationship between the chemical potential and the reaction (or production) rate. In practice, experienced operators would be able to select a reasonable chemical potential with which to start running the kiln and would accumulate the necessary information — build and improve on a mathematical model of the process — over a few weeks or months of operation. When the relationship between chemical potential and production rate has been established, the actual choice of chemical potential becomes a policy decision which should depend upon the current relative values of energy and productivity. When demand for the product is high it would be reasonable to go for productivity and ignore the extra fuel costs, while in a period of low demand productivity has less value so advantage should be taken of the best fuel rate available.

In principle, then, the chemical potential most suitable to current circumstances can be selected. In effect this means that the fuel/air ratio has been decided because it is through that ratio that the chemical potential is controlled. Chemical potential also influences the reaction rate and hence the production rate which determines the feed rate of concentrate. All the information is available for calculation of the necessary materials/energy balance, from which the fuel supply rate and the analysis and temperature of the gases leaving the kiln can be obtained, always provided that good estimates of heat losses from the kiln can be made.

Once this has been accomplished it should be possible to operate the kiln by adjusting the rates at which the concentrate, fuel and air are supplied and adjusting the rotation speed to be appropriate for the driving rate. The temperatures and compositions of calcine and gas should be as specified or calculated, but it may be prudent to measure some or all of these so that corrective action may be taken should things go wrong. If the temperature of the calcine is seen to be falling, for example, it might seem to be not unreasonable to increase the fuel-supply rate slightly. This would be satisfactory if the fall were due to a lowering of the ambient temperature or a rise in the relative humidity of the combustion air or even a fall in the calorific value of the fuel. If, however, the concentrate supplied had a coarser size distribution than normal and assuming the reaction to be exothermic, a fall in temperature could be due to a reduced reaction rate. The feed rate would require to be reduced to ensure a sufficient degree of roasting. The fuel and fuel/air ratios would then have to be recalculated. In a sophisticated control system all of these possible causes of a fall in temperature would be monitored and appropriate adjustments made *before* the fall in temperature was actually observed, but the facility would remain for making small adjustments in both fuel input and feed rate to compensate for any small errors in the set values arising from errors in the measurements on which the basic calculations depend.

This rather synthetic example has been dealt with at some length to illustrate that a good understanding is necessary if even a simple process is to be brought under a reasonable degree of control. It may have been noticed that the word "computer" has not been mentioned. In principle, computers are not necessary except as a convenience — a very great convenience — which may make it possible to take into account many parameters simultaneously and perform lengthy calculations very quickly. If such speed really is needed, then the computer becomes a necessity and it must be provided with all the information that a human calculator would need though perhaps in a different form. In the example, a computer would need additional parameters to describe the composition of the concentrate which might also change in a real case, another one or two to evaluate the particle size and its distribution, and a mathematical relationship between particle size and reaction rate. As this last is temperature dependent, with a range of temperature in the kiln, that relationship would not be a simple one. Given all that information, however, a computer can go through the arithmetic of a new heat balance in a few seconds at any time to produce an up-to-the-minute set of operating conditions. The example also shows that a single observation is not enough on which to decide what remedial action to take. If several observations are made at the same time (e.g. temperature and composition of calcine) the computer may be programmed to deduce the reason for any deviation from a normal state before making adjustments.

This general approach is applied to very large complex processes including the basic oxygen steelmaking (BOS) process (see page 323). In that case the input is molten iron from the blast furnace, of known composition and temperature. The energy comes entirely from the oxidation of the silicon, manganese and phosphorus, most of the carbon and some of the iron from the metal charged. The product is steel at a higher temperature than the iron charged, slag and a large volume of carbon monoxide which is at about the same temperature as the steel. A materials balance shows how much flux must be included with the charge and how much oxygen must be blown to complete the necessary oxidation reactions. A heat balance shows how much surplus energy is available from the reactions, from which may be deduced the quantity of scrap steel which must be charged to absorb that energy and so keep the temperature at the required level in the most sophisticated equipment, a computer when fed with the appropriate information not only calculates the details of the charge but also activates the mechanical procedures of weighing, charging and blowing. At the end of the blow the steel should have the expected composition and should be at the required temperature but it is prudent to measure its temperature and have it analysed before it is deoxidized and cast. When the computer is fed with the actual values of temperature and composition, any necessary remedial action is calculated and put into effect.

The need for some correction should be expected. Analytical accuracy may be good but sampling of large masses of materials is not easy and the effect of small errors can be great. If the carbon content of a 200-ton charge is believed to be 2.98% when it is actually 3.01% there is about 60 kg more carbon charged than had been declared to the computer so that the heat input has been underestimated by about 700 MJ. If all of that carbon were retained at the end of the heat the carbon content of the steel would be 0.03% high – which could be significant in low carbon steel. There are many other opportunities for errors to arise in the analytical, weighing and metering operations. While these will usually be largely self-cancelling there will always be some residual error which will occasionally be large. The energy balance can only have been made using estimates of heat losses to the gases and through the crucible walls. Whatever formulae are used to make such estimates, they will be another source of uncertainty which will normally be diminished, however, as the programming of the computer is modified in the light of experience during the early months of its use. (See example 16 in Appendix 1.)

Apart from controlling the purely technical aspects of production, computers have been used, so far to a much greater extent, for progressing work through factories and for determining the best order in which work ought to be done or stocks used up. This has been applied more to the finishing stages of metal production than to the extractive stages, in the rolling mills, for example, where computer control is used to match rolling schedules to the order books in order to reduce the quantity of off-cuts produced which are too small to be sold. With the aid of a computer the total length of a bar can be measured as it is leaving the rolls and the best selection of lengths which will give least waste calculated before the end of the bar reaches the guillotine, the appropriate order number for each length of bar being recorded. On the extractive side, a similar technique could be used to select ore mixes from current stocks according to specified criteria which might include the requirement that usable mixes should be retained in stock for a specified period ahead. The computer might be programmed to take into account materials in transit or even just on order.

Once a computer has been programmed, it can be left to do what may appear to be a decision-making job until changing circumstances dictate a change of programme. The success of an operation so controlled depends upon how cleverly the programme has been written. It is likely to have been written today, not by the manager who formerly made the decisions but by a new specialist in the mathematical modelling of industrial processes. His function should be to interpret the requirements of the management into the kind of programme that will cause the computer always to produce the desired effect. It is obviously desirable that the rising generation of engineers and metallurgists whose work will inevitably be done with the aid of these machines should acquire some familiarity with the language in which the machines are instructed by their operators.

CHAPTER 12

EXTRACTION PROCESSES

EXTRACTION metallurgy is like a crossword puzzle with its clues across and its clues down. There are the processes which have been devised on the basis of chemical, physical and engineering principles; and there are the procedures, which are actually adopted for the separate metals, which are sequences of processes arranged in particular ways dictated mainly by economic circumstances (but partly also by tradition). To complete his study of extraction metallurgy the student must examine both the processes and the procedures and see for himself how they are woven together. A good process is one that is economical in its use of capital, labour, energy and chemical reagents. The correct procedure in any case is that sequence of processes by which the metal can be produced in the desired quality at the lowest overall cost from the ores available.

In this chapter the processes are classified and each type of process is discussed briefly mainly from the standpoint of the underlying chemical theory which has been expounded in the previous chapters. While some examples are given of uses to which processes are put, consideration of the procedures is held over until the next chapter.

Calcination

Calcination is the thermal treatment of an ore to effect its decomposition and the elimination of a volatile product – usually carbon dioxide or water. The necessary temperature can be calculated from the free energy–temperature relationship of the reaction concerned. For example for

$$CaCO_3 = CaO + Co_2; \quad G_T^o = 177,100 - 158 \text{ J mol}^{-1} \, T \quad (7.8)$$
$$\text{and} \ (7.13b)$$

when the CO_2 pressure is 1 atm, $\Delta G_T^o = 0$ and $T = 1123$ K or $850°C$, so that a kiln temperature of $1000°C$ would be ample to provide a rapid temperature rise in the mineral particles to decomposition temperature. As explained on page 193 the rate of decomposition is probably controlled by the rate of heat transfer into the particle, so that even higher kiln temperatures would be expected to increase production rates but at some cost in fuel. Since the solid residual product is likely to be of porous and permeable texture the escape of the gaseous product is not inhibited and its pressure at the decomposition front is likely to be little higher than atmospheric.

Most carbonates decompose at lower temperatures than calcium carbonate – $MgCO_3$ at $417°C$, $MnCO_3$ at $377°C$ and $FeCO_3$ also about $400°C$. Hydrates (e.g. bauxite) also decompose at relatively low temperatures and kilns may be run at about $700°C$ for the calcination of typical ore minerals provided that no other purpose is involved.

Calcinations may be carried out in shaft or rotating kilns applying as far as possible the counterflow principle for efficient heat transfer.

Roasting

The main purposes of roasting are as follows:

1. Oxidizing roast to burn out sulphur from sulphides and replace them in whole or in part with oxides.
2. Volatilizing roast to eliminate other elements with volatile oxides such as As_2O_3, Sb_2O_3 or ZnO which can be recovered as "fume" – a suspension of fine solid particles condensed out of the vapour phase (see pages 358, 359).
3. Chloridizing roast for conversion of certain metals to chlorides (see page 330). These may be carried out under oxidizing or reducing conditions.
4. Sulphating roast to convert certain metals from sulphides to sulphates, usually prior to leaching.

5. Magnetizing roast, usually a controlled reduction of haematite to magnetite, to enable magnetic separation to be effected.
6. Reducing roast of oxide to metal prior to leaching or smelting.
7. Carburizing roast to prepare calcine for chlorination (see page 330).
8. Sintering or blast roasting often with the primary purpose of modifying the physical condition of the ore (usually an oxidizing roast), but also used for purpose (1).

Roasting involves chemical changes other than decomposition, usually by reaction with the furnace atmosphere, and depends on diffusion of chemical species through the product to the reaction front in each particle (see page 195). A roast may effect drying and calcination in passing – e.g. the first stage in the roasting of CuS is calcination to $Cu_2 S$ with the elimination of sulphur which escapes as a gas to the surface of each particle and burns there to SO_2. In the second stage, however, oxygen must diffuse into the particle and SO_2 out of it as $Cu_2 S + 2O_2$ → $2CuO + SO_2$, so that a draught must be maintained to keep the partial pressure of the gaseous products low outside the particles in order to assist diffusion.

Primitive roasting in "heaps" and "stalls" was replaced first by the use of shaft furnaces and simple air furnaces, and later by rotating kilns and more complicated machines like the Wedge multi-hearth roaster. In this furnace the counter-current principle was applied to both heat and mass transfer as the ore progressed down the shaft and across each of a series of hearths, where it was continuously turned over by rotating rakes which also advanced the ore across the hearths. Access of furnace gases to the surfaces of finely ground ore particles was scarcely adequate in many cases, however, and a high proportion of the reactions took place only as the ore fell from one hearth to the next below. The natural development was *suspension roasting* in which the ore was dried and preheated on two top hearths and then allowed to fall through the central section of the furnace against the oxidizing furnace gases. The modern development of this is *flash roasting* in which the preheated ore is injected through a "burner" with preheated air rather like pulverized fuel (Fig. 66). This process is most appropriate for the roasting of sulphides which oxidize exothermally and require no additional fuel.

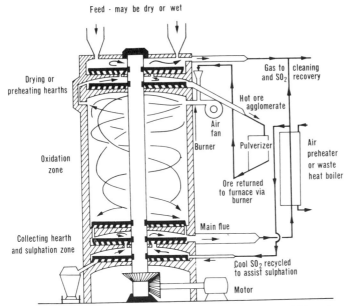

Feed - may be dry or wet

Drying or preheating hearths

Oxidation zone

Collecting hearth and sulphation zone

Gas to cleaning and SO$_2$ recovery

Hot ore agglomerate

Air fan

Burner

Pulverizer

Air preheater or waste heat boiler

Ore returned to furnace via burner

Main flue

Cool SO$_2$ recycled to assist sulphation

Motor

FIG. 66.   Schematic representation of a Wedge-type multi-hearth furnace modified for the flash roasting of sulphide concentrates. The upper hearths may be used for drying out slurry or for preheating concentrates. The lower hearths can be used for sulphating if required.

This is called *autogenous* or *pyritic roasting*. In flash roasting the benefits of counterflow operation are partially lost.

The use of fluidized beds for roasting fine concentrate is obviously attractive. If a gas is passed upwards through a bed of solid particles of small and preferably regular size in the range 2–0.02 mm diameter, the behaviour of the bed will depend upon the velocity of the gas. At very low flow rates the gas permeates the bed without moving the particles at all and the pressure drop across the bed is proportional to the flow rate. An increase in gas velocity to a critical value causes the bed to expand as the effective weights of the particles become balanced by the drag forces of the gas stream upon them. Over a short range of velocities the particles remain individually suspended each with a downward velocity relative to the gas stream approximately equal to its terminal velocity. As the gas velocity is increased further the bed continues to expand as the distance between the particles increases and the pressure drop across the bed decreases. This "particulate" fluidization is not the condition used for fluidized bed reactors but is approximately the condition which is found in ore dressing equipment like hydraulic classifiers, jigs, etc. A further increase in velocity of the gas causes a rather less stable condition to arise in which the expansion of the bed ceases and the pressure drop across the bed becomes independent of gas velocity. The bed appears like a boiling liquid as "bubbles" burst on its surface. It flows like a liquid, and solids of density less than the bed density measured as the weight of the particles divided by the volume of the expanded bed can be floated upon it. This is called "aggregative" fluidization and the bed seems to consist of large numbers of fairly stable but loose clusters of particles dispersed in an atmosphere of gas laden with individual particles, which is called the "lean phase". This lean phase tends to form pockets or "pseudo-bubbles" which rise to the surface and cause the appearance of boiling. In this condition gas–solid contact is very complete. The gas flow is very turbulent so that the solids are thrown around in all directions with high energy but the *relative* velocity between gas and solid particle is not outside the laminar flow range. Both heat and mass transfer rates in fluidized beds are very high. Effective thermal conductivities of fluidized beds of about 100 times that of silver have been measured. The mechanism of this effect is not understood. Uniform temperatures in a bed are assured by convective transfer as the particles move about rapidly and randomly, and heat transfer between particles seems to be by conduction during collisions which break through surface films of gas. Heat transfer from the gas may be facilitated on these same occasions. The specific surface area of the particles is, of course, very high so that even if much of the transfer is across a gas film by an inherently slow process the rate of transfer need not be low. In the case of mass transfer the transfer coefficient (mass per unit area per unit of concentration gradient) is not as high as in a fixed bed but again because of the available area the mass transfer rate is very high between gas and solid. A fixed bed of similar specific surface would be *very* impermeable.

Roasting in a fluo-solids reactor (Fig. 67) may be carried out autogenously if the reactions are sufficiently exothermic. Otherwise, fuel must be used which may be pulverized coal fed with the concentrate or gaseous

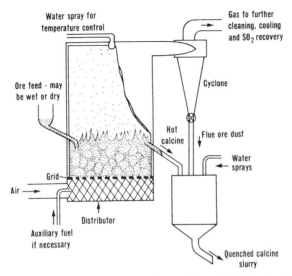

FIG. 67. Schematic representation of a fluidized bed for roasting sulphide ore. Obviously a wide variety of feed and discharge arrangements is possible, appropriate to different operating conditions.

fuel introduced with the air. The concentrate must be fed continuously either dry or as a slurry, displacing the column to discharge over a weir remote from the feed point. Some solid particles are inevitably swept out with the gas stream and must be recovered in dust catchers or cyclones. The gases may also be passed through a waste-heat boiler and further treated to remove sulphur dioxide before being discharged. These steps are necessary, of course, after any roasting process but in this case the dust losses are likely to be greater than from conventional roasting kilns and the sulphur dioxide concentration in the gas is also rather higher, other things being equal. Because reaction rates are higher than in traditional process the heat loss per unit of production is lower. This leads to higher temperatures being developed if steps are not taken to remove the excess heat. This may be done in several ways. If the process requires fuel, less will be needed if it is being conducted in a fluidized bed and in some cases lean ores with insufficient sulphur content to make fully pyritic roasting possible in traditional types of kiln can be treated without any fuel ad-

dition in a fluo-solids reactor. Richer sulphide concentrates usually produce an excess of heat and the temperatures are controlled by adding
water either in the slurry or as a spray (as shown). Much of the excess heat
is then recovered in the waste-heat boiler. Alternatively the heat may be
abstracted by means of cooling coils built into the turbulent bed. This
has the advantages that it controls the temperature of the solids more
directly and that there is no interference with the chemistry of the
process or dilution of the gases. The heat is recovered, however, in a
more degraded, less valuable form. One other way of cooling the bed is
to recycle a part of the calcine. This can serve an additional purpose.
Calcine is not necessarily completely roasted despite the speed of the
reactions. If this is the case it may be classified after quench cooling and
the coarser fraction, which is most liable to be inadequately converted,
may be returned, wet, for further oxidation and at the same time to
help to control the temperature.

Incomplete calcination of particles is partly due to the range of
particle size fed and partly due to some particles finding their way to
the discharge weir more quickly than others. In some designs of reactor
attempts are made to classify the particles within the bed so that larger
particles will remain in the bed longer than small ones. This can be done
by injecting the air at different velocities at different parts of the bed, so
imposing pressure differentials and inducing flow patterns which will
delay the passage of larger particles.

The earliest applications of fluidized bed techniques to extraction
metallurgy were to sulphating where close temperature control is necessary
particularly if differential sulphating is being carried out — that is at such a
temperature that one metal present will form the sulphate while others
remain as oxide so that separation by leaching (see page 293) can then be
effected. Extensive developments can be expected including multi-stage
treatments, but agglomeration of the particles by sintering puts an upper
temperature limit on the processing of any particular material.

While roasting may appear at first sight to be rather a simple,
unsophisticated process, it does usually require quite close control,
sometimes of temperature, sometimes of furnace atmosphere, sometimes
of rate of throughput of materials and occasionally of all three at once
(see page 273). In the case of a magnetizing roast of haematite to
magnetite the composition of the furnace atmosphere is of greatest

importance. Supposing that the reduction is to be effected by carbon monoxide, obtained by burning a fuel with a small deficiency of air, the principal reaction will be:

$$3Fe_2O_3 + CO = 2Fe_3O_4 + CO_2; \Delta G_T^o = -33{,}230 - 53.9T \text{ J mol}^{-1}.$$

At $1000°C$, $K_{1273} = CO_2/CO = 1.5 \times 10^4$, corresponding to a CO/CO_2 ratio of 6.6×10^{-5}. In combustion gases there will be about 75% of nitrogen so that this ratio is equivalent to the concentration of carbon monoxide being 0.0016%. This concentration must be exceeded if the reaction is to proceed and must be exceeded generously if the reaction is to be fast. The further reduction of magnetite to wustite must however, be avoided. The reaction is—

$$Fe_3O_4 + CO = 3FeO + CO_2; \Delta G_T^o = 29{,}616 - 38.3T \text{ J mol}^{-1}.$$

For this reaction $K_{1273} = CO_2/CO = 6.11$, corresponding to about 3.5% of carbon monoxide in the furnace atmosphere. This must not be exceeded if the formation of wustite is to be avoided at $1000°C$: If the ore were at only $800°C$ the carbon monoxide could safely be raised to 5.4% but if the temperature rose to $1200°C$ the upper limit of carbon monoxide would fall to only 2.5%. In practice it is the upper limit that is important. It should be approached as closely as practicable in order to enhance the rate of reaction in general and to make it unlikely that the concentration of carbon monoxide will fall near to the lower limit in any part of the kiln. In this particular roasting process the purpose is to render the iron oxide magnetic so that it can be separated from the gangue. Complete conversion to magnetite is not necessary. Depending upon the particle size, the proportion of iron in the particles and the strength of magnetic field available, a partial conversion will suffice — perhaps in the range 10—20%. The residence time of the ore can be selected so that the largest particles are sufficiently converted and the material is then passed through at the appropriate rate. It will be seen that all three factors mentioned above interact with each other. Although the composition of the gas seems to be of prime importance, it depends upon the temperature selected and the production rate depends upon them both. The higher the temperature, the faster the reaction at any selected reduction potential but as the temperature is raised the maximum permissible concentration of carbon monoxide in the gases falls so that their capacity for reduction is reduced and the maximum production rate possible does not rise as might have been expected.

In the case of a volatilizing roast to remove antimony as its oxide from a sulphide ore, it is necessary to maintain an oxygen pressure high enough to oxidize the antimony to the trioxide Sb_2O_3, which is volatile but not so high as to form the tetroxide Sb_2O_4, which is not very volatile. Free energy data for the relevant reactions are not available and in practice the formation of the higher oxide is avoided by keeping the temperature below 500°C. A higher temperature could probably be used in a slightly reducing atmosphere with a CO/CO_2 ratio of about 1/100 but control would have to be close to avoid reduction to the metal. Another volatilization process, the removal of zinc from lead smelting slags by fuming, requires that the metal be reduced and vaporized under a bed of coke and the gaseous zinc oxidized in the atmosphere above the hearth – another process requiring considerable skill for its execution. These processes have the advantage that the products are removed from the site of the reactions. The reverse reactions do not occur so that even quite small chemical potential differences from equilibrium values can be expected to drive the reactions at a reasonably fast rate.

The oxidation of sulphides may be complete, as in the case of the sintering of zinc sulphide to the oxide prior to its reduction in a retort, or only partial as when copper–nickel concentrates are prepared for matte smelting. In the former case too high an oxygen pressure or too long a time in the kiln is only a waste of money, but in the latter case the quality of the calcine would also be affected.

When a sulphide is roasted to sulphate for leaching, the overall reaction probably proceeds in two stages. The first produces oxide and sulphur dioxide and, since it is exothermic, can cause the temperature to rise to beyond 1000°C. The second is sulphation which can proceed only below the decomposition temperature of the salt being formed which is about 850°C for $CuSO_4$ and $ZnSO_4$, 710°C for $Fe_2(SO_4)_3$, 900°C for $NiSO_4$ and 980°C for $CoSO_4$ when $pSO_2 = 1$ atm or rather less when, as is usual, other gases are also present. In practice the sulphation stage for copper is carried out at about 680°C. The two stages can be carried out together in a fluo-solids reactor in which very good temperature control can easily be maintained. The thermodynamics of sulphation reactions can conveniently be presented in "predominance" or "stability" diagrams which are like the Pourbaix diagrams of the corrosion technologists. Consider the reactions involved in the roasting of cobalt sulphide to cobalt sulphate. The lower sulphide is now considered to be Co_9S_8 and the relevant reaction is

$$2/25Co_9S_8 + O_2 = 18/25CoO + 16/25SO_2 \, ; \Delta G^{\circ}_T = -294{,}393 + 43.98T$$

from which

$$\Delta G^{\circ}_{950} = -252{,}612 \text{ J mol}^{-1}.$$

Hence

$$K_{950} = \frac{p^{16/25}_{SO_2}}{p_{O_2}} = 7.76 \times 10^{13}$$

and

$$\log K = 13.89 = 16/25 \log p_{SO_2} - \log p_{O_2}.$$

Rearranging,

$$\log p_{SO_2} = 21.7 + 1.56 p_{O_2},$$

from which it is seen that $\log p_{SO_2}$ is a linear function of p_{O_2}. This relationship can be represented as a straight line with a slope of 1.56 and is included as line A in Fig. 68 where it is the boundary between the domains of stability of the sulphide and oxide phases. As a boundary it extends until intersected by other lines bounding domains of stability of other phases.

One other such line is the boundary between the oxide and sulphate phases, which is determined by the reaction

$$CoO + SO_2 + \tfrac{1}{2}O_2 = CoSO_4 \, ; \Delta G^{\circ}_T = -384{,}400 - 50.21T \ln T + 663.6T$$

from which

$$\Delta G^{\circ}_{950} = -81{,}030 \text{ J mol}^{-1}$$

so that

$$K_{950} = p^{-1}_{SO_2} \times p^{-1/2}_{O_2} = 2.85 \times 10^4.$$

Hence

$$\log p_{SO_2} = -4.46 - \tfrac{1}{2} \log p_{O_2}.$$

This is again a linear relationship between the logarithms of the sulphur dioxide and oxygen pressures and it is included in Fig. 68 as line B. The rest of the diagram has been constructed in a similar manner. It is valid

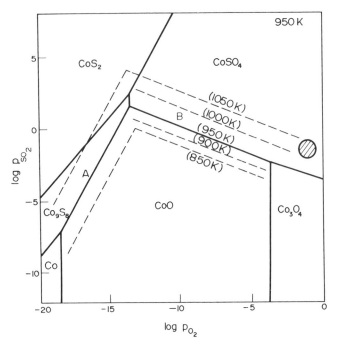

FIG. 68. Predominance diagram for the system Co–O–S at 950 K with additional lines at 850–1050 K. The hatched circle indicates the range of roaster gas compositions. (After Ingraham.[72])

only at the single temperature 950 K but similar diagrams can be prepared at other temperatures. To illustrate the effect of temperature on this diagram, however, broken lines are included for the above two reactions at 850 K and 1050 K. There is also a hatched area showing the range of p_{SO_2} and p_{O_2} values encountered in roasting kilns. It is apparent that if the temperature is raised above 1050 K ($777°C$) cobalt oxide, rather than cobalt sulphate, is the stable phase. Similar diagrams for the Cu–O–S and Fe–O–S systems at 950 K are shown as Figs. 69a and 69b. From these it is

FIG. 69. (*Opposite*) Predominance diagrams for the systems (a) Cu–O–S and (b) Fe–O–S at 950 K. The hatched circles indicate the range of roaster gas compositions. (After (a) Ingraham[73] and (b) Davenport and Biswas.[74])

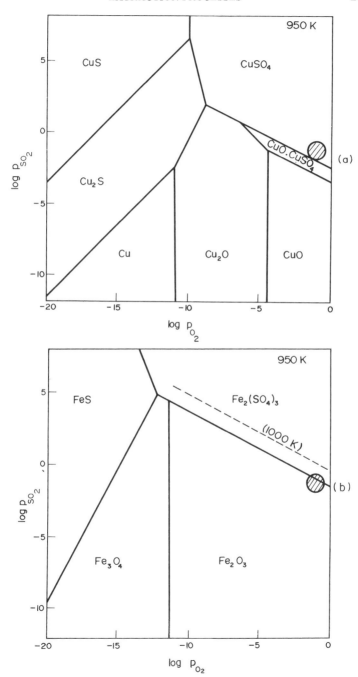

clear that at that temperature copper will be sulphated as well as cobalt but iron will remain as an oxide, Fe_2O_3, so that a good separation of copper and cobalt from iron by leaching after roasting at $677°C$ should be possible. At a rather higher temperature the copper will also be retained as an insoluble oxide so that a separation of cobalt from copper and iron would be possible, following roasting at about $750°C$. Control of the temperature must be very good if this separation is to be made efficiently. Sufficiently good control may be possible in a fluo-solids reactor, but in practice there are better ways of separating copper from cobalt.

Another application of roasting which requires close control is chloridizing, particularly when differential chloridizing is being practised to produce the lower, volatile chloride of one metal, which can be distilled away from less volatile compounds of other metals (see page 333).

Smelting

Smelting is essentially a melting process in which the components of the charge in the molten state separate into two or more layers which may be slag, matte, speiss or metal. Constituents of the charge, sometimes including values, may also appear in the furnace gases. Fluxes may be included in the charge to facilitate the formation of the slag which is usually the least fusible phase. Smelting does not necessarily involve any refining but the opportunity may be taken to adjust the composition of the slag, the oxygen potential or the temperature so that unwanted elements are collected preferentially into the slag, speiss or vapour phases. Sulphur in iron smelting is collected in the slag, copper in lead smelting into matte and cobalt in copper smelting into speiss. It should be understood that at the high temperatures involved in smelting, favourable partition of selected elements between phases can be effected but not complete separations. Smelting operations often incorporate a roasting stage in the same furnace — for example, the reduction roast in the stack of the iron blast furnace or the oxidation of sulphide concentrates in the flash smelting of copper–nickel ores. More generally it is unavoidable that the furnace atmosphere will react with the charge during melting. Allowances must be made for the chemical and thermal effects of these reactions which should be made to work for the smelter rather than against him. In principle it is better to carry out the two functions of roasting and smelting separately, but in practice the economic advantage of requiring only one piece of plant instead of two prevails. The principal smelting processes are for the production of matte and metal.

Matte smelting is usually carried out on concentrate which has been roasted to reduce its sulphur content to such a level that, when it is melted with a suitable flux, a high-grade matte will be produced along with a slag

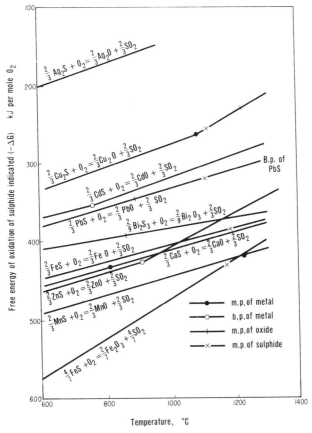

FIG. 70. Standard free energies of oxidation of some metallic sulphides, the standard states being the pure condensed oxide and sulphide and gaseous O_2 and S_2 at 1 atm pressure. (After Hopkins.[75])

containing most of the gangue materials. The matte contains sufficient FeS to protect the more valuable sulphides from oxidation, FeS being one of the most readily oxidized of the common sulphides (see Fig. 70). Most of the control is in the roasting. In practice, however, some or even all of the roasting reactions may take place within the smelting furnace, as has been mentioned above. The slag must have a low enough viscosity to facilitate the separation of droplets of matte which become entrained in it. This consideration usually determines the operating temperature. Losses of values to smelting slags can rise to 3% of the metal in the ore and slags may be held in settling furnaces to allow additional time for the separation to take place. Magnetite crystals separating from an otherwise fluid slag have the effect of raising its viscosity. Additions of pyrites may be made to reduce the magnetite and restore the fluidity of such slags.

Matte smelting may be carried out in blast furnaces, reverberatory furnaces or electric arc furnaces. Blast furnaces are the least commonly used today. They are most appropriately used for smelting lumpy ores. Finely ground concentrates must be sintered before smelting in a blast furnace. This is a roasting operation as well as an agglomerating process but is usually too expensive to justify except, perhaps, to provide part of the feed for a furnace fed principally on lump ore. Blast furnaces are fired with coke but a sufficiently oxidizing combustion gas is maintained to effect some oxidation of sulphides in the stack and to prevent the reduction of iron oxide from the slag. Fine concentrates are usually smelted in reverberatory furnaces. They are usually calcined before charg-

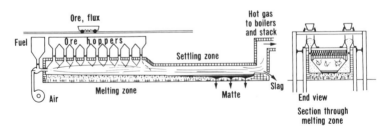

FIG. 71.   Schematic representation of a copper reverberatory smelting furnace. The end view shows the suspended magnesite roof and the protection offered to the side walls by the ore feed. Fuel may be oil or pulverized coal. Facilities must be available for charging converter slag.

ing but there is a trend toward feeding raw concentrate to reverberatories after adjusting the Cu–Fe–S ratio by differential flotation. These furnaces are built principally of magnesite-based refractories. The concentrate is fed into a long, narrow melting zone where it is heated by flames from the gas, oil or pulverized coal burners set in the end wall. Matte and slag run to the lower end of the long hearth where they are tapped off almost continuously. Flash smelting is a pyritic smelting process which combines the operations of flash roasting and smelting, fine concentrate being burned like pulverized fuel with auxiliary fuel blown in the form of pulverized pyrites (FeS_2) to create enough heat to melt out matte and slag into the settling hearth at the bottom of the furnace (Fig. 72). This is an obvious case in which to use oxygen instead of air to increase the combustion temperature and raise the sulphur dioxide concentration in the furnace gases. Electric arc smelting is also practised on copper concentrates and calcine where the price of electricity is favourable. Although no air is blown into these furnaces, they operate under a small negative pressure and enough air is drawn through the brickwork to oxidize some of the sulphides and produce a furnace gas rich in sulphur dioxide. The sulphur dioxide from all of these processes must be collected and converted into some marketable

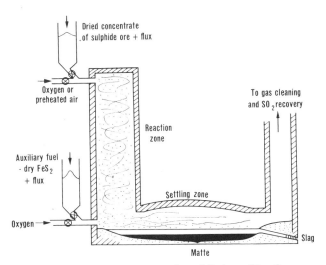

FIG. 72. Schematic representation of flash smelting furnace.

form – elementary sulphur, sulphuric acid or liquefied SO_2. If it cannot be used or sold it is reduced to the element which can be stored.

There is obviously always some interaction between the burden and the furnace atmosphere which must affect the quality of the product. There is a small excess of oxygen in the combustion gases which diminishes as they pass through the furnace, reacting with sulphides and slag constituents as they do so. If the excess oxygen is about 1%, $p_{O_2} \simeq 10^{-2}$ atm so that from Fig. 60 it can be seen that the iron in the slag will be oxidized approximately to the composition of magnetite at the surface where it is in contact with the furnace gas. It would be expected that any FeS present would reduce this Fe_3O_4 and indeed under oxidizing conditions very little sulphide is found in the slag. The oxygen diffuses from the magnetite-rich surface to the slag–matte interface where it oxidizes the iron in the matte, raising the O/Fe ratio in that phase above unity. Some oxidation of sulphur might be expected here also but this may be restrained by the kinetics of nucleation of SO_2. The thermodynamics of this complex system have not yet been worked out but in practice the O/Fe ratio in the matte can rise so high that magnetite is precipitated. This appears first in the cooler parts, on the hearth bottom, where the solubility of Fe_3O_4 is least. In extreme cases, however, it can appear almost anywhere, even the slag–matte interface, indicating that both phases have become saturated with respect to Fe_3O_4. Magnetite is difficult to redissolve and its deposition can make continuous operation of a furnace very difficult. It is necessary to avoid or, at least, delay its formation by maintaining a high steady temperature, and by keeping the excess oxygen in the combustion gases as low as possible. The oxygen pressure is lower in blast furnaces than in reverberatories and the magnetite problem is less severe. The mattes from these processes show a corresponding difference in oxygen content, which can be so low in a blast furnace matte that some metal is produced.

Smelting for metal involves reduction, usually by carbon as coal or coke, but sometimes by ferrosilicon, and may be performed in a blast furnace, hearth, or reverberatory furnace or in an electric arc furnace. The blast furnace acts as a gas producer as it burns coke to CO_2 which reacts further with the carbon to produce CO. The ascending gases preheat the solid charge descending the stack and reduce metal oxides to "sponge" metal. The stack reactions amount to drying, calcination and reduction roasting. The metal melts and the slag forms from gangue and flux in the bosh just above the combustion zone, and the two phases separate in the crucible below, from which they are tapped at regular intervals. Matte may also be produced here as in lead smelting where judicious additions of pyrites will bring copper and some other impurities down in that form. Little real refining is done at this stage beyond controlling the amount of some impurities (sometimes desirable ones) which come into the metal. In ironmaking, the silicon and manganese in iron are held within limits by adjusting the charge and the temperature in the hearth, and the iron is partly refined with respect to sulphur through slag composition and hearth temperature. The reduction of iron from the slag in a lead blast furnace is inhibited by maintaining the CO/CO_2 ratio in the furnace atmosphere oxidizing toward iron through reducing toward lead, i.e. with a value about 1.0 (see Fig. 45). In tin smelting this does not work because tin oxides enter both acid and basic slags in a very stable and irreducible form and a two-stage process had been developed, the first yielding a fairly pure tin and a slag rich in tin and iron oxides while the second is a further treatment of that slag to produce a tin–iron alloy (hardhead), containing about 5% of iron, and a slag low in tin. Obviously further treatment of the alloy is necessary.

The temperature of smelting is usually determined by the free-running temperature of the slag rather than the melting point of the metal, or sometimes by the minimum temperature at which reduction will take place. In the case of tin it is the reduction temperature which decides, and that the reduction temperature for tin-bearing slags ($1250°C$), not that for SnO_2 ($500°C$).

{ Hearth smelting is used where very reducing conditions are either unnecessary or undesirable. Tin is now smelted in reverberatories in both stages to minimize the reduction of iron at the high temperature which is necessary. Some lead concentrates were formerly mixed with coke and smelted in an "ore-hearth", blown with air through tuyères as in a blacksmith's hearth. Mechanical rabbling of this hearth maintained reaction rates high and permitted drainage of slag and metal to a well at one end of the hearth. The functions of roasting and reduction were both performed in this furnace. It may appear that a sulphide was being reduced by carbon, but the process in fact proceeded in two stages.

{ Electric arc smelting is most appropriately applied where the necessary temperature for fusion or reduction is higher than can be conveniently reached with carbonaceous fuel, i.e. over about 1500°C. This is sometimes a batch process with ore coke and flux all melted together in a hearth type of furnace. Pre-heating and partial pre-reduction of the ore may, however, be carried out in a separate furnace such as a rotating kiln continuously feeding the smelting furnace which may then be operated semi-continuously like a blast furnace (see Fig. 73). Where electricity is cheaper than metallurgical coke, iron is smelted electrically in small quantities in this way. The more important applications of arc smelting are to the production of ferro-alloys, where the temperatures necessary to effect reduction are very high, and some contamination by carbon is of no consequence. Silicon may be used as the reducing agent, if high carbon in the ferro-alloy is to be avoided.

Where carbon is to be avoided altogether, high-grade concentrates can

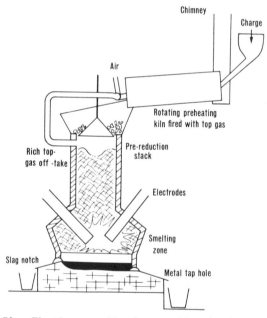

FIG. 73.   Electric arc smelting furnace with preheating kiln fired by top gas. Electricity would normally be supplied in three phases rather than through two electrodes as shown.

be smelted by the alumino-thermic or "thermit" process. If aluminium and a metal oxide are thoroughly mixed and ignited with, for example, a cartridge of magnesium powder, the reaction $3MO + 2Al = 3M + Al_2O_3$ takes place (with a few exceptions) with the evolution of heat. In some cases the heat evolved is enough to melt out the whole charge. If necessary, an irreducible oxide may be added to lower the slag melting point, or the charge may be preheated to raise the combustion temperature of the reaction. In other cases it is accepted that the metallic and slag phases cannot readily be separated by liquidation. Then the mixed products of the reaction are leached with a suitable solvent to dissolve away the slag and release the metal either as a fine powder if its melting point has not been exceeded or as small pellets if partial coalescence of liquid globules has commenced. This type of process has been developed to embrace the use of reducing agents other than aluminium, such as Mg, Ca and even Na. These are capable of reducing not only oxides but also halides.

It will be appreciated that the reaction does not always go to completion as implied in the equation above. In the case of reduction of an oxide of iron with aluminium, the yield of iron is rather low and FeO is found in the slag. If an excess of aluminium is used to reduce this FeO from the slag, much of it enters the iron. The product of any thermit smelting process is liable to some degree of contamination by the reducing agent unless the latter is volatile (like Mg or Na) or is completely insoluble in the metal being produced. It will be appreciated that thermit smelting is essentially a batch process and also that it is primarily a reduction process, in which the products are (sometimes) conveniently separated by liquation.

Hydrometallurgy

Large tonnages and immense values of metals are produced in whole or in part by hydrometallurgical techniques. Much of the copper and almost all of the aluminium, gold and platinum metals are involved as well as part of the base metals lead, zinc and nickel. The basic steps are dissolution or *leaching* of the values from the ore, elimination of unwanted elements from the pregnant liquors, concentration of the values and their deposition from solution either in compounded form by chemical precipitation or by reduction of the metal either chemically or electrolytically. Precipitated compounds may be treated further by any appropriate method. The essential conditions are that the valuable mineral must be soluble in water or in some other cheap reagent like sulphuric acid or that it can be rendered so soluble. Pre-treatment is usually necessary to render the valuable mineral accessible to the solvent and chemically capable of being dissolved in it — i.e. crushing and grinding, and possibly roasting to sulphate or chloride may be necessary. Leaching is also used, and to an increasing extent, as an intermediate process to be applied to the crushed products of smelting operations. The purpose of such a leaching operation might be to free a metallic product from a soluble halide type slag, or to take the metal into solution prior to pre-refinement by wet methods.

Leaching may be carried out in several ways. *Leaching in situ*, i.e. in the mine itself, has been used at least once with economic success on a large caved-down deposit of a very lean copper ore in Ohio, U.S.A. The circumstances must be exceptionally favourable, however, the rock being well broken up and of permeable texture and the ore body of suitable shape. It is sometimes worth while to divert surface water so that it percolates through old mine workings and takes values in solution down into a sump from which they can be pumped and recovered. In another technique known as the five-hole method a number of drill holes are bored into the upper surface of an ore body and an additional hole to another point near to its lowest level. Water is fed through the shorter holes and pregnant liquor is pumped from the deepest point. This can be used only on a very soluble oxidized ore overlying an impermeable rock. It is frequently necessary to shatter the ore body with explosives to render it sufficiently permeable. Normally such a technique would be used on a deposit which is considered to be too lean to justify mining conventionally, is at no great depth and is associated with other larger workings, so that drilling and pumping costs are low and little additional processing equipment is required. True *in situ* leaching of undisturbed ore bodies is not likely ever to be successful.

Heap leaching was used in Spain and now is in the U.S.A. In Spain the ore treated, having a very high FeS_2 content, would have required very heavy fluxing if smelted. Run-of-mine ore (< 200 mm) is stacked in the open on impermeable clay pads with drainage culverts built into them. The heaps are sprayed regularly with the oxidizing and acid tailing solution from the precipitation plant, rich in ferric sulphate. They must be well ventilated to provide plenty of air to keep down the temperature in order to inhibit the oxidation and sulphation of FeS_2. In the first part of each of many cycles this solution seeps into the ore lumps, reacts with the copper minerals and dissolves some of the copper. It is then allowed to dry out and the dissolved salts creep to the surface by "reversed capillarity". The next spraying washes this salt down to the drains and so to the recovery plant. This mechanism is faster than diffusion but it still took 2 years to exhaust a heap of 100,000 tons of sulphide ore at Rio Tinto in Spain. Even this time would have been much longer had it not been for the fortuitous presence of a micro-organism known as *Thiobacillus ferroxidans* which, in effect, catalyses the oxidation of sulphides. It derives its energy from the oxidation of ferrous ions to ferric and in this way ensures that the oxidizing agent necessary for the conversion of sulphides of copper to the soluble sulphate is maintained in good supply. Fortunately this organism is usually found to be thriving naturally in areas where it is required. Heap leaching is also used on oxidized ores, in which case the reactions are faster and the bacillus is not required. *Dump leaching* is a similar process applied on the scale of millions of tons and over periods of several years to mine wastes of copper concentration too low to justify conventional treatment.

More often, ore is crushed to about −6 mm and leached in large concrete tanks by *percolation leaching*. These tanks hold about 10,000 tons of ore. Leaching liquor is fed into each tank, held for several hours and drained out. Each tank is percolated by a series of perhaps a dozen solutions, each more acid than the preceding one, the last of them being fresh reagent. Likewise the solution is passed to fresher ore at each move until it is strong enough to be sent to the recovery plant. In this way a counterflow process is operated discontinuously. A tank load of about 10,000 tons would be leached out in about 5 days. The most successful applications of percolation leaching have been to oxidized ores requiring only simple solution without oxidation. The largest operation is in Chile where the ore was a basic copper sulphate which dissolves readily in sulphuric acid which is continuously regenerated in the associated electrowinning plant.

When further grinding is necessary to make accessible all the mineral particles, the ore becomes too fine for the percolation method to work well. Flotation concentration may be preferred as the next step. *Agitation leaching* is, however, a further possibility, in which the ore is suspended in the leaching solution in a vat and stirred mechanically or with compressed air jets. The contact time is reduced to about 2 days. Separation of the pregnant liquors from fine ore requires classifiers and thickeners or a filter press. Agitation is more efficient in that it leaches out the metal salts faster, but it is in most aspects more costly than percolation leaching, and is not very commonly practised.

Leaching reagents should be cheap and water soluble. Common reagents are sulphuric acid and ferric sulphate, ammonia and ammonium carbonate. The oxidizing ferric salt is needed to attack sulphide minerals, the straight acid is used on oxide minerals, or oxidized ones. The ammoniacal solutions are used on native copper or on copper and nickel pre-reduced to the metallic state. Aluminium oxide is leached with sodium hydroxide, gold and silver with sodium cyanide and uranium ores with sodium carbonate solution. Following sulphation, water itself may be sufficient.

The chemistry of leaching is fairly simple in most cases but may involve the formation of little-known complexes such as $Ni(NH_3)_6SO_4$ or $NaAu(CN)_2$. Differential leaching may be possible under carefully chosen conditions. In heap leaching, for example, the solution of iron is minimized if the temperature of the heap can be kept low. In the pre-refining of aluminium, bauxite is leached with NaOH solution, under pressure, at $150°C$. The high temperature is needed to ensure rapid and complete dissolution of alumina (see page 355). Leaching under pressure makes possible the use of elevated temperatures at which dissolution is faster than under normal conditions. The high pressure also allows gaseous reagents (particularly oxygen) to be used at higher activities than are available in solutions at ambient pressures and temperatures. This accelerates dissolution and modifies the reaction equilibrium position favourably. Pressure leaching with oxygen atmosphere can be used in the cyanidation of gold (see page 368) and for the sulphation of nickel sulphide. Probably the most important application is to the leaching of mixed nickel–copper–cobalt sulphide ores with ammonia and oxygen to bring the ammines of these metals into solution while leaving associated iron in an insoluble condition. The net reaction will be something like—

$$NiS + 2O_2 + 6NH_3 = Ni(NH_3)_6 SO_4$$

but conditions may be modified to provide a lower ammine if this is required (see page 302).

The recovery of the metal from the leach solution may be effected by electrolysis, which is dealt with in the next section, by chemical precipitation or by reduction to the metallic state either by another metal or by hydrogen. It is first necessary to separate the liquor from the gangue residue and to carry out any preliminary purification that is required.

The amount of purification and concentration required by a leach liquor before the metal can be recovered varies greatly from one case to another. Suspended solids must usually be removed by settling or using a bowl classifier or even by filtration. The almost ubiquitous iron may often have to be removed by adjusting the pH with lime or soda ash to about 3.5 and bubbling through air to precipitate ferric hydroxide without taking out other heavy metals. Other metals present in minor proportions may often be easier to remove at this stage rather than later and they may be precipitated as hydroxides or even sulphides fairly specifically at appropriate pH values. Finally undesirable anions may be removed, for example chloride prior to electrolysis, using cement copper as the reagent to form the insoluble cuprous chloride.

Concentration of the metal being collected can be effected using the solvent extraction (or liquid–liquid extraction) technique. This was developed for the extraction of uranium from very dilute solutions and until recently has been used only for some of the less common metals. It is now being used in copper extraction, however, and is likely to be applied to other base metals in the near future. It is also used in the separations of the platinum metals. Solvent extraction has the potential for separating a metal from a relatively impure solution at the same time as it concentrates it but the effectiveness in this function depends on which metal and what impurities are involved. Modern reagents can be very specific in their action.

The "solvent" is usually a complex organic compound with a replaceable hydrogen atom. Some of these are organic derivatives of phosphoric

acid but the most popular now appear to be chelating compounds marketed under proprietary trade names "LIX" and "KELEX" – usually with a number appended. LIX reagents are hydroxy-oximes and KELEX are hydroxy-quinolines and their mode of operation is illustrated below, the example being one of the hydroxy quinoline type of reagents. This reaction will go to the right when the hydrogen ion concentration is low and to the left when it is high.

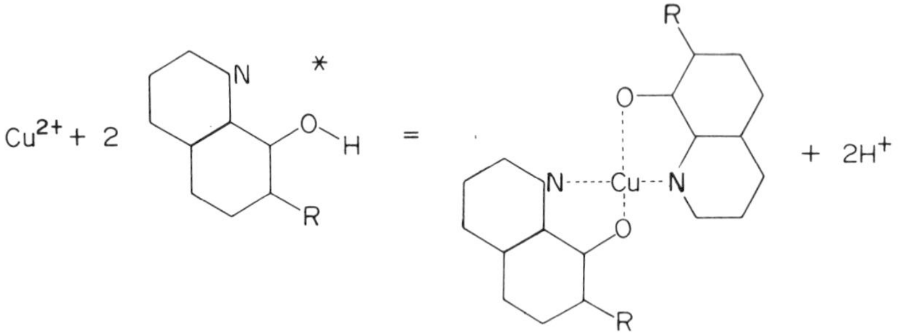

In use the reagent is dissolved in a carrier which is usually based on kerosene. This is agitated with a similar volume of leach solution, partially purified if necessary and with a sulphuric acid content of about 0.1%. Any copper in the leach solution is chelated and enters the kerosene phase. The kerosene is allowed to rise and separate from the aqueous layer. It is then agitated with a smaller volume of aqueous solution containing, this time, about 15% sulphuric acid. The reaction is reversed. The copper escapes from the complex and enters the aqueous solution. If the volume ratios of the solutions have been selected for the greatest effect, the concentration of copper in the concentrate liquor may be up to 50 times that in the leach solution, and this is a suitable feed for electrolysis tanks. The flow of materials is presented diagrammatically in Fig. 74.

Large-scale continuously operated mixer/settler units are in use as well as various types of columns in which the two immiscible phases pass in opposite directions. In the perforated plate column some form of pulsation is needed to cause the lighter phase (the kerosene) to be injected

*Alkyl hydroxy quinoline – marketed as KELEX 100 by Ashland Chemical Co.

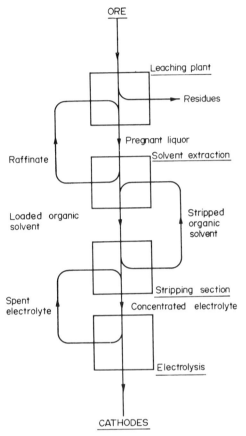

FIG. 74. Flow sheet for an extraction procedure involving
leaching, solvent extraction and electrowinning.

through the perforations as suitably sized droplets to maintain a large
interfacial area in the system. It will be appreciated that the reagent is continuously being regenerated for re-use.

Ion exchange is another technique which serves the same purpose as
does solvent extraction. It is based on a similar principle. A familiar
example of the technique is the softening of water by zeolites. Ion
exchange depends on the peculiar properties of certain synthetic resins

which can be treated with mineral acids to produce "ion exchange sites" which may be indicated at RX where X = NO_3, Cl or SO_3H. In reactions with solutions containing suitable cations the H in the sulphonic group could be replaced by one of them, e.g. $CuSO_4 + 2R \cdot SO_3H = 2(R \cdot SO_3)Cu + H_2SO_4$. In a reaction with a solution containing suitable anions the NO_3 or the Cl could be replaced. The major application is to the recovery of uranium which may occur in leach liquor as a complex anion in sulphuric acid solution so that the reaction with $R \cdot NO_3$ is—

$$[UO_2(SO_4)_3]^{4-} + 4R \cdot NO_3 = R_4 \cdot UO_2(SO_4)_3 + 4[NO_3]^-.$$

When all the available exchange sites have been occupied, the collected ions can be released by washing with a strong solution of X ions – in this case nitrate – the resin being simultaneously re-activated. In this particular case, the separation of uranium from other metals present as cations is elegantly effected. The extension of the technique to other metals is controlled only by cost.

Chemical precipitation can be brought about by heating, by altering the pH, by dilution, or by adding appropriate reagents. Ammines can be decomposed by driving off the ammonia by steam distillation to precipitate the nickel and cobalt which had been complexed. In a few cases a final separation of a metal in salt form may be made by crystallization. This is all simple inorganic chemistry practised on a large scale. Examples can be found in Chapter 13 in the sections on beryllium, aluminium, platinum and uranium.

Reduction by another metal is called cementation. The principle applications of its use are to the precipitation of copper and gold. Cement copper used always to be made from heap leach liquors before solvent extraction became available for their concentration. The liquor containing copper sulphate is run through long troughs containing scrap steel. The iron displaces copper from solution according to

$$CuSO_4 + Fe = FeSO_4 + Cu.$$

Copper forms a curd-like precipitate which is heavily contaminated with the residues of the steel. It must be fire-refined before it can be sold, whereas electrolytic copper requires little more than a re-melt before casting. Modern equipment is available for cementing copper on a continuous basis. The cementation of gold is effected by adding zinc dust to a purified and deaerated cyanide leach solution. The product, like cement copper, is a slime, of gold and zinc, which needs further processing. Obviously it is the more noble metals which can be reduced most easily by this method.

The reduction of metals from solution by hydrogen[78] is chemically similar to cementation but is technically more difficult. Its principal applications are to the recovery of nickel and cobalt from leach solutions, particularly after pressure leaching with ammonia. It is also used on a smaller scale for making metal powders.

Considering the case of the reduction of nickel, the reaction may be written

$$Ni^{2+} + H_2 = Ni + 2H^+$$

for which

$$K = \frac{a_{Ni} \times a_{H^+}^2}{a_{Ni^{2+}} \times p_{H_2}}$$

where a_{Ni} = 1. The equilibrium conditions are defined by $\Delta G_T^\circ = RT \ln K$ where ΔG_T° can be written as $(nFE^\circ{}_{Ni} - nFE^\circ{}_{H_2})$ for this reaction (see page 140). The condition for equilibrium can therefore be written

$$E_{H_2}^\circ - E_{Ni}^\circ - \frac{RT}{nF} (\ln a_{Ni} + 2 \ln a_{H^+} - \ln a_{Ni^{2+}} - \ln p_{H_2}) = 0. \quad (12.1)$$

By convention, $E_{H_2}^\circ = 0$; by definition $-\log a_{H^+} = \text{pH}$; collecting terms together, when $T = 298$ K and $n = 1$,

$$\frac{RT}{nF} \log_{10} e = 0.059.$$

The standard electrode potential for nickel $E_{Ni}^\circ = -0.236$ so that equation (12.1) becomes, at 25°C,

$$0.0295 \log a_{Ni^{2+}} + 0.059 \text{ pH} + 0.0295 \log p_{H_2} = +0.236. \quad (12.1a)$$

If it is the purpose of the process to reduce the nickel ion concentration as far as possible it will be obvious from equation (12.1a) (and the Law of Mass Action) that this can best be achieved by reducing the activity of hydrogen ions, that is increasing the pH, and by increasing the hydrogen pressure. Equation (12.1a) puts this on a firm quantitative footing. If, for example, the temperature is $25°C$ and the pressure of hydrogen is 1 atm, and if the concentration of Ni^{2+} ions is 0.01 molal (at which concentration $_{aq}a_{Ni^{2+}} = 0.004$) then there is a critical pH which must be exceeded if further reduction of nickel from the solution is to be possible. This is obtained as

$$pH_{critical} = (0.236 - 0.0295 \log 1 - 0.0295 \log 0.004) \div 0.059$$
$$= 5.2.$$

If the pH is kept at 5.2 and the hydrogen pressure is increased to 100 atm, the lowest activity of Ni^{2+} ions attainable would be 4×10^{-5}. With a $pH = 10$ and a pressure of 100 atm, $_{aq}a_{Ni^{2+}}$ could in theory be reduced to 10^{-14}. If, however, the case of the reduction of zinc is considered, the value of E_{Zn}^{o} is -0.76 at $25°C$ and at $pH = 10$ and $p_{H_2} = 100$ atm the limiting value of $_{aq}a_{Zn^{2+}}$ would be 5790. This means that no reduction of zinc would be possible under these conditions.

In practice the calculated equilibria are not easily approached. Even the reduction of copper by hydrogen is kinetically difficult and does not occur anywhere near the calculated equilibrium condition. In the case of nickel, reduction is carried out in an autoclave at about 200 atm pressure, at $160°C$ to hasten the reaction, and at a pH of about 11. The high pH is reached without precipitating hydroxides by using ammonia to produce soluble ammines. An NH_3/Ni molar ratio of 2 is preferred so that the reaction becomes

$$Ni(NH_3)_2^{2+} + H_2 = Ni + 2NH_4^+$$

which has no effect on the pH of the solution as the reaction proceeds. It is also necessary, in the cases of nickel and cobalt among others, to "seed" the solution with a fine powder of the metal, these particles to act as nuclei for the deposition of new metal. It is an advantage too if a catalyst is added to the solution. The usual catalyst is anthraquinone which acts on the surface of the growing particles, apparently increasing the proportion of potential growth sites which are active. This not only increases the rate of deposition but also produces a denser and more spherical particle.

In operation the autoclave is supplied with hydrogen at a sufficient rate to maintain the pressure as set. A vigorous agitation is maintained which ensures that the gas is thoroughly entrained into the solution. The kinetics are those of a first order reaction, the rate being dependent only upon the pressure of the hydrogen. When the supply rate falls off, the end of the reaction is indicated. The possibility of reducing only one metal out of a mixed solution and so effecting separations has obvious attractions.

To an increasing extent pre-refining is being carried out to produce pure oxide or salts suitable for direct reduction to pure metal. Typically, a roast and leach are followed by a series of stages involving precipitations and filtrations which are straight-forward applications of classical inorganic chemistry – the "group separations" on a large scale. Even H_2S is used as a reagent in these separations. Under industrial conditions it is no more unpleasant or dangerous to use than many other chemicals in common use. These are often referred to as "wet" methods. They are often carried out on a relatively small scale in simple equipment – steam-heated vats, stirrers and filter presses – and no further description will be offered here.

One of the greatest difficulties met in hydrometallurgy is the inhibition of corrosion, often combined with erosion. Tanks and ducts may be made of concrete lined with lead, rubber or mastic. Transport should be by gravity as far as possible and air-lifts may be used instead of pumps for regaining height.

Electrolysis

Electrolysis is used both for extraction and for refining and in either case the electrolyte may be an aqueous solution or a mixture of fused salts. In electrolytic extraction or electrowinning processes the metal is usually in solution in the electrolyte from which it is plated on to the cathode, the anode being an insoluble conductor. In electro-refining processes the anode is the impure metal, the electrolyte is a solution of high electrical conductivity and constant concentration, and the cathode is of pure metal built up on a pure "starting sheet" or sometimes on a blank of another metal. Occasionally pure metal is deposited on a pool of molten metal covering the conducting hearth of the cell or floating on the electrolyte, in contact with the cathode.

The voltage necessary to operate an electrolytic cell may be expressed as the sum of the five components which are discussed below.

1. The e.m.f. for the net chemical reaction involved according to—

$$E_R = E^\circ + \frac{RT}{nF} \Sigma \ln a_i \qquad (12.2)(6.73a)$$

which can be calculated from the free energy change for the net reaction and values of the activities of reactants and products (see page 140). The simple net reaction for the deposition of copper from $CuSO_4$ solution is—

$$CuSO_4 + H_2O = Cu + H_2SO_4 + \tfrac{1}{2}O_2$$

or

$$Cu^{2+} + H_2O = Cu^0 + 2H^+ + \tfrac{1}{2}O_2$$

for which the standard free energy change is the sum of the free energies of the partial reactions—

at the cathode

$$Cu^{2+} + 2e^- = Cu^0 ;$$
$$\Delta G^\circ_{298} = -64,410 \text{ J}$$

and at the anode

$$H_2O = H_2 + \tfrac{1}{2}O_2 ;$$
$$\Delta G^\circ_{298} = 285,920 \text{ J},$$

and

$$H_2 = 2H^+ + 2e^- ;$$
$$\Delta G^\circ_{298} = 0,$$

the standard free energy of ionization of hydrogen being zero because E° for the standard hydrogen electrode has been arbitrarily chosen as zero. The free energy of the net reaction is therefore $\Delta G^\circ_{298} = 221.6 \text{ kJ}$ so that

$$E^\circ = -\frac{\Delta G^\circ}{nF} \simeq -1.2.$$

The negative e.m.f., like the positive free energy value, indicates that the reaction would occur spontaneously only from right to left, and the value -1.2 V is a measure of the e.m.f. which would have to be opposed to prevent this reverse reaction from taking place. The application of 1.2 V would therefore bring the cell into the equilibrium condition in which no net reaction occurred while the application of a higher voltage would bring about the deposition of copper, the positive direction *within the cell* being from the metal in the state in which it appears on the left-hand side of the equation to the metal as it appears on the right — i.e. in this case, from solution to cathode. This potential is subject to an adjustment, using equation (6.73a), for the deviation of the activities from the values in the standard solutions in which electrode potentials are measured, as has been discussed on page 142. Activity coefficients of acid and salts in solution are generally much lower than unity at the concentrations found in electrolytic tanks. At unit molality, which is of the order of concentration met in practice, $_{aq}\gamma_{CuSO_4} = 0.04$ while $_{aq}\gamma_{H_2SO_4} = 0.13$. Values of these coefficients can be obtained only from standard reference books (see Bibliography) in which the range of data available is still rather restricted. E_R can be seen, however, from equation (10.1) to be a variable quantity rising as $_{aq}a_{CuSO_4}$ falls and as $_{aq}a_{H_2SO_4}$ increases, i.e. as deposition proceeds. The change is slow, however — a ten-fold decrease in the copper activity adding only 0.0295 V to E_R (see page 142) while a ten-fold increase in the activity of the acid has the same effect, so that the increase in E_R would be only 0.06 V if both these changes occurred together. Such a large variation in concentration is unlikely within any single practice. Clearly no great error is incurred if it is assumed that E_R is constant and has a value equal to the standard value. In electro-refining, where the thermodynamic state of the metal being dissolved is almost identical with that of the metal being deposited, $\Delta G° \simeq 0$ and therefore $E_R \simeq 0$. The cell voltage would normally be about 0.3 V in this case.

2 and 3. These are e.m.f.s E_A and E_C due to polarization at the anode and cathode respectively. That at the cathode is due to the concentration gradient in the Cu^{2+} ions caused by the depletion in the solution just off the cathode surface. Its effect may be reduced by agitation. That at the anode is sometimes called the "oxygen over-voltage". It is partly due to a local surplus in the sulphuric acid

concentration and partly due to the difficulty of nucleating oxygen bubbles. This e.m.f. can be reduced by agitation and by incorporating a catalyst or a reducing agent to take up the oxygen without forming gas bubbles.

4. An e.m.f. E_Ω is required to overcome the ohmic resistance of the electrolyte. This can be kept low by adjustment of its composition to ensure a high concentration of conducting ions. It can be kept low by designing the cell so that the electrical paths are wide and short – leaving space for the growth of the cathodes of course – since, as in solid conductors, resistance R is proportional to the length of the path l and inversely proportional to its cross-sectional area s,

i.e. $$R = r \cdot \frac{l}{s} \qquad (12.3)$$

where r is the specific resistivity of the electrolyte in ohms per cm cube. r depends on the ionic species present and the least contribution is made by hydrogen ions and the next smallest by hydroxyl ions so that low resistances are most easily obtained in fairly strong acid or alkaline solutions (in water). These need not be compatible with low values of E_A. E_Ω increases with current in accordance with Ohm's law.

5. The contact potential E_K at each electrode, in connections between electrodes and bus-bars, must be included in the grand total.

Thus the total e.m.f. required in a cell is—

$$E = E_R + E_C + E_A + E_\Omega + E_K \qquad (12.4)$$

which may be substantially higher than E_R, perhaps two or three times as high, subject to some control in the design of the cell. In the case of the deposition of copper, for example, a potential drop of about 2.3 V might be expected per cell although E_R is only 1.2 V; and in the extraction of aluminium the value of E_R is only 1.2 V for the net reaction but $E_C \simeq 0.55$, $E_A \simeq 0.40$, $E_\Omega \simeq 3.1$ V and $E_K \simeq 0.8$ V making a total cell e.m.f. of about 6 V.

It will be appreciated that the cell potential actually used is determined by the magnitude of the current which is required to provide the desired rate of production. The higher the selected current, the more severe will be the polarization at the electrodes and the greater the voltage drops across the electrolyte, the bus-bars and the contacts. Only E_R is independent of the value of the current.

Faraday's law is, of course, valid in these operations but it is found that the current used exceeds that theoretically necessary to deposit the metal being extracted and this leads to the term "current efficiency" which is the ratio of the weight of metal actually deposited to that which should have been deposited in accordance with Faraday's law. This apparent deviation is largely due to side reactions which cannot always be avoided. Co-deposition of hydrogen at the cathode is one such reaction which can be and must be controlled by limiting the cell voltage, as is discussed later. Some hydrogen may be occluded in the deposited metal but it should not be evolved as a gas. Electrolytic reduction of ferric iron to ferrous within the cell is not so easily prevented. Even if the iron is reduced chemically before the electrolyte is passed into the tanks, some of it is re-oxidized at the free surface with air or by oxygen at the anodes and is then available for reduction at the cathodes. When iron in copper leach solutions is very high, this reaction can depress the current efficiency in the tank house below 80% compared with a more normal value of 95% when the iron is low. There is obviously good reason to control roasting prior to leaching so that iron is rendered insoluble (see page 285). The re-solution of cathode material is another cause of low current efficiency. This may occur because acidity has been allowed to build up to too high a level and it will certainly be a consequence of using current-reversal practice to improve the quality of the deposits (see below). Current can also be lost through leakages to earth or between bus-bars. In the humid conditions prevailing in tank houses these losses probably cannot be eliminated but they can be minimized by maintenance of a high standard of cleanliness in the plant.

The current, together with the resistance of the electrolyte, determines how much heat $(I^2 R)$ is produced. When the electrolyte is an aqueous solution, its resistance should be kept low so that a high current can be passed without there being an appreciable rise in temperature. The resistivities of acid solutions are lower than those of alkaline solutions. Neutral solutions have the highest resistivities and should be avoided. The cell resistance. depends also on the distance between the anodes and cathodes and this should not be excessive. The resistivities of electrolytes fall with rising temperature and tanks are commonly run at about $40-50°C$ to take advantage of this. When the electrolyte is a solution of molten salts or oxides a higher resistance is necessary so that for any particular current the I^2R heating effect will suffice to maintain a high cell temperature. Obviously a correspondingly high cell voltage will be required. In some circumstances supplementary heating may be required.

Electrolytic deposition from aqueous solutions is restricted to those metals whose ions are discharged with a lower cell e.m.f. than is required to discharge hydrogen ions. The net reaction involved in the dissociation of water is—

$$H_2O = H_2 + \tfrac{1}{2}O_2$$

which can be split into the partial reactions—
at the cathode,

$$2H^+ + 2e^- = H_2 \quad : \Delta G°_{298} = 0.0 \text{ kJ}$$

and at the anode,

$$H_2O = H_2 + \tfrac{1}{2}O_2 \quad : \Delta G°_{298} = 285.9 \text{ kJ}$$

and

$$H_2 = 2H^+ + 2e^- \quad : \Delta G°_{298} = 0.0 \text{ kJ}.$$

The free energy of the net reaction is clearly

$$\Delta G°_{298} = 285.9 \text{ kJ}$$

so that

$$E° \simeq -1.5 \text{ V}.$$

In the case of the deposition of zinc the net reaction is—

$$Zn^{2+} + H_2O = Zn^0 + 2H^+ + \tfrac{1}{2}O_2$$

for which $G°_{298} = 431$ kJ so that $E° \simeq 2.3$ V. It would appear, therefore, that copper would readily be deposited electrolytically with the application of 1.2 V, but that attempts to deposit zinc by application of the necessary 2.3 V would lead only to a vigorous evolution of hydrogen. Fortunately, however, the nucleation and deposition of hydrogen does not occur at the minimum thermodynamic value of E°, for kinetic reasons. In other words there is a high hydrogen "overpotential" or "over-voltage" by virtue of which the deposition of hydrogen at the cathode is inhibited until the cathode potential is at a much lower value than the arbitrary 0.0 V assigned to the standard hydrogen electrode, but measurable only when the hydrogen evolution is catalytically assisted by the nature of the platinum black surface of the special electrode. At more normal surfaces, zinc and cadmium, cobalt and nickel, tin and lead can all be·deposited in preference to hydrogen despite their inferior positions in the electrochemical series (see Table 4). In practice the cell e.m.f. is usually so much higher than E° that some evolution of hydrogen accompanies deposition of the metal. This should be kept to a minimum by maintaining the cell e.m.f. as low as is practicable and keeping potential catalysts out of the system.

An exceptional case is the deposition of sodium from aqueous solutions on to a mercury cathode. In this case the hydrogen over-voltage on the extremely smooth surface of the liquid metal is very high. At the same time the activity of sodium in dilute solution in the mercury is

TABLE 4. THE ELECTROCHEMICAL SERIES AND NORMAL ELECTRODE POTENTIALS OF COMMON METALS (EUROPEAN CONVENTION)[83]

Metals electropositive to hydrogen	Metals electronegative to hydrogen but which can be deposited from aqueous solutions under the hydrogen over-voltage	Electronegative metals which can be deposited only from molten salts
E°	E°	E°
Au + 1.68	H 0.00	Nb − 1.10
Pt + 1.4	Pb − 0.13	U − 1.40
Ag + 0.80	Sn − 0.14	V − 1.50
· Hg + 0.80	Mo − 0.20	Al − 1.67
Cu + 0.34	Ni − 0.25	Be − 1.70
Bi + 0.23	Co − 0.28	Ti − 1.75
Sb + 0.10	Cd − 0.40	Mg − 2.38
W + 0.05	Fe − 0.44	Na − 2.71
H 0.00	Cr − 0.56	Ca − 2.8
	Zn − 0.76	
	Mn − 1.05	

very low and the two effects are additive. This phenomenon has industrial applications.

Metals which cannot be electrolysed from aqueous solution may be deposited from fused salts or "igneous melts". These include aluminium, magnesium and the alkalies and alkaline earths — all very reactive metals which require some protection from the atmosphere when produced. Metallic aluminium is collected at the cathode at the bottom of the cell and is protected by the electrolyte but in the magnesium process the metal floats on the electrolyte and elaborate cell design is needed to keep the magnesium and the anodic chlorine apart. In the aluminium cell oxygen is released at the carbon anode which reacts forming CO and CO_2 in approximately equal proportions, so depolarizing the anode, which is consumed and must continuously be replaced. One type of electrode is continuously replaceable. This device, which is known as the Söderberg electrode, is an aluminium tube (of rectangular section) filled with a paste of carbon and tar at the top, which bakes hard as it is fed down into the cell. Fresh sections of the tube are welded on top so that the anode is always serviceable.

Volatile matter from the tarry bond leads to some difficulties in the operation of the associated gas-cleaning plant, which is now required to work with high efficiency. Consequently there is a trend toward the use of carbon block electrodes which are prepared and fired outside the cell. There are several of these per cell and they must be replaced from time to time as they are oxidized away. The electrolyte must be easily fused and should contain no metals of lower electrode potential than that of the metal being extracted. The anode and cathode materials must be chosen as refractories, special consideration being given to the avoidance of contamination of the metal. At least part of the heat required in these processes will arise in the electrolyte acting as a resistor and this must be considered when the composition of the electrolyte and the cell dimensions are being chosen.

In electrolytic extraction plants, electrodes are usually arranged in parallel in each tank while the tanks are arranged in series as indicated in Fig. 75a. Anodes are shaped as shown in Fig. 75b with lugs, one of which rests on the positive bus-bar running along one side of the tank while the other rests on an insulated support. Cathode blanks have a similar shape but depend from a rigid bar which spans the tanks and

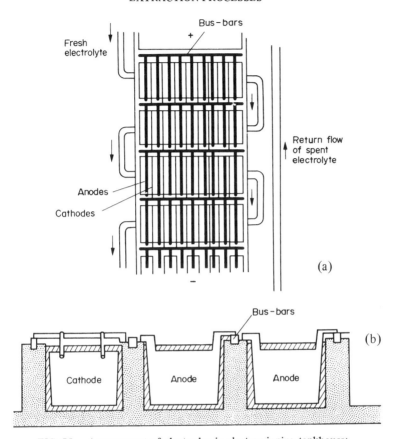

FIG. 75. Arrangement of electrodes in electrowinning tankhouse: (a) electrodes in parallel within each cell but cells in series. Several such "strings" of cells might be arranged in parallel; (b) detail of electrode—bus-bar contacts.

makes contact with the negative bus-bar running along the other side of the tank. That same bus-bar provides the anodic current to the adjacent tank. Tanks typically hold about forty to fifty anodes and cathodes, each with an area of about 1 m^2. If the electrical supply is at 100 V and the potential drop in each cell is 2 V there would be fifty tanks in series. The d.c. supply can be generated specifically for the purpose or converted from a.c. using motor-generators. Silicon diode rectifiers are now preferred, however, as being more efficient than converters, cheaper to maintain and amenable to supplying reversed current if required to do so. The current must be carried outside the cells by heavy bus-bars to minimize I^2R losses. The electrolyte may be allowed to rise in temperature to about $50°C$ where its resistivity is lower than at $20°C$, for the same reason. Aqueous solutions usually cascade by gravity from one tank to the next in a group, the flow rate being arranged so that the electrolyte in the lowest tank has a sufficiently low metal content and a high enough acidity for it to be returned to the leaching plant. The flow rates through the two plants must be planned to be compatible. No attempt is made to reduce the metal ion concentration to a very low value. Pregnant leaching solution may have to be purified with respect to certain ions before being electrolysed to improve process efficiency or the purity of the product. Temperature, pH and current density also affect the quality of the cathodic material with respect to composition and physical character.

Current densities usually lie in the range $100-300$ A/m^2, the highest value practicable being used in order to achieve the maximum production rate. This is important because cathodes are not removed or "pulled" until they have been in the tank about a week and this accumulation of valuable material represents a considerable capital charge on the process. High current densities, however, give irregular deposits. Sometimes the metal becomes loose and spongy and may even fall to the bottom of the cell. In other cases there is excessive "treeing" or formation of dendritic growths which can lead to short-circuiting in the cells and must be avoided. Colloids, gums and glues may be added to improve the physical quality of the deposit at high current densities. A recent development is the use of periodic reversal of the direction of the current which has the effect of electropolishing the high spots off the cathode surfaces before dendrites can develop. (During the reversal the cathodes are, of course, temporarily anodes.) The cycle is something like 200 seconds forward followed by 20 seconds reverse. Considerably improved net current densities and so higher productivities are claimed for this technique[79] but obviously at considerable cost in energy consumption which must be offset by savings in other directions. Other methods which have been suggested to achieve similar results include ultrasonic agitation of the electrolyte and fast circulation of the electrolyte, each being costly to operate in its own way.

Metal extracted electrolytically can usually be produced at a high level of purity and often needs little further refining other than re-melting to get rid of occluded hydrogen and provide the opportunity to make a sound ingot of suitable shape. To achieve a suitable degree of purity, however, it may be necessary to eliminate from the solution, by chemical means, any metals which are electro-positive to the product metal. This may be done by simple precipitation or cementation methods or by solvent extraction. In this last case the electrolyte produced is very acid and attacks the antimonial lead anodes which are normally used. If contamination of the product by the lead taken into solution must be avoided it becomes necessary to look for more suitable anodes. In copper practice titanium anodes coated with an alloy of platinum group metals are now sometimes used. The solvent extraction reagents sometimes contaminate the electrolyte and enter the cathode where they show as friable patches of mixed metal and organic material on the surface and adversely affect the efficiency of the process. The phenomenon is known as "organic burn".

The optimum rate of deposition of metal is not necessarily the highest rate. Whereas the rate of deposition is proportional to the current passing through the cell, the energy consumed, per unit of metal produced, increases as the deposition rate rises. This is because the energy is the product of current and potential and because a rise in current can only be achieved by increasing the potential also (assuming the characteristics of the cell are not to be altered). If Ohm's law were applicable the energy would be measured as I^2R and would be proportional to the square of the production rate. Because E_R is not dependent upon the current and because E_A and E_C change rather unpredictably as the current is altered the relationship between energy consumption and production rate is not readily predictable but the cost of energy per unit of production must rise while at the same time the standing charges per unit of production must fall. In some circumstances there can be a production rate at which the total cost per unit of production has a minimum value. (See example 18 in Appendix 1.)

In refining processes the arrangement of cells and tanks is very similar to that in electrowinning. The cell potential is lower so that the number of tanks in series is often larger. The electrolyte is circulated to minimize polarization of the electrodes and a portion of it may have to be bled off for purification and adjustment of its composition. The behaviour of the impurities in the anode is, of course, important. More noble metals do not enter the solution but fall away as an insoluble *anode slime* which may fall to the bottom of the tank or may be collected in a fabric bag hung round the anode. These slimes may contain gold, silver and platinum metals whose values may justify the choice of an overall process which includes electrolytic refinement. The slimes may also contain sulphur, selenium and tellurium which are of little value. Metals only slightly less noble than the principal metal go into solution with it and will normally be deposited with it unless steps are taken to ensure that they are precipitated chemically. This may be done within the cell by including suitable reagents in the electrolyte as is the case with As and Sb in copper refining. The precipitate becomes part of the anode slimes. Alternatively the chemical treatment may be carried out external to the cell in which a "diaphragm" must separate the anode compartment from the cathode compartment. The diaphragm is a screen of canvas which does not impede the flow of electric current. The electrolyte levels are maintained so that only a small flow of solution can occur and that from the cathode to the anode compartment. The "anolyte", the solution from the anode compartment, is drawn off and taken to the chemical treatment plant which produces the purified "catholyte" solution which is fed back into the cathode compartment, the whole process being continuous. As already explained the catholyte will seep into the anode compartment by gravity but cannot become contaminated by untreated solution.

Metals which are significantly electronegative to the principal metal will accumulate in the electrolyte, from which they must be periodically removed.

There are some interesting applications of electrolysis both to winning and to refining which do not always conform with the general descriptions given above. Three such processes are now considered briefly.

Nickel can be produced to better than 99.5% purity in a diaphragm cell from impure nickel sulphide anodes. These are 95% Ni_3S_2 with copper, iron and cobalt as the main impurities. Although an electrowinning process, this technique has the characteristics of a refining process and the product is essentially a refined product. The anodes are contained in synthetic fabric bags and as the nickel is electrolytically leached out a large volume of sludge collects around the skeletal residue of elemental sulphur. The electrical resistance increases as the anode is consumed. The cell voltage must be increased to maintain the production rate, as a result of which there is an appreciable and variable rise in the temperature of the anolyte. The rather high potential also causes some decomposition of water at the anode by the partial reaction

$$2H_2O \rightarrow O_2 + 4H^+ + 4e$$

the four electrons being in excess of those made available by the dissolution of nickel. Consequently some additional nickel salt must be introduced into the catholyte to take up these electrons in the cathodic reaction. The purification of the anolyte is by removal of copper and arsenic using H_2S and precipitation of the iron and cobalt as hydroxides. The sludge is over 95% sulphur with undissolved sulphides of nickel, copper iron and cobalt along with some precious metals.

In cobalt extraction, deposition from acid solutions is not possible. Cobalt hydroxide is precipitated with lime and the slurry electrolysed. Cobalt is dissolved continuously at the anode and deposited at the cathode, the concentration in the solution always being very low.

The refining of aluminium is elegantly performed using a dense solution of the impure metal in copper as the anode, with a less dense electrolyte containing, however, enough barium salt to raise its density above that of the pure aluminium which floats as cathode on top.

Fire Refining

Most metals, however they have been extracted and refined, are brought into the liquid state for their final adjustment of composition. In some cases this is little more than simple melting to allow cathodically entrained hydrogen to escape by diffusion. Sometimes volatile impurities are drawn off under reduced pressure (see page 342). In other cases, which include important metals like steel and lead, most of the impurities present after the primary extraction are removed by selective chemical reactions which render them insoluble in the molten metal. The most common reagent for this purpose is oxygen (or air), but sulphur, chlorine, carbon, lime and zinc are among others used in particular processes. Some of these have already been mentioned in Chapters 8 and 9.

Oxidation may be effected by gaseous air or oxygen, or through an oxidizing slag maintained high in FeO or some other easily reduced oxide or salt. Litharge and sodium nitrate are used in the refining (softening) of lead, for example. In the traditional processes slow oxidation was effected by "flapping and tossing" the molten metal with rabbles to increase the surface exposed to air. Modern trends are toward oxygen lancing, which can be carried out so that the oxygen is fed into the slag, and so through it by diffusion, or in a faster jet which will pierce the slag and oxidize the metal directly (see page 324). Oxidation by means of an oxidizing slag involves transfer of metal as well as oxygen to the bath, with an increase in the metallic yield of the process, particularly if the oxygen is fed in as ore or mill-scale or in some other solid form. Oxygen lancing burns up metal to produce oxide for the slag. Some heat is produced, but metal is expensive fuel. The advantages of lancing lie in improved control and productivity.

Cupellation is a special application of oxidation to the refining of noble metals. It is preceded by the collection of the noble metal in a bullion or lead alloy. The lead is then oxidized by melting in air or by jetting air on to the molten alloy, and the PbO and any other oxides form a slag, most of which runs away down a sloping hearth to a tapping point leaving behind a pool of noble metal alloy called *Doré* metal which can be separated into its constituents by "wet" methods. The hearth of the cupel must be sufficiently concave to retain the pool of *Doré* metal. It usually absorbs a part of the very reactive slag and is lined with a suitable refractory to take this up and leave the metal clean. Bone ash is the traditional material for this purpose but other refractories can be used. The hearth should be easily replaceable so that a fresh hearth can be made available quickly when necessary.

It will be obvious that only impurities less noble than the metal being refined can be removed by oxidation. More precisely, an impurity can be removed by oxidation only if its oxide can be brought into a state in which its free energy is more negative than that of the oxide of the metal being refined. Thus, as discussed on page 168, phosphorus can be oxidized out of iron in the presence of a basic slag in which P_2O_5 dissolves with a chemical potential much lower than that of FeO even though the standard free energy diagram for oxide formation shows phosphorus to be more noble than iron at the operating temperature.

Despite exceptional cases of this kind, the more noble metals often cannot be removed by oxidation, and in steelmaking the removal of copper, nickel, tin and antimony is not possible by the ordinary methods and indeed a major difficulty facing that industry is the gradual accumulation of these so-called "residuals" which are continually being returned in scrap and never escape the system.

When noble metals are valuable (gold, platinum) other means are found of removing them, such as electrolytic refining (see page 313), the preferential formation of a sulphide (see page 375) or preferential solution in another metallic phase (see page 375).

Removal of an impurity can never be quite complete and always involves some contamination by the reagent used. In the oxidation of carbon from iron by oxygen, for example, the reaction may be written—

$$C \text{ (in metal)} + O \text{ (in metal)} = CO \text{ (gas)}$$

for which the reciprocal of the equilibrium constant is—

$$m = {}_\%a_c \times {}_\%a_o = 0.0020 \text{ at } 1600°C, \qquad (12.5)$$

the pressure of the CO formed being very nearly 1 atm. The activities are referred to the 1 per cent by weight solutions as standard states and Henry's law assumed valid. Clearly, a very low carbon content can be attained only by increasing the oxygen content, possibly to the limit of its solubility. Similar considerations apply in all other cases, "purified" copper, for example, being high in oxygen after being refined of its iron and zinc. The best reagent would be one whose solubility in the metal being refined was extremely low. The solubility of oxygen is usually low, but its effects are often very great.

Obviously the excess of the first reagent must in turn be refined out if a really pure metal is required. The deoxidation of copper is carried out by "poling" with greenwood logs which decompose with evolution of reducing gases (hydrogen) which gradually reduce the concentration of oxygen but introduce some hydrogen, most of which escapes by diffusion, however, as the metal freezes and cools. A lower residual oxygen content can be obtained by melting under charcoal or pouring the metal through a bed of hot charcoal to obtain the oxygen in equilibrium with carbon instead of hydrogen.

In steelmaking, deoxidation is effected by adding elements the free energies of formation of whose oxides are more negative than that of iron oxide and whose presence as an excess is not undesirable in the steel. More precisely, the free energy of oxidation of a deoxidizing agent in solution in iron (and in the presence of other solutes) must be more negative than the free energy of solution of oxygen in the same iron solution. In other words, the thermodynamic state of all of the species involved must be taken into consideration. Inspection of Fig. 45 indicates that ΔG° for the formation of SiO_2 is more negative than ΔG° for the formation of FeO so that silicon would be expected to deoxidize iron because, for the reaction

$$2FeO + Si = 2Fe + SiO_2,$$

ΔG is always negative. But the deoxidation reaction is more correctly written

$$2\underline{O} \text{ (in solution in Fe)} + \underline{Si} \text{ (in solution in Fe)} = SiO_2 \text{ (pure solid)}.$$

The standard free energy of this reaction, the standard state of each solute being a 1% solution, is given by $\Delta G^{\circ}_T = -594{,}400 + 230T$ J mol^{-1}, from which it can be calculated, using equation (6.46), that

$$K_{Si(1873)} = \frac{a_{SiO_2}}{\%a_{\underline{O}}^2 \times \%a_{\underline{Si}}} = 3.6 \times 10^4$$

and that if the solutions of $\underline{Si}$ and $\underline{O}$ are sufficiently dilute

$$(\%\underline{O})^2 \times \%\underline{Si} = 2.7 \times 10^{-5}. \tag{12.6}$$

If the oxygen concentration is 0.01%, the equilibrium silicon concentration is 0.27% and silicon will be deoxidizing toward that particular solution but only at concentrations higher than 0.27%. If the oxygen concentration is as low as 0.005%, however, it will require more than 1.08% of silicon to effect further deoxidation. Such a high concentration of silicon would have deleterious effects on most grades of steel so that silicon could not be used as a deoxidizing agent for steel which already had so little oxygen present. In practice carbon is always required in steel and its re-introduction after all impurities (and particularly phosphorus) have been oxidized to acceptable levels rapidly reduces the oxygen content in accordance with equation (12.5).

Raising the carbon concentration to 0.15% (mild steel) reduces the oxygen from about 0.05% toward the equilibrium concentration which is 0.015%, but if the carbon required is 0.3% (medium carbon steel) the oxygen will be further reduced toward a lower equilibrium concentration of 0.007%. Because the reaction involves the nucleation of carbon monoxide bubbles these low concentrations of dissolved oxygen are probably not reached. Further reduction of the dissolved oxygen is achieved by using manganese, silicon and aluminium as deoxidizers. Manganese is a weak deoxidizer but it performs a similar function with respect to sulphur in steel, removing it from solution and converting it into a harmless form. Aluminium is a very powerful deoxidizer,

$$K_{Al(1873)} = (\%Al)^2 \times (\%O)^3 = 3 \times 10^{-15},$$

so that 0.05% aluminium in solution in steel is in equilibrium with only $10^{-4}\%$ of oxygen. The weaker deoxidizers are added first and aluminium and sometimes the even more powerful zirconium are added later as "scavengers". Except for carbon monoxide, the products of deoxidation are solids or liquids which should, ideally, be allowed to float out of the melt. Apart from the time required, this is not possible to achieve because, as the steel cools to its solidification temperature, the solubility products decrease so that precipitation of oxides continues until the metal is solid. At 1500°C, for example, data for the silicon reaction show that

$$K_{Si(1773)} = 3.1 \times 10^{-5}$$

so that $(\%O)^2 \times \%Si = 3.2 \times 10^{-6}$. A solution containing 0.27% Si and 0.01% oxygen at 1600°C will, when cooled to 1500°C, contain 0.0034% oxygen at equilibrium and much of the difference will have been precipitated as SiO_2 during cooling. In practice the conditions are much more complex with other oxides and manganese sulphide precipitating at the same time.

When really pure iron is required, oxidation is taken as far as possible, the excess oxygen is reduced with some carbon and the metal then treated with hydrogen to remove the last of the carbon. It is then finally melted under vacuum (see page 334) to remove hydrogen and other volatile impurities.

Gaseous impurities can be removed by vacuum methods of by bubbling an inert or non-reactive gas through the melt. In this latter case, as each bubble comes into equilibrium with the metal it will dissolve out the gaseous impurity to the point that its partial pressure in the bubble will be in accord with Sievert's law — i.e. wt% $X = K\sqrt{p_{X_2}}$ for a diatomic gas X_2 dissolving monatomically. As each bubble passes, wt% X will be reduced slightly but since it is a proportion which is removed each time, only an infinitely large number of bubbles can cleanse the metal completely of the gas X_2.

It will be clear that purity is a relative term and that the absolute elimination of any element is, even in theory, not possible. In practice, side reactions with containing refractories cause contamination. "Pure" iron will pick up oxygen from siliceous refractories, rather less from alumina, less again from zirconia — but never none.

Fortunately the achievement of extreme purity is seldom necessary. For the electronics industry, however, impurities in silicon and germanium are reduced to the order of 1 part in 10^{14} by the elegant technique of zone refining. The metal, in the form of a long rod, is melted, over a very short length only, by induction heating or by an electron beam, the molten section being retained in place by surface tension. The molten zone is then made to traverse slowly the length of the rod. As it does so segregation occurs, the metal freezing at one end of the pool always being purer than the liquid from which it is separating, just as in dendritic segregation. Repeated treatment of one rod, always in the same direction, gradually sweeps the impurities to one end, leaving the end at which each pass commences increasingly pure. This must be carried out under high vacuum. It would be a costly process on a large scale but probably represents the ultimate in refinement technique.

The principle underlying this most sophisticated of refining techniques is exactly the same as that underlying the crudest of all — liquation — in which a very fusible metal can be freed·of such impurities as raise its melting point by heating it on a sloping hearth to a temperature just above the melting point of the pure metal. Then only pure metal can melt and run down the hearth. Every slight increase in temperature allows a slightly less pure fraction to escape. Applications to refining are not numerous but include the refinement of tin and silver. There are a few cases where a similar technique is used in ore-dressing for removing fusible sulphide from gangue, e.g. antimony and bismuth.

Converter Processes

Converters are used for oxidizing impurities out of blast-furnace iron in steelmaking and for the oxidation of sulphur from copper and nickel mattes by blowing air or oxygen either through or over the surface of the molten charge. The reactions involved are strongly exothermic and the processes are autogenous. The matte or metal is charged in the liquid state but fluxes and any scrap are charged solid and heat must be available to melt them. The reaction vessels, called "converters", are of compact cylindrical form with a low surface : volume ratio which minimizes the heat loss through the shell. Reaction rates are high and process times short. This also conserves energy. The products are either metal or matte, usually at a higher temperature than the material charged, and a gas which must usually be cleaned and processed. Other processes in which air or oxygen are injected into molten metal, such as the softening of lead or the manufacture of steel from 100% scrap, do not produce enough heat to be self-sustaining and so must be carried out in furnaces in which the energy deficit can be made up by burning fuel. The development of lances and tuyères which can deliver fuel and oxygen in a wide range of ratios may be expected to extend the range of application of converters. The energy in matte smelting comes from the oxidation of sulphides (FeS, Cu_2S, etc.) and in steelmaking from the oxidation of carbon, silicon, phosphorus and manganese in the blast-furnace iron. The composition of the charge must be such that it will provide exactly the correct quantity of heat for the process.

Several types of converter are shown in Fig. 76. They can be rotated on trunnion mountings into several working positions for charging, blowing, sampling, pouring and fettling. Some special types may also be rotated axially. Facilities may be incorporated for separating the slag

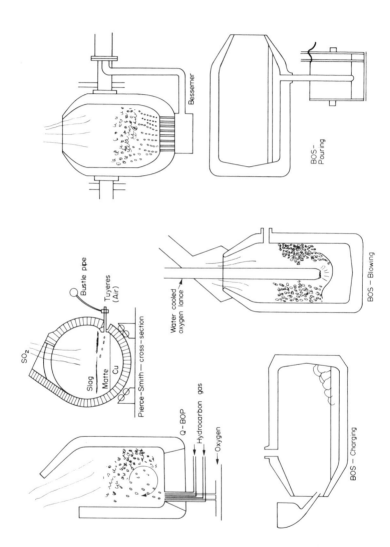

FIG. 76. Schematic diagrams of various converters.

from the matte or metal or for pouring off the metal and leaving the slag behind. Converters may be top blown, bottom blown or side blown. In top blowing the air or oxygen is delivered through a single water-cooled lance which may, however, have several jets. In side-blown converters there is a row of tuyères which may direct air on to the surface of the charge or may be submerged during blowing so that the air passes into the molten matte or metal. In bottom-blown converters the air comes through an assembly of tuyères in the base of the vessel and passes through the charge. There are usually a very large number of small tuyères but in one recent design there are only about four of them. The refractories around these tuyères suffer severe erosion and it must be possible to replace them easily at regular intervals. The linings require less frequent replacement — perhaps once per week for linings and once per shift for bottoms — but are subject to big differences among the various applications.

In steelmaking, the converter was invented by Bessemer in about 1850. His original idea was to blow air on to the top of the iron but the heat available was not sufficient to raise the temperature up to the melting temperature of the steel made. The Bessemer converter was bottom blown with air, the heat being derived from the oxidation of silicon, manganese and carbon, in that order, along with some iron which also entered the slag. Thirty years later the discovery was made that phosphorus could also be oxidized from the iron under a basic slag in an "after-blow", after the carbon had gone. This basic process was used from 1880 until 1960 for making steel from iron obtained from ores containing a high proportion of phosphorus. The process also produced a phosphatic slag of great value as a fertilizer but unfortunately the steel had a high nitrogen content and methods were sought to eliminate this embrittling element using modifications to vessel geometry, and by blowing mixtures of oxygen and steam and oxygen and carbon dioxide instead of air. These were all technically successful but the use of pure oxygen blown on to the surface of the metal was eventually seen to be economically preferable to the others. It gave high production rates and low refractories consumption and proved to be adaptable to a wide range of grades of iron. Now known variously as the L–D, BOS and BOF or basic oxygen steelmaking process it has been adopted everywhere as the standard method of converting iron into steel and has been developed so far that batches of up to 300 tons of steel can be produced at 40-minute intervals from one vessel. This requires that oxygen be supplied at a rate of the order of 1500 m³ per minute from a lance which reaches to within 1 m of the metal surface. The physical interaction between gas and metal becomes very important and the lance height and oxygen velocity must be adjusted so that droplets of metal are thrown up into the slag to form with it an emulsion with a large interfacial area which ensures a very fast reaction rate. The lance height must be a proper compromise such that there is a balance between the amount of metal blown off the surface and into the slag and the amount of oxygen in the slag available to decarburize it. If the lance is low the iron is transferred rapidly to the slag. If it is high the slag is strongly oxidized. The lance height is adjusted during the blow, initially low to oxidize iron in order to make slag and later at a higher position to maintain the oxygen in the slag high enough to ensure the rapid removal of phosphorus to a low level.

There have been several variations of this process, including low-pressure jetting of oxygen so that oxidation is almost all through the slag. This allowed phosphorus to be removed ahead of carbon but the reactions are slow and some simultaneous bottom blowing is needed to provide agitation. In another process, lime is blown with the oxygen until the slag is formed. The most recent development is a return to bottom blowing, but with jets shrouded with an annular stream of a reducing gas such as methane which delays the interaction of oxygen and iron at the nose of the tuyères sufficiently to protect the refractories around the tuyères from severe erosion. Only about four tuyères are used and these are so placed as to cause the maximum of turbulence in the bath. Remarkable claims have been made for this "Q–BOP" process.[80] Production rates are said to be even higher than in the BOS process. Phosphorus and sulphur can be removed to lower levels and there is less loss of iron and manganese. The erosion of refractories is slight and the tuyères can be used for argon flushing (see page 335) to reduce hydrogen and nitrogen in the melt. This process may replace the BOS process in time.

The conversion of matte has traditionally been carried out in a side-blown Pierce–Smith converter because it is essential that air should enter through the matte layer in the later stages of the operation, rather than pass through the metal which, because it does not react with the air, would be chilled by it. The use of oxygen is not usually necessary because the oxidation of sulphide is adequately exothermic but a top-blown process using oxygen has been employed for the treatment of nickel matte where the attainment of a temperature in excess of 1600°C is thermodynamically desirable.

The conversion of copper matte is carried out in two stages. First, iron sulphide is converted to the oxide and sulphur dioxide and iron oxide is slagged off. Successive portions of crude matte are added and treated in a similar manner until the vessel has a full charge of cuprous sulphide or "white metal". This is then blown to oxidize the sulphur to produce "blister copper" which contains only about 0.1% of sulphur but nearly 1% of oxygen. Converter slags contain some copper which must be recovered.

Gases leaving converters are laden with dust and fume which must be removed before they are released to the atmosphere. Gases from the BOS process carry fine ferric oxide fume which is probably produced by the oxidation of iron vapour and this is extremely difficult to collect. The sulphur dioxide from matte converters must obviously be recovered but the heat and carbon monoxide in the BOS gas are also valuable (see page 251). Converter processes are designed to be in thermal balance, with the product at the desired finishing temperature. They can be brought under sophisticated computer control on the basis of thermal balances as is discussed on page 272.

Distillation Metallurgy

In only a few cases are metals amenable to extraction or refinement by distillation at normal pressures. To be eligible for such treatment a metal must have a low normal boiling point (i.e. at 760 mm Hg). The vapour pressure p (mm Hg) of a metal is approximately related to its temperature by the equation—

$$\log p = -L_e/2.303RT + C \qquad (12.7)$$

where L_e is the latent heat of evaporation from the solid or liquid phase concerned and C is a constant involving the entropy of evaporation. This equation can readily be derived from equation (6.47). The normal

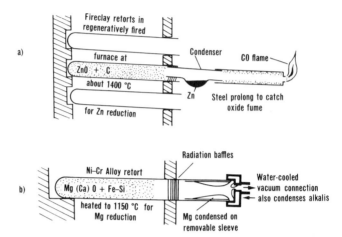

FIG. 77. Schematic representations of horizontal retorts for (a) the reduction of zinc oxide with carbon and (b) the reduction of magnesia in dolomite with ferrosilicon (Pidgeon process).

boiling point can be obtained by putting $p = 760$ and evaluating T ($^\circ$K). The metals with the lowest boiling points are Hg (357°C), Cd (765°C), Zn (905°C and Mg (1107°C), along with the alkalis Cs (690°C), K (774°C) and Na (892°C), and As (610°C) which so often turns up in extraction operations. Of these sodium and potassium are always reduced electrolytically from hydroxide or chloride fusions but the others can be extracted by heating the oxide with a suitable reducing agent and condensing the metal from the gases evolved. In the case of mercury it is usually the sulphide that is roasted and no reducing agent is needed.

Traditional retorts were steel or fireclay tubular vessels which were heated in batteries in retort furnaces. Each might lead to its own condenser or through a main to a common condenser. This horizontal retort provided batch operation only (Fig. 77). Continuous vertical retorts replaced horizontal retorts in many plants. These developed differently for different metals. To accommodate fine mercury ore, a shaft furnace has been developed down which ore slides in a zig-zag manner on inclined shelves, while elsewhere a rotary kiln was used to give mercury ores what amounts to a calcine or "volatilizing roast", the pro-

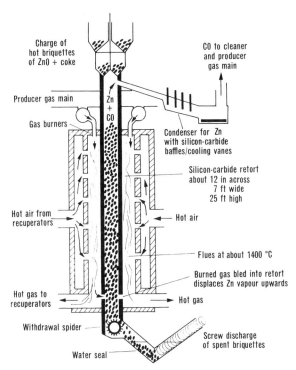

FIG. 78.   Schematic representation of a vertical retort for the
reduction of zinc calcine briquetted with coke (end view).

duct being the metal. Zinc oxide sinter is briquetted with coke to produce
suitable feed for zinc retorts which rather resemble gas retorts (Fig. 78).
Zinc vapour and carbon monoxide are drawn off at the top of a high narrow
chamber. Spent briquettes – gangue and coke ash – are withdrawn
mechanically at the bottom. Heat is supplied from gas burned in the flues
on either side. These retorts are built in batteries like coke ovens. More
recently, zinc and lead ores are being smelted in a blast furnace, the lead
being collected as bullion from the hearth, iron entering the slag, and zinc
coming off with the top gas.

Condensers also vary. Zinc from a horizontal retort is easily
condensed in an air-cooled chamber. The zinc vapour is protected by

carbon monoxide from re-oxidation and need not be cooled below 400°C at which temperature equation (12.7) applied to zinc shows that its vapour pressure is so low that metal losses in the gas are of negligible value. Mercury on the other hand must be cooled right down to near room temperature because its vapour pressure is appreciably high even at 100°C. More elaborate water cooling is needed' here and especially as the gases to be cooled are fuel gases, not merely retort gases.

Zinc vapour from the blast furnace is accompanied by a gas which is slightly oxidizing toward it and the more so as its temperature falls so that very rapid cooling is essential to prevent re-oxidation. In practice this is preceded by a temporary reheating of the gases by partial combustion of their CO content to throw the equilibrium safely on the reducing side. Shock cooling is effected in a lead spray condenser in which the zinc dissolves in liquid lead at 550°C so quickly that its reoxidation during cooling is negligible (see page 392). In one process, magnesium is produced by electro-roasting magnesia with coke. The gases emerging from this unit must also be shock cooled with a blast of cold reducing gases. In another process it is retorted in a batch process in a closed retort (Fig. 77b). The reducing agent is ferrosilicon, so that no gas other than magnesium vapour is formed and there is no danger of re-oxidation in the closed vessel. The metal is allowed to condense on a cooled receiving plate at one end of the retort situated outside the furnace. Despite their higher boiling points, in some cases the alkali and alkaline earth metals can be prepared in a similar way, using aluminium as the reducing agent.

Volatile metals can also be refined by distillation at normal and reduced pressures. Reduced pressure enables lower temperatures to be used and reduces the degree of contamination by oxygen either from the atmosphere or from the refractories. This is particularly useful for the more highly reactive metals but it is also used for refining mercury. Zinc can be separated from lead and cadmium by fractional distillation using a separate column for each impurity. The fractionating column (Fig. 79) is a tower containing a large number of refractory trays each of which has a hole or holes through which liquid metal can drain down to the next lower tray thus exposing a large area of metal surface from which the lower boiling point species can evaporate. The lower end of the column is built into a furnace: the upper end extends beyond it and

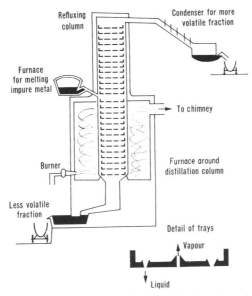

FIG. 79.   Schematic representation of a fractionating column for
separating, say, cadmium from zinc.

runs cool enough to condense the higher boiling-point metal before the
vapours pass over to the condenser for the more volatile metal.

Distillation occurs incidentally in many processes, sometimes causing
severe losses of metals or severe pollution of the atmosphere with oxide
fume. At steelmaking temperatures manganese is quite volatile and is
lost in the furnace gases.

Distillation or vapour-phase separations can also be made when a
compound rather than the metal itself is volatile. The volatilizing roast
has already been mentioned and it is of particular value for the
separation of antimony as the volatile trioxide after the oxidation of a
sulphide ore. This can be carried out in a simple shaft furnace, or in a
rotating kiln, the antimony oxide leaving with the flue gases, to be
collected in a bag house while the gangue is either slagged or discharged
as a clinker at the lower end of the kiln. In the next section the
separation of halides by vapour-phase transportation is discussed. In nickel

metallurgy the low-boiling nickel carbonyl is used in one important process for separating nickel from associated impurities (see page 380).

Halide Metallurgy

The use of halides in extractive processes has made possible the large scale production of some of the rarer metals usually associated with the atomic energy industry – uranium, beryllium, niobium, zirconium, titanium and vanadium. Reduction of the oxides of all these metals is difficult. All are refractory and can be reduced by carbon only at high temperatures but in most cases carbides are formed, the decomposition of which is not easy. Another difficulty is that several of these metals take a large amount of oxygen into solid solution and the removal of this oxygen to very low concentrations (activities) is very difficult and can indeed only be effected properly by reduction with calcium. This leads to the possibility of using liquid calcium and solid metallic oxide, which is likely to be slow, or gaseous calcium at a higher temperature with liquid oxide, but the solid calcium oxide which is the deoxidation product is often difficult to separate. The reduction of chlorides by magnesium with the formation of a liquid slag of magnesium chloride is generally a much easier operation.

Other applications of halogen-based processes are feasible but halogens are costly reagents and they are seldom regenerated cheaply in the process. Except for the electrolytic extraction of aluminium and magnesium, halide metallurgy has not yet been used for common metals.

Halides generally have low melting points. They are miscible in the liquid state and when cooled form very low melting-point eutectics. Most of them have high vapour pressures and quite low boiling points so that distillation and even fractional distillation are practicable. Chlorides are somewhat hygroscopic and for this reason the more expensive fluorides are sometimes preferred. When necessary, purification can be carried out by any standard process of inorganic chemistry prior to the reduction of the halide to the metal.

Metallic halides are easily produced by several methods, but the halogen alone is not usually capable of reducing oxides. These may be roasted in a stream of chlorine, in the presence of either carbon or hydrogen to effect reduction. Alternatively the oxide may be pre-reduced to metal or even to carbide which is then roasted in chlorine. If reduction conditions are such as have differentiated between different oxides in a concentrate, subsequent chlorination will convert only those which have been reduced. Some metals, like vanadium and niobium, are commonly smelted to ferro-alloys and these are sometimes used as a starting-off point for conversion to chlorides. Anhydrous hydrogen chloride is an alternative to chlorine in these treatments, probably dissociating to hydrogen and chlorine in the process. Conversions to fluoride may be by reactions with anhydrous hydrogen fluoride or through aqueous hydrofluoric acid.

The free energy of the formation of some chlorides is shown on Fig. 80. From this it appears that most chlorides are stable up to very high temperatures and cannot easily be decomposed thermally. Both $TaCl_5$ and $NbCl_5$ have been thermally decomposed on to hot wires, however, at about $2000°C$ and under reduced pressure, though this is not the normal method of their reduction. Fluorides are similar in this respect but some iodides, notably those of titanium and zirconium, are easily decomposed by heat under vacuum and this provides a very simple means of producing these metals in a very pure state. One of the least stable chlorides is CCl_4 indicating that carbon is not a reducing agent for chlorides. Hydrogen reduces a number of chlorides particularly if a high H_2/HCl ratio can be maintained at an elevated temperature. Most chlorides cannot be reduced by hydrogen so that one of the expensive and highly reactive metals must be used for their reduction — aluminium, magnesium or an alkali metal. In practice magnesium is very suitable for reduction by a thermit process in which metal halide and reducing agent are heated together sealed in a refractory lined steel "reaction bomb" until the reaction occurs such as, for example—

$$UF_4 + 2Mg = U + 2MgF_2 : \Delta H = -351 \text{ kJ.}$$

The heat of this reaction is adequate to fuse both products and permit their complete separation. The corresponding reduction of the oxide,

$$UO_2 + 2Mg = U + 2MgO : \Delta H = -138 \text{ kJ,}$$

produces less heat and could not form a molten slag even if the metal melted, unless a flux were provided. (A further advantage of halide reduction in this case is that the metal is produced free of oxygen) The

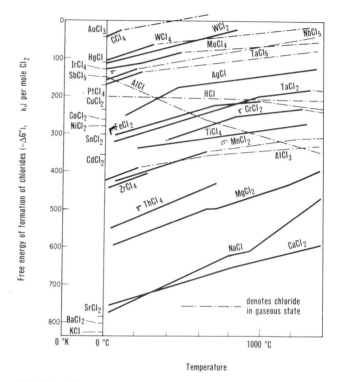

FIG. 80.   The standard free energies for formation of some
chlorides, the standard states being the pure condensed phases and
gases at one atmosphere pressure as appropriate. (Based on similar
diagram by Villa,[76] with additional data incorporated from various
sources including Kubaschewski and Alcock[30] and Smithells.[77])

halides of magnesium, calcium, sodium, lanthanum, etc., which appear at
the foot of Fig. 80 can be reduced only by electrolytic means from melts.

    There is a wide range of interesting processes possible when metallic
halides are used.[81]

    Ores may be concentrated by differential chlorination of impurities
which can then be volatilized off. For example, volatile iron and manganese
chlorides are formed when chlorine and hydrogen are passed over a
tungsten ore at $850°C$. If the hydrogen pressure is maintained high

enough, however, tungsten chloride will be reduced to the metal which will remain *in situ*. Similarly lead and zinc can be removed from ores containing vanadium, leaving the vanadium behind.

Differential oxidation of their chlorides can be used to separate those very similar metals zirconium and hafnium. At 800°C—

$$ZrCl_4 + O_2 = ZrO_2 + 2Cl_2 : \Delta G^{\circ}_{1073} = -143.1 \text{ kJ}$$

and

$$HfCl_4 + O_2 = HfO_2 + 2Cl_2 : \Delta G^{\circ}_{1073} = +204.4 \text{ kJ}.$$

If the mixed chloride vapours are held at 800°C in an oxygen/chlorine atmosphere in which $O_2/Cl_2 = 2$ the zirconium will oxidize and condense but the hafnium will remain in the gaseous phase.

When two or more valencies are possible some interesting separations can be effected. The chlorination of ferro-niobium, for example, produces a mixture of $SiCl_4$, $AlCl_3$, $TiCl_4$, $FeCl_3$, $TaCl_5$ and $NbCl_5$. If these are cooled to 200°C the $FeCl_3$, $TaCl_5$, and $NbCl_5$ condense. Iron can be removed from these by hydrogen reduction at 350°C when $FeCl_2$ remains solid but $TaCl_5$ and $NbCl_5$, being only slightly reduced, are volatilized. Further hydrogen reduction of these chlorides at 500°C reduces and condenses the niobium as $NbCl_3$ or at 600°C as the metal, leaving Ta in the vapour phase.

An application of changing valency to refining is found in the case of aluminium. $AlCl_3$ gas reacts with molten metal at 1200°C to form $AlCl$ which is also volatile. At a reduced temperature, however, about 600°C, the sub-chloride reverts, depositing pure aluminium and regenerating $AlCl_3$ for re-use. The purity of the refined aluminium depends, of course, on none of the impurities behaving in a similar manner.

These few examples serve to indicate the flexibility and variety to be found in halide processes. Operating temperatures are often quite low and established chemical engineering methods are often applicable as, for example, the use of fractionation columns. The major operating difficulty is probably the control of corrosion. Where elementary halogens are needed costs are high and the extension of halide metallurgy to low-priced metals is likely to be restricted to rather special conditions. The choice of halogen for a particular application must always depend on a number of considerations, some technical and some economic in nature. It is quite futile to generalize or try to lay down a set of rules by which the choice ought to be made. This would only lead to a justification of all the decisions made in recent years without necessarily providing clear guidance as to how the next decision will be made. Chlorine is the cheapest halogen and the least unpleasant to use. The major disadvantage attending its use is that chlorides are often hygroscopic but much of the work is carried out in closed systems so that is not always important. Chlorides are less stable and more volatile than fluorides while iodides are less stable and more volatile than chlorides. Bromine has properties intermediate between chlorine and iodine but it is the most expensive and least pleasant to use and finds few applications. In practice chlorides are used where there is no good reason for using another halogen. Uranium and sometimes beryllium are extracted using fluoride routes. In the former case the reason is associated with the thermal requirements of the thermit reduction process. Iodides are used when titanium and zirconium are refined using the van Arkel process where high volatility makes vapour-phase transportation very easy and low stability in each case means that thermal decomposition is possible.

Vacuum Metallurgy [82]

The first and most obvious application of vacuum to extraction metallurgy is to the removal of dissolved gases like hydrogen and nitrogen from refined products like steel. This, it might appear, could easily be carried out by subjecting a ladle of the metal to low pressure within a vacuum chamber for a short time prior to casting into moulds. Unfortunately the removal of gases under such conditions involves the transport of the gas, as atoms in solution, to the free surface and, particularly, across a diffusion layer at the surface. This makes the process so slow that more elaborate techniques have to be used. The simplest of these is to teem the metal through a nozzle in the bottom of the ladle through an evacuated space into another ladle or into a mould. As the metal enters the vacuum, dissolved gases are released with some sputtering which extends the exposed surface and hastens the further release of gas. The time of exposure to reduced pressure is, however, rather short so that this simple technique is of limited value. The removal of gases can be improved by entraining an inert gas such as argon into the stream of metal as it passes through the nozzle. This causes a more vigorous sputtering, the argon carries away some of the dissolved gases (as in flushing) and the area exposed is greatly increased so that in the time available as each drop of metal falls into the ladle below, a higher proportion of the gases present can escape. Two-stage variations of stream degassing have also been developed in which an evacuated tun-dish is interposed between the ladle and the mould which is situated in a second evacuated chamber[89] (see Fig. 81). Degassing by purging with argon under vacuum is more commonly practised, the ladle being placed in an evacuable chamber and the inert gas being injected through a porous plug in the ladle bottom. The argon stirs the metal very effectively and removes dissolved gases in the manner discussed on page 320. The main disadvantage of this process when it is applied to steel is that the temperature falls rather rapidly, making it difficult to reach the teeming bay with the steel still sufficiently hot. The use of arc or induction heating is possible but expensive. To start with steel at a higher temperature would increase the quantity of gas to be removed as well as increasing the time available for its removal and the small advantage would be costly in terms of energy consumed. A third method which is currently in favour is the vacuum lift technique (see Fig. 81). An evacuable vessel has a long nozzle extending downwards from its base. This nozzle is dipped into molten metal in an open ladle and a portion of the metal is drawn up into the evacuated space. By raising and lowering the vessel, the steel is made to flow into and out of it as often as is needed to achieve an adequate degree of degassing. This is under the control of the operator and may require rather a long time. Arc heating can be used in the upper vessel to maintain the temperature of the metal. Since the evacuated space is quite small and only a small part of the metal is being treated at any time the capacity of the pumps need not be as great as in some other processes. In each case the object is to extend the surface area as far as possible and to introduce as much turbulence as possible to assist the transport of the atoms of the gaseous element to the surface. For many purposes these relatively simple arrangements are sufficient to bring about considerable improvement in metal quality.

More complete degassing can be effected by melting under vacuum. This is most simply done by including the whole of an induction melting furnace in the vacuum chamber. This can be followed by casting

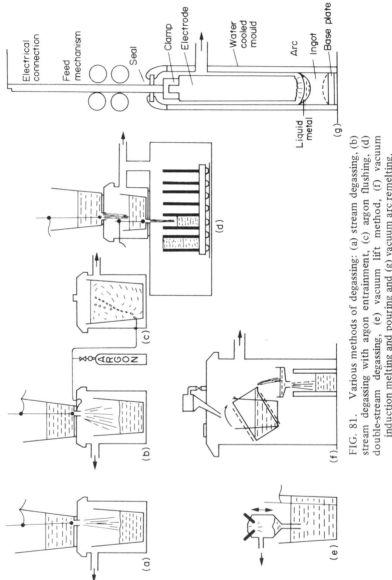

FIG. 81. Various methods of degassing: (a) stream degassing, (b) stream degassing with argon entrainment, (c) argon flushing, (d) double-stream degassing, (e) vacuum lift method, (f) vacuum induction melting and pouring and (g) vacuum arc remelting.

within the same evacuated chamber so that oxidation of the metal during pouring is prevented. Induction melted charges are well stirred by currents within the melt so that the rate of loss of volatiles is determined mainly by diffusion from the stirred bulk of the metal across a boundary layer to the surface at which the maximum rate of evaporation is that given by the Langmuir relationship which is discussed below. In a poor vacuum the net evaporation rate will be lower than the Langmuir value because a proportion of the escaping atoms return to the surface after collisions or if the vacuum is very poor, because of the formation of films of compounds. like metal oxides on the surface. Where the vacuum is good and the surface clean so that diffusion controls the rate of loss of volatiles, purging with argon can be used to accelerate the removal of gases. In principle vacuum induction melting should give the best control over gas removal and temperature and the control can be held for a long time to reach any desired equilibrium. Much has been written of this most elegant process for producing steel of superlative quality but only a very small tonnage is made annually by this means, most of it for special applications within the aerospace industries.

Vacuum arc remelting involves the re-melting of either ingots or compacts made from powder, pellets or swarf, under vacuum, using an electric current to strike an arc between the metal in its original form as cathode and a pool of purified metal over a water-cooled anode at earth potential. The main purpose is sometimes only to re-melt and consolidate reactive metals like titanium or zirconium without the possibility of their being oxidized but a major application is to the improvement of certain qualities of steel, particularly by removing hydrogen and redistributing inclusions. Some reduction of oxygen may also be effected. An ingot cast for the purpose is cleaned by machining off the surface. It is placed in a water-cooled mould just a little larger than the ingot itself, suspended from a heavy vertical rod which passes through a vacuum seal and carries the electrical current. An arc is struck between the bottom of the ingot and a starter plate on the bottom of the mould. Metal passes across the arc gap as tiny droplets, as in the arc welding process, filling the bottom of the mould. As the ingot is consumed it is lowered to maintain a constant arc gap until its complete transfer to the new secondary ingot has been effected. The exposure time for each droplet is short but the specific surface is large and the removal of hydrogen is efficient. The process has been applied to the manufacture of large mill rolls in which normal ingot structure is rather coarse, the normal segregation pattern is inappropriate, and the risk of hairline cracking due to hydrogen in alloy steel qualities is unacceptable. Whereas steel ingots up to 10 tons can be treated in this way other metals are usually dealt with on a smaller scale.

Electron beam melting involves bombarding a pool of metal in a water-cooled crucible with a beam of high-energy electrons. Impure metal is fed into the beam where it melts and the liquid is drawn in a finely dispersed state by the electric field into the pool. As the crucible is retracted an ingot can be built up. In either of these last two cases the starting metal may be a power compact or similar physically unsatisfactory form requiring consolidation. These methods were developed for the rarer and more refractory metals but the vacuum arc process is used to a limited but increasing extent for the production of high-quality steel.

In general it is hydrogen which is removed by the processes described. Other elements normally gaseous at the temperature of the molten metal are not necessarily drawn off into the vacuum. Nitrogen in steel is not normally reduced below the low concentration usually found and may even be increased slightly when argon purging is employed because argon usually contains some nitrogen as an impurity. The passage of the nitrogen into the steel rather than out of it is, however, due to the free energy of solution of nitrogen in iron being slightly negative under the conditions encountered. The standard free energy of solution is given by

$$\tfrac{1}{2}N_2(1 \text{ atm}) = \underline{N}(1\% \text{ in Fe}); \Delta G_T^{\circ} = 3600 + 23.9T \text{ J g-atom}^{-1}. \quad (12.8)$$

From this it can readily be shown that at $1600^{\circ}C$, if the concentration of nitrogen is 0.002%, the free energy of solution becomes zero when the nitrogen pressure is 0.002 atm or 1.5 torr so that it should be possible to remove nitrogen below 0.002% by reducing the pressure below that value. While this is not a low pressure, it cannot be maintained easily during vigorous degassing or during purging and it should be appreciated that it is the pressure exerted by a head of only about 2.3 mm of steel. A further important factor is that if any aluminium or titanium has been added to the steel as deoxidizers, these will combine with nitrogen to form non-metallic inclusions rather than allow it to escape. An element in solution in the steel will evaporate only if the phase change lowers its chemical potential. The case of oxygen is more obvious. The dissociation pressure of FeO at $1600°C$ can be read off Fig. 45 as about 10^{-8} atm, which is lower than can be applied on an industrial scale, so that the removal of oxygen even from its saturated solution in iron by vacuum alone is not possible and to draw oxygen off a more dilute solution would require an even lower pressure to be applied.

Oxygen can be removed from steel in the presence of sufficient carbon because the oxides of carbon are volatile and have more strongly negative free energies of formation than FeO (or the solution of oxygen in iron). The appropriate reaction is

$$\underline{C}(1\% \text{ in Fe}) + \underline{O}(1\% \text{ in Fe}) = CO_{(g)}$$

for which

$$K_C = \frac{p_{CO}}{\%a_C \times \%a_O} = 500 \text{ at } 1600°C.$$

If $p_{CO} = 1$ atm, the product $\%a_C \times \%a_O$ is equal to $\%\underline{C} \times \%\underline{O}$ which has the value 0.0020. This is the solubility product for carbon and oxygen in iron at $1600°C$. If $\%C = 0.1$, the equilibrium concentration of oxygen is obviously 0.020%. If, however, the effective value of p_{CO} is reduced to 10^{-2} the solubility produce becomes 2.0×10^{-5}. If $\%C = 0.1$ the equilibrium concentration of oxygen is only $2.0 \times 10^{-4}\%$. In an induction melting furnace a pressure as low as 10^{-5} atm (10^{-2} torr) might be used, in which case the corresponding equilibrium concentration of oxygen would be $2.0 \times 10^{-8}\%$. The attainment of such a low concentration of oxygen is not possible in practice partly for kinetic reasons and partly because of side reactions involving refractories.

Because the nucleation of gas bubbles is kinetically difficult (see page 206) the rate of formation of carbon monoxide is restricted by the area of liquid/gas interface available at which it can be precipitated. If there is only the top surface of the metal in the crucible or ladle, the reaction is likely to proceed only very slowly. When carbon in steel exceeds about 0.2% or when the oxygen is very high the formation of carbon monoxide may occur as a boil (see page 206), the bubbles being generated at crevices in the refractory lining. As these bubbles rise through the melt they grow rapidly and may collect other gases as they pass through the metal. When the carbon and oxygen are too low to generate a boil, argon may be bubbled through the melt to improve stirring and create more surface at which carbon monoxide can be deposited.

At the low oxygen potentials which can be reached under vacuum even refractory oxides like silica and alumina become thermodynamically unstable and begin to dissociate. In the case of silica the reaction is

$$SiO_2 = \underline{Si} + 2\underline{O}$$

where both the silicon and the oxygen dissolve in the metal. If the metal is iron at $1600°C$ the equilibrium constant is

$$\underline{K}_{Si(1873)} = \%\underline{Si} \times (\%\underline{O})^2 = 2.7 \times 10^{-5} \qquad (12.6)$$

so that if the oxygen in the iron were reduced to $10^{-4}\%$ the calculated equilibrium concentration of silicon in the iron would be 2700%. Obviously the implicit assumption that Henry's law remains valid is not justified but it is clear that significant contamination of the metal by silicon can be expected if deoxidation to such a low level is attempted in a silica-lined vessel. The activation energy of silica reduction is very high, however (about 400 kJ mol^{-1} SiO$_2$), so that there is no possibility of the refractories collapsing. In the case of alumina the appropriate reaction is

$$Al_2O_3 = 2\underline{Al} + 3\underline{O}$$

where the products again dissolve in the metal. If the solvent is iron at $1600°C$ the equilibrium constant will be

$$\underline{K}_{Al(1873)} = (\%\underline{Al})^2 \times (\%\underline{O})^3 = 1.8 \times 10^{-14}. \qquad (12.7)$$

In this case if the oxygen concentration in the iron were to be reduced to $10^{-4}\%$ the equilibrium concentration of aluminium in the iron would be 0.13%. If the oxygen were further reduced to 10^{-5} and $10^{-6}\%$ the corresponding (calculated) equilibrium concentrations of aluminium in the iron would be 4% and 134%, respectively. Once more significant contamination of the steel appears to be possible particularly at very low oxygen concentrations. The only other oxide commonly used for lining furnaces is magnesia. Magnesium is virtually insoluble in molten iron and is a gas at $1600°C$ so that its dissociation can best be examined in the reaction

$$MgO = Mg(g) + \tfrac{1}{2}O_2 .$$

In this case

$$\underline{K}_{Mg(1873)} = p_{Mg} \times p_{O_2}^{\frac{1}{2}} = 2 \times 10^{-10} . \qquad (12.8)$$

If the total pressure over pure magnesia at $1600°C$, due entirely to gaseous oxygen and magnesium, is 10^{-6} atm it can be shown that these will share the space in the ratio 3 : 1 so that $p_{O_2} = 7.75 \times 10^{-7}$ atm and $p_{Mg} = 2.25 \times 10^{-7}$ atm. In atomic terms there is 6 times as much oxygen as magnesium in the atmosphere. At the slightly lower pressure of 6.4×10^{-7} atm the ratio of O : Mg atoms in the gas would be 1 : 1 as in the oxide whereas at still lower pressures there would be a preponderance of magnesium in the atmosphere under equilibrium conditions. It would appear that at the pressures found in vacuum-induction melting furnaces the loss of magnesium would be restricted but in practice the atmosphere contains gases from other sources. Within a total pressure of 10^{-6} atm the p_{O_2} value might be only 10^{-8} atm in which case the equilibrium p_{Mg} value would be much higher at 2×10^{-6} atm. Magnesia would then decompose quite rapidly but as it did so the partial pressure of the oxygen would be increased so that the dissociation reaction would be to some extent self-stifling. The important point is that the magnesium would not dissolve in the iron (though it could do so in other metals). The magnesia refractories would not be destroyed by decomposition because the reactions are much too slow but the accumulation of magnesium and its oxide in the pumping equipment could be troublesome. The choice of refractory for any particular application must be such that the quality of the product is satisfactory without the costs being higher than necessary. In vacuum arc and electron beam melting contamination of the melt by contact with refractories is negligible because the duration of such contact is extremely short.

In systems other than steel, elements other than carbon may be used for deoxidation. These must form oxides, usually sub-oxides like SiO or

ZrO, which are more volatile than the elements themselves – and more volatile than the metal being deoxidized. This sort of treatment can be applied to the deoxidation of refractory metals like molybdenum using zirconium or tantalum as deoxidizers, but a small excess of the reagent would always be left in the metal.

Apart from elements which are normally gaseous, other impurities can be removed from a melt at high temperature and reduced pressure if their partial pressures over the melt exceed by a reasonable margin that of the metal being purified. Elements of inherently high volatility do not necessarily exhibit a high partial pressure under such conditions. Oxygen in iron is one example already discussed which can be removed only as a volatile oxide of lower free energy than iron oxide. Similarly phosphorus, though normally volatile, cannot be removed from iron because of its high affinity for that metal. Its negative free energy of solution in iron is high (see equation (7.23)) and it forms a compound with iron at lower temperatures. Other similar cases, though none involve an element so inherently volatile, are those of sulphur, silicon, aluminium, arsenic and antimony which all form compounds with iron at low temperatures and are either difficult or impossible to remove by vacuum unless some suitable volatile compound can be found. On the other hand, zinc, magnesium, bismuth, tin, copper and manganese are rapidly drawn off even under modest vacuum (*ca.* 1 torr) and in the case of tin and copper this may become very important as they are particularly difficult "residuals" in steel (see page 344). Nickel and cobalt cannot be removed because they have low volatilities. Other metals of low volatility present no problem as they can usually be taken out by oxidation.

In this context the terms high and low volatility must be used with reference to the volatility of the metal being purified. In principle, the individual components of a solution evaporate at rates corresponding to their several vapour pressures over the solution. Langmuir[84] derived the expression

$$w_i = p_i \left(\frac{M_i}{2\pi R T} \right)^{\frac{1}{2}} \tag{12.9}$$

from the kinetic theory of gases for w_i, the mass of gaseous molecules of species i striking unit area of surface in unit time, where p_i is its partial pressure, M_i its molecular weight, R the gas constant and T the absolute temperature. It is argued that if the solid or liquid surface is in thermodynamic equilibrium with the gaseous phase the same mass of matter must leave the surface as impinges upon it and that the equation also shows the maximum mass of i which can evaporate from the surface per unit area in unit time at the temperature and pressure under consideration.

If, in a binary solution, the mole fraction of the solute i is x_i that of the solvent j will be $(1 - x_i)$. If x_i is small enough, Henry's law applies to the solute and Raoult's law to the solvent, and the respective partial pressures over the solution will be $\gamma_i^\circ x_i p_i^\circ$ and $(1 - x_i)p_j^\circ$ where γ_i° is the Henry's law constant and p_i^υ is the vapour pressure of the pure species i. If the solution is exposed to a high vacuum the evaporation rates of i and j in g per cm^2 per second are given by

$$w_i = 44.33 \; \gamma_i^\circ x_i p_i^\circ \left(\frac{M_i}{T}\right)^{\frac{1}{2}} \qquad (12.9a)$$

and

$$w_j = 44.33 \; (1 - x_i)p_j^\circ \left(\frac{M_j}{T}\right)^{\frac{1}{2}}, \qquad (12.9b)$$

respectively, where the vapour pressures are in atmospheres and the coefficient 44.33 incorporates $(2\pi R)^{-\frac{1}{2}}$, where R is in ergs K^{-1} mol^{-1}, with the conversion factor for pressure in dynes cm^{-2} to standard atmospheres. If the concentration x_i is to become smaller

$$w_i/x_i > w_j/(1 - x_i).$$

i.e.

$$w_i/w_j > x_i/(1 - x_i).$$

If $x_i = 0.01$, for example, and $(1 - x_i) = 0.99$ then w_i/w_j must exceed $1/99$ if x_i is to decrease by differential evaporation but that ratio must be greatly exceeded if rapid economic removal of i is to be effected. If the ratio $w_i/w_j = 1$, for every gramme of i removed there is a corresponding loss of 1 g of j. If the total quantity of i present is very small an equal or even greater loss of the solvent being refined can possibly be tolerated.

Consider the evaporation rates relative to that of iron of copper, nickel, manganese and aluminium in binary solutions in molten iron at $1600°C$, calculating in each case the value of x_i at which the ratio w_i/w_j falls to 1.0.

If

$$\frac{w_i}{w_{Fe}} = \frac{\gamma_i^{\circ} x_i p_i^{\circ}(M_i/T)^{\frac{1}{2}}}{(1 - x_i) p_{Fe}^{\circ}(M_{Fe}/T)^{\frac{1}{2}}} = 1.0, \qquad (12.10)$$

since $M_{Fe} = 55.85$ and $p_{Fe}^{\circ} = 7.9 \times 10^{-5}$ atm (0.06 torr) at $1600°C$,

$$x_i = \frac{5.9 \times 10^{-4}}{\gamma_i^{\circ} p_i^{\circ} \sqrt{M_i} + 5.9 \times 10^{-4}} \qquad (12.11)$$

If $i = Cu$, $M_i = 63.5$, $\gamma_i^{\circ} = 8.5$ and $p_i^{\circ} = 8.32 \times 10^{-4}$ atm and $x_{Cu} = 0.0103$.

If $i = Ni$, $M_i = 58.71$, $\gamma_i^{\circ} = 0.66$ and $p_i^{\circ} = 4.7 \times 10^{-5}$ atm and $x_{Ni} = 0.71$.

If $i = Mn$, $M_i = 54.93$, $\gamma_i^{\circ} = 1.0$ and $p_i^{\circ} = 0.053$ atm and $x_{Mn} = 0.0015$.

If $i = Al$, $M_i = 26.98$, $\gamma_i^{\circ} = 0.063$ and $p_i^{\circ} = 2.23 \times 10^{-3}$ atm and $x_{Al} = 0.44$.

The range of calculated values of x_i is very wide. These are mole fractions of the solutes, below which it is not possible to reduce their concentrations without beginning to lose the solvent at the same rate as the solute is being removed. Clearly the removal of manganese and copper down to very low concentrations without incurring appreciable losses of iron appears to be quite practicable while the removal of either aluminium or nickel appears to be impossible. The choice of unity for the ratio w_i/w_{Fe} in these calculations is, of course, quite arbitrary and in industrial practice it would be made on economic grounds. A reduction of the value from 1.0 to 0.1 would, however, lead to a calculated critical value of x_{Ni} of about 0.2 which is still not low enough to make vacuum refining a practical proposition in that case.

    It will be noticed that it is necessary that the value of p_i° must not be low if an element is to be removed by vacuum but it is even more important that γ_i° shall not be low. Any tendency toward compound formation with the solvent shows as negative deviation from Raoult's law, as a low value of γ° and in low effective volatility. Multi-component systems can be considered in a similar manner but with the interaction coefficients included along with γ° values in the equation (12.11). Interaction between manganese and carbon in steel would be expected to slow up the evaporation rate of manganese because of its tendency to form carbides. Evaporation rates are very dependent on the hardness of the vacuum used. In the above calculations it is implicitly assumed that none of the atoms which escape from the surface will return to it. In practice, a proportion of these atoms do collide with molecules in the low-pressure atmosphere at a short distance from the surface and some of these will return to the melt. Working on a large scale it is not easy to maintain a very high vacuum, so some inefficiency compared with the theoretical case is inevitable. Another important consideration is that when vacuum treatment is being used for any purpose, all elements present are affected, more or less, and some losses may have to be made good after the treatment has been completed.

While the possibilities are very important both in ferrous and in non-ferrous applications of vacuum techniques it should be realized that, apart from the degassing of steel and the use of vacuum to protect reactive metals during re-melting operations, the tonnage of metals produced or refined under vacuum is still quite small. Vacuum-induction steelmaking is practised on only a very small scale. The removal of the residuals copper and tin from steel using vacuum is feasible but not yet necessary. The dezincing of lead after desilvering can be carried out using vacuum distillation but the established method using chlorine gas continues to be preferred.

There are four principal types of vacuum pump used on the industrial scale: mechanical, steam ejector, vapour booster and diffusion.

Mechanical pumps include the standard oil-sealed rotary pump which is really efficient only down to a pressure of about 1 torr. The mechanical booster or Roots pump is made to fine tolerances and runs dry. This can be used down to 10^{-3} torr but must be backed by a rotary pump.

Steam ejector pumps have no moving parts and are cheap to run in plants which can generate steam from waste heat but they require a lot of head room. They can be used down to about 10^{-2} torr, which is adequate for degassing operations, but for that pressure an elaborate five- or six-stage system backed with a cold-water spray condenser is needed.

The vapour booster pump combines features of the steam ejector and the classical diffusion pump. It requires to be backed by a rotary pump, and can bring pressures down to about 10^{-3} torr. It has the unusual feature that it works more efficiently on hydrogen than on air. This is probably the most popular type for use on induction and arc melting furnaces.

Diffusion pumps using oil of low vapour pressure as the working fluid are most efficient in operation at pressures below 10^{-3} torr and must be backed to about 10^{-1} torr. They can bring pressures down to 10^{-6} torr if necessary but have a poor capacity actually to move gas. Their use should be restricted to situations where there is only a small amount of gas being evolved and the main purpose of the vacuum is protective.

The choice of pumping system must be determined first by the pressure at which it is necessary to work and secondly by the capacity required to move gas. The third technical consideration is contamination — mainly contamination of the pump and working fluid by the gas being drawn through it and by any solid which may accompany this gas. Possible contamination of the melt by the working fluid of the pump may occasionally have to be taken into account. Other factors of importance will mainly be economic.

Vacuum metallurgy is in its infancy and may yet become a very important branch of the subject as the advantages to be gained from levels of purity hitherto thought to be quite impossible of achievement come to be investigated.

Electro-slag Refining[85]

Electro-slag refining has been developing in the steel industry since about 1965 as a method of improving the composition and the structure of steel ingots at the same time. The additional cost of such a process must be justified, however, and it is not usually incurred unless steel of an exceptionally high quality is required. Electro-slag refining (or ESR) serves a similar purpose to the vacuum arc re-melting already discussed. It is better able to remove rather than re-distribute non-metallic inclusions but less suited to the removal of hydrogen and other gases. It is an adaptable process, however, and could in principle be combined with vacuum re-melting. It could be used with metals other than steel and might be applied with electrodes made of compacted swarf, powder or pellets. It has even been suggested that it could be applied to the refining of molten steel.

The metal to be refined should already be of high quality and will usually be in the form of a specially cast cylindrical ingot, longer but narrower than the size required after refining. As is shown in Fig. 82, it is lowered into a mould, molten slag is poured in and current is passed

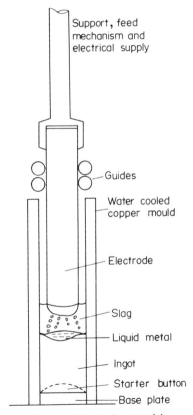

FIG. 82. Electro-slag re-melting.

to a starting button on the water-cooled base plate at earth potential. As in electric arc welding, metal is transferred from the consumable electrode of unrefined metal to build up on the hearth. The tip of the electrode is surrounded by the slag which carries the electric current and acts as a resistance heater, as well as a sink into which any oxide, silicate, or sulphide inclusions can dissolve. As the tiny droplets of metal pass through this slag the conditions for the transfer of inclusions to the slag are very favourable. As the process is continued a new ingot builds up in the mould and as the electrode becomes smaller the gap between them must be kept constant. The new ingot may be allowed to fill a long water-cooled mould in a conventional manner or the base of the mould may be retracted slowly, much as in continuous casting, so that the metal is deposited and solidified continuously within a short mould. A large ingot may be built up using several consumable electrodes.

The electrolyte or slag consists of between 70 and 80% of CaF_2 along with CaO and Al_2O_3. The fluoride confers a high electrical conductivity on the slag while the lime and alumina lower the melting temperature from $1418°C$ to about $1230°C$, but increase the electrical resistance and modify the chemical properties and viscosity. The atmosphere over the slag may be controlled, particularly to eliminate any reaction with moisture, in the air, which would introduce hydrogen into the steel, or with oxygen which brings iron oxides into the slag. The current may be a.c., d.c. or a combination of these. Direct current of the correct polarity can assist the transfer of sulphur and oxygen into the slag by providing the electrons for

$$S° + 2e \rightarrow S^{2-}$$

and

$$O° + 2e \rightarrow O^{2-}.$$

Power is supplied at 30–50 V, the current being typically several thousands of amperes depending upon the cross-section of the mould. It is controlled jointly by the applied voltage and the resistivity of the slag and should be such that the heat produced (I^2R) can be removed rapidly by the water-cooling system, so that only a shallow pool of molten metal is present at any moment. In this way piping of the ingot can be avoided. Thus the available cooling rate determines the production rate and that rate must fall as the diameter of the mould is increased.

The equilibria approached in ESR are determined by the composition of the slag which is generally very basic. Consider the silicon–oxygen reaction—

$$\underline{Si} + 2\underline{O} = SiO_2 \, ; \, \Delta G_T^{\circ} = -594,400 + 230T \text{ J mol}^{-1} \qquad (12.12)$$

from which

$$K_{Si} = \frac{_ra_{SiO_2}}{\%a_{\underline{Si}} \times \%a_{\underline{O}}^2}$$

At $1600°C$, $K_{Si(1873)} = 3.6 \times 10^4$, so that, if the product is pure silica $(_ra_{SiO_2} = 1)$, the $\%\underline{O}$ in equilibrium with 0.3% Si is 0.0096. If the silica is taken into solution in a very basic ESR slag in which $_ra_{SiO_2}$ is very low, typically about 10^{-3}, the $\%\underline{O}$ in the equilibrium with 0.3% Si at $1600°C$ would be only 0.0003. In the ESR process, however, the temperature would be higher than $1600°C$. At $1800°C$ the value of $K_{Si(2073)} = 899$, from which it can be shown that $0.0019\% \underline{O}$ will be in equilibrium with $0.3\% \underline{Si}$ at that more realistic temperature under an ESR slag. This is considerably lower than can be reached in steel which is conventionally deoxidized with 0.3% of silicon but similar improvements can be shown in steel deoxidized by other methods. It is unlikely that such low concentrations of dissolved oxygen are actually achieved but some reduction in oxygen is probably effected while the ESR process is performing its principal function which is the removal of the larger inclusions of oxides and sulphides, which remain after conventional casting, by dissolving them in the slag.

It is claimed that ingot quality produced by this process is good and surface finish excellent. Such inclusions as remain are extremely small and well dispersed. The process is expensive to operate but is already applied to the refining of steel for such special purposes as the manufacture of ball bearings and for rolling mill rolls.

Ingot Production

After the final refining operation, metal must usually be cast into some convenient form. This may be an ingot for sale for re-melting in a foundry, in which case it need only be made to a standard shape and size. Alternatively it may be cast for subsequent forging, rolling or extrusion into semi-finished goods — billets, plates, or tubes, for example. In this case its size and shape will have been selected to suit the necessary working programme.

Ingots for stock used to be cast into open pig beds in sand but moulding machines are now used in which metal moulds are presented in turn to receive the flow from a tun dish. The filled moulds are cooled in air or with water spray as they move round a conveyor, until the metal is solid. The ingot is then ejected and the mould cooled down, dressed and returned to the casting position. In general it is necessary only to skim any surface oxide off the metal and to give the mould a dressing of lime wash or graphite, to ensure that the ingots are satisfactory.

When the ingot is to be worked into billets or slabs, its shape will be determined by the form of the billet to be made from it — square section for rolling to billets, rectangular for slabbing or octagonal with fluted faces for making large forgings. Round ingots are not favoured because they are prone to cracking but they are made for vacuum re-melting, and for extrusion into round bar or tube and continuously cast metal (see later) is also often cast to circular section. Ingots for working are usually cast into metal moulds. These may be very heavy cast-iron moulds which rely upon their high heat capacity to absorb the heat from the cast metal or water-cooled copper moulds. The former are used mainly for steel ingots.

When the metal is nominally pure there is no problem about segregation during freezing but when an alloy is being produced, segregation can become very important. In either case there could be a very coarse crystalline structure in the cast ingot which would not be compatible with good mechanical properties. In an ingot produced for re-melting neither segregation nor coarseness of grain is of any consequence but when an ingot is being made for forging or rolling excessive segregation and coarseness should be avoided by the exercise of tight control over pouring temperature and cooling rate.

Variations in chemical composition within the primary crystals or dendrites is called micro-segregation. This is found in all cast alloys but is more noticeable in some than in others. It can usually be eliminated by annealing for several hours at as high a temperature as is practicable for the particular alloy. This must be below the liquidus of the alloy. Steel ingots are "soaked" at temperatures up to about 1300°C (depending upon composition and quality). If the temperature is much higher the liquidus temperature may be exceeded as a complex oxy-sulphide eutectic melts in some of the original dendrite boundaries, so damaging the steel beyond repair.

Variations in chemical composition between one part of the ingot and another is called macro-segregation and this includes variations in the amounts of non-metallic inclusions. Macro-segregation cannot be cured by heat treatment and must therefore be prevented as far as is possible. It is at its worst in steel ingots. At the high temperature of molten steel the solubilities of both oxygen and sulphur are much higher than in the solid steel so that as the metal cools from its casting temperature at about 1600°C precipitation of oxides and sulphides is inevitable. As the metal freezes both of these elements concentrate in the liquid phase from which precipitation continues until solidification is complete. A second important factor is that steel ingots are so large that convection currents can develop in the molten interior during the later stages of solidification. These currents redistribute the inclusions, generally transferring them upwards, as relatively pure metal crystals sink toward the bottom. At the same time inclusions which are still liquid may coalesce and rise quickly toward the top. The final distribution of inclusions varies considerably. It depends on the ingot size and shape and on the composition of the alloy. The general trend is, however, that the upper axial zone contains most of the inclusions and an above average proportion of most alloying elements while the lower axial zone is clean and is likely to show negative segregation of alloying elements. The top of an ingot is also likely to have shrinkage cavities or "pipe" because of the difference in volume between the liquid and solid metal. These are likely to contain slag and it is usually necessary to discard this part of the ingot which may amount to 20% of the whole. If directional cooling can be imposed on the metal so that it freezes from the bottom end upwards the slag and the shrinkage cavity can be concentrated into a smaller volume at the top. This can be effected through mould design, by using insulation over the metal in the mould or by actually heating the metal at the top of the mould either by using "exothermic compounds" or by applying a flame or an electric arc. Commonly moulds are made wider toward the top but are surmounted with a narrower extension within which most of the defective material should be found. This extension is often useful for handling the ingot in the forge.

The above description applies to fully deoxidized or "killed" steel. A high proportion of mild steel with C $<$ 0.2% is made in "rimming" quality. The oxygen is eliminated from the molten steel by allowing it to combine with some of the carbon during solidification in the mould. The evolution of CO gives the steel another name — "effervescing" steel. The gas sweeps inclusions to the top of the mould. This reaction is stopped when half the metal has solidified by "capping" the mould with a cold steel slab. The resultant macrostructure is a rim or envelope of almost pure iron columnar crystals with clean blowholes lying parallel between them and a core containing most of the metalloids and impurities. Most of the larger inclusions are at the top along with large, clean, gas-filled cavities. The blowholes are closed up by rolling. The crop-off should be only about 5%. Rimming steel is cheaper than killed steel but is not suited to all purposes.

It is important that the surfaces of ingots for forging or rolling should be free from defects which could become extensive laminations or seams on the surface of the rolled product. It is difficult to avoid some splashing during direct teeming into moulds from the top. The simplest device is to set a piece of wide tube in the bottom of the mould to divert the first splashes. A more expensive but more effective technique is called "up-hill" casting in which the metal is poured through a funnel and gate to the bottom of the mould so that it rises into the mould quietly and without splashing. Ingots can be surface dressed before being forged if necessary and in some plant overall "flame-washing" (with an oxy-gas torch) is carried out as a routine treatment.

An increasing proportion both of steel and of non-ferrous metals and alloys is now continuously cast. Sometimes these are only semi-continuous but the principle is the same. The metal is fed through a tun dish into a water-cooled copper mould of simple section. Initially this mould has a false bottom which can be withdrawn when the casting starts. The rate of withdrawal, the rate of pouring and the heat-abstraction rate must be exactly in balance so that the level of the liquid and the position of the solidus surface remain stationary as long as pouring continues. In practice the mould oscillates through a few centimetres, going down slowly with the solidified metal and then suddenly being knocked back up. In this way there is less wear on the mould than if the hot metal were sliding through it continuously. There are several advantages to the method of which the least is probably the continuous nature of the operation, because the product is usually cut into lengths immediately prior to further treatment. The cross-section is usually quite small so that the cooling rate is high. Consequently the primary grain structure is not so coarse as in conventional ingots and the segregation is limited to micro-segregation and that is not severe. The amount of crop-off becomes negligible and the surface finish is generally excellent. The ingot is sawn into blooms which need little or no soaking so that soaking pits and cogging mills are not required. The blooms can be cut to whatever exact length is required to produce the billet, bar or plate that is to be made, so minimizing waste. The main disadvantage of the process is its lack of flexibility. It is best used where long runs of standard-sized products are being made.

CHAPTER 13

EXTRACTION PROCEDURES

IN THIS chapter a brief description is given of extraction procedures for thirty elements which are currently manufactured in greater or smaller quantities. The procedures described are "typical" but to some degree "synthetic" in that in a few cases there may be no plant actually using the sequence of processes suggested. The alternative of describing how particular plants solve their problems leads to untypical descriptions which can be misleading. In general, procedures are more complex than indicated if only because of the large amount of recirculation of intermediate products which is usually necessary in order to maintain high yields and work up by-product values in drosses, slags, flue dusts, slimes and liquors. The impact of the subsidiary procedures involved is only touched on in these descriptions.

No attempt has been made to analyse the economic aspects of these procedures. They are very complicated and differ from one country to another, and even from one mine to another in the same ore-field. Common metals are often produced by simple traditional methods involving low capital and reagent costs but high energy costs and much labour. When extraction is practised in industrially advanced countries or on a very large scale the capital costs rise as improved technology leads to more efficient use of energy and labour. The extraction of the newer or rarer metals involves the most sophisticated techniques in which reagent costs are often very high, capital costs high and labour charges low. These metals are usually produced by non-traditional procedures largely because the traditional methods are inadequate to deal with such very reactive metals. The new methods may, however, find some applications toward improving techniques used for the "old" metals as economic conditions change and as ores available become yet

354

leaner. In particular "wet" methods of pre-refining may be developed which will require simpler, cheaper plant and restrict the expensive pyro-metallurgical operations to the very last stages.

Aluminium

Ores: Bauxite, a mixture of $Al_2O_3 \cdot H_2O$ and $Al_2O_3 \cdot 3H_2O$ containing 50–60% Al_2O_3 with Fe_2O_3 as the principal impurity, is the only important ore. Silica is preferred as low as possible, 10% being typical.
Reserves: Bauxite reserves are good in large isolated deposits. These are controlled to some extent by the International Bauxite Association. Many clays and rocks and the mineral alunite $(K_2O \cdot 3Al_2O_3 \cdot 4SO_4 \cdot 6H_2O)$ could be used as sources of Al and methods of extraction are known but are not at present economic to use.
Occurrence: Large isolated deposits mainly in tropical regions. The principal mines are found in Queensland, Australia, Jamaica, Guinea, Guyana and Surinam. Other countries where the production of bauxite is significant include Greece, France and Yugoslavia, Haiti and the Dominican Republic, India and Indonesia, the U.S.A. and the U.S.S.R.
Ore dressing: In the past ore was calcined before shipment. Now pre-refining is usually carried out near the mine on ore that is crushed to about −500 μm.
Pre-refinement: Crushed ore is treated by the Bayer process to produce pure Al_2O_3. The ore is digested in a solution of about 30% NaOH at a temperature between 150°C and 230°C in large pressure vessels at a pressure sufficient to suppress boiling. This is about 4 atm at the lower temperature but up to 30 atm at the highest temperature used. The reason for the range of temperature is associated with the proportions of monohydrate and trihydrate in the ore. The trihydrate dissolves readily at 120°C but the monohydrate requires either a higher temperature or more concentrated NaOH or both. As the temperature is raised, the rate at which silica is taken into solution is increased and as the concentration of NaOH is raised, the subsequent precipitation of aluminium trihydrate becomes more difficult and costly. In the past the ores have contained principally the trihydrate and only part of the monohydrate was dissolved and collected. Today many bauxites contain a high proportion of the monohydrate all of which must be collected. The

original Bayer process has been modified to handle these. Silica dissolves in NaOH as sodium silicate but this combines with sodium aluminate to form an insoluble alumino-silicate which forms a part of the "red mud" of insoluble material, which is mainly iron oxide. Obviously alumina equivalent to the silica going into solution is being lost. The solution of $NaAlO_2$ with excess NaOH is separated from the insoluble mud by settling and filtration. The mud may be worked to recover the alumina in it if its value is high enough. The solution is hydrolysed by dilution in the presence of freshly precipitated $Al(OH)_3$ "seed" to nucleate the reaction

$$NaAlO_2 + 2H_2O = NaOH + Al(OH)_3 \text{ (precipitated)}.$$

The precipitate is settled and filtered off, calcined to $a\text{-}Al_2O_3$ at 1200–1350°C and shipped to the smelter which will be near to a source of cheap energy. The Al_2O_3 should be almost free of both Si and Fe both of which would appear as impurities in the aluminium after reduction. The clarified solution must be concentrated with respect to its NaOH content for re-use in the digesters. This is done by evaporation of the water used to dilute the solution for hydrolysis. The energy required for this evaporation is a major item of cost and the design of a process must be such as to minimize this cost. It may even be preferable to lose some Al to the red mud rather than incur increasing evaporation costs.

Extraction: Al_2O_3 is reduced electrolytically by the Hall–Heroult process. It is dissolved in a molten mixture of synthetic cryolite (Na_3AlF_6), AlF_3 and CaF_2 (about 87 : 5 : 8) at 970°C and electrolysed between a consumable carbon anode and a carbon cathode which is the lining of the cell. The Al_2O_3 concentration varies between about 8% and 1.5%. When it falls too low, a high resistance develops at the anode and the Al_2O_3 in solution must be replenished by breaking the crust over the cell and stirring in fresh calcined solid. The anode carbon reacts with oxygen and is consumed. The metal collects at the bottom of the cell and is withdrawn at intervals by suction. Preparation of the anodes is an important part of the process. Special low-ash cokes are used because the ash contributes most of the impurities in the metal. The process is continuous in batteries of cells or "pots" connected in series. Individual cells can be taken out for re-lining as required. The cells operate at about 6–7 V and the current densities are up to about 10,000 A m^{-1} of cathode surface with total currents of about 10^5 A. The current efficiency is about 90%.

If the cell reaction were

$$Al_2O_3 = 2Al + 1\tfrac{1}{2}O_2 \qquad (13.1)$$

at 970°C, since $\Delta G^{\circ}_{1243} = 1279$ kJ the reversible cell potential needed to dissociate the alumina would be $E^{\circ}_{1243} = -2.21$ V. If the combination of the oxygen with carbon at the anode is included in the electrochemical reaction the free energy is much smaller. For

$$Al_2O_3 + 1\tfrac{1}{2}C = 2Al + 1\tfrac{1}{2}CO_2, \qquad (13.1a)$$

$\Delta G^{\circ}_{1243} = 686.0$ kJ and $E^{\circ}_{1243} = -1.18$ V and for

$$Al_2O_3 + 3C = 2Al + 3CO \qquad (13.1b)$$

$\Delta G^{\circ}_{1243} = 616.7$ kJ and $E^{\circ}_{1243} = -1.07$ V subject to small adjustments for deviations of the activities and fugacities from unity. In practice the gases leaving the cells contain between 30 and 50% of CO but it is now thought that CO_2 is the principal anodic reaction product and that much of the CO is produced in side reactions between the CO_2 and metallic aluminium and sodium dissolved in the electrolyte. This would impair the current efficiency as the aluminium so oxidized would have to be reduced a second time. Whatever the detailed mechanisms involved it will be seen that almost half the energy required to decompose the Al_2O_3 is supplied by the oxidation of carbon and the rest by electricity. In practice the potential drop per cell is at least 6 V, the difference being due to overpotentials at the anode and cathode and to the potential needed to drive the very high current against the resistances in bus-bars and electrolyte. This last is very important because the energy (I^2R) dissipated there is required to maintain the temperature of the electrolyte. Since the current is determined by the total resistance of the line of pots in series it is essential that the resistance in any one pot should not deviate far from its prescribed value, otherwise its temperature will wander. The resistance of the electrolyte is determined by its composition and temperature along with cell geometry. The overpotentials depend on the magnitude of the current, rising as the production rate is increased.

The minimum energy required to produce 1 kg of Al from Al_2O_3 can be calculated from the free energy value given above to be 6.70 kWh. If half of this is provided by oxidation of the carbon of the anode, about 3.55 kWh have to be supplied by electrical energy and, if the cell voltage is 5 times the reversible potential for the effective reaction (13.1a), the total requirement of electrical energy is 16.75 kWh per kg Al, subject to some variation with practice. In addition about 0.5 kg of anode carbon is consumed. It is obviously an advantage if a supply of cheap electricity is available for this process and most smelters are sited near to hydroelectric power stations. In recent years there have been some cases of smelters being built alongside conventional and nuclear power stations. If a coal-fired station has an overall efficiency of conversion of 30% it will burn about 7.5 kg of coal to provide 16.75 kWh of electrical energy. Only in special circumstances can the price of this coal be low enough to permit the process to be carried out profitably.

Alternative processes for the treatment of ores other than bauxite have been under consideration for many years. Of some importance is the recent Alcoa smelting process which is claimed to require only about 10 kWh of electrical energy per kg Al. The alumina is converted to $AlCl_3$ which is electrolysed. The chlorine is recycled. Apart from the saving in energy this new process avoids the problems associated with emissions of fluorine which give a lot of trouble in the Hall–Heroult process.

Refining: The product of the Hall–Heroult process is pure enough for most purposes and may be cast directly or, after alloying, into slabs or billets suitable for working. A small proportion of metal may be refined by the method mentioned on page 315.

Antimony

Ores: Stibnite (Sb_2S_3) is the principal ore mineral. A complex copper antimony sulphide called tetrahedrite is found in ores of Cu, Pb and Ag and a high proportion of Sb is obtained, as a by-product, from these ores.

Reserves: Apparently adequate.

Occurrence: Widespread in small amounts, but production is concentrated in Bolivia, Mexico, Union of South Africa, U.S.A. and

Algeria. The greatest reserves are probably in China, but Sb from that country comes on the world market spasmodically.

Ore-dressing: Various gravity and flotation techniques are used on lean ores. Rich ores and concentrates are "liquated" by heating in crucibles with perforated bottoms through which the molten sulphide can drain free from gangue at about 550°C. This may also be done on a sloping hearth but more oxidation and fuming occurs.

Extraction: An oxidizing roast converts Sb_2S_3 to volatile oxide Sb_2O_3 at 450°C with restricted air, or non-volatile Sb_2O_4 at 550°C — the former being desired when separation from gangue has to be effected. This process is then called "fuming". The oxide formed is reduced with charcoal under a flux, basically of Na_2CO_3, in crucible, small hearth or blast furnaces. Iron can be used as the reducing agent for the direct reduction of lean sulphide ore in a small blast furnace. A matte is formed from the FeS produced along with other sulphide impurities, these being "fluxed" with Na_2S derived from sodium salts incorporated in the charge. This constituent reduces the specific gravity of the matte and thereby facilitates separation of matte and metal layers. The crude Sb so produced is high in Fe which must be removed by liquation prior to refining in the usual way. At all stages in these processes Sb_2O_3 fume is formed and must be collected in bag-houses for re-introduction to the system.

Refining: Classical fire refining is used with additions of sulphur to remove Cu and Fe followed by oxidation of As and poling under a protective flux. The "starred" pattern is achieved by freezing under special flux covers which remain fluid until the metal is solid. This former indicator of quality is of no significance today. Pb is difficult to separate and is normally retained if present for use in making antimonial lead.

Arsenic

Ores: Orpiment (As_2S_3) and realgar (AsS) have been mined from veins and used as yellow and red pigments for centuries. Today all As comes from the flue dusts of extraction plants handling ores of Pb, Cu, Co, Zn, Sn, Au, etc., as the element is commonly associated with sulphide ores of all kinds.

Reserves: The problem is often how to get rid of As rather than where to find it.

Occurrence: In association with sulphide ores of many kinds all over the world.

Ore dressing: Flue dusts are often recycled through sinter plants, etc., in such a pattern that while the major metal is being recovered minor elements like As are gradually concentrated in the dusts. When a suitable concentration has been reached the dust is withdrawn from circulation for further treatment.

Extraction: As_2O_3 is sublimed or fumed at about $1100°C$ from the dust in a reverberatory and is further purified by a series of fractional sublimations, and condensations in brick chambers called "kitchens". Only small amounts of elementary As are prepared. As_2O_3 is reduced with charcoal in iron retorts, the element subliming at $616°C$ and being collected either in amorphous or "metallic" form in a water-cooled condenser. Most As_2O_3 is marketed unchanged or after conversion to another compound.

Beryllium

Ores: Beryl ($3BeO·Al_2O_3·6SiO_2$) occurring in granitic pegmatites is the only source of Be. Principal suppliers are India, Argentine and Brazil but it is also worked in South Africa, Rhodesia, Morocco and the U.S.A.

Ore dressing: Mainly by hand picking the largest crystals out of crushed ore.

Extraction: Pre-refinement or chemical concentration to pure BeO or BeF_2 is the first requirement. The ore is sintered with alkali or even fused to render Be soluble and other elements are precipitated by standard "wet" methods. Pure hydroxide or chloride is converted to $(NH_4)_2BeF_4$ which is crystallized out of solution and ignited to BeF_2. This is reduced with Mg under a slag containing excess BeF_2 at $1300°C$. At this stage Mg is the main impurity. An alternative route is by producing pure $BeCl_2$ by fractional distillation and electro-depositing the Be from a fused solution of its chloride in NaCl. Vacuum arc melting is a necessary final stage.

Bismuth

Ores: Bismuth occurs native in appreciable quantities and as bismuthinite $(Bi_2 S_3)$ and bismuth ochre $(Bi_2 O_3)$ usually associated with other sulphide deposits. Most Bi is obtained as a by-product of the extraction of Cu, Pb or Sn.

Reserves: Production rate probably tied to that of Pb ultimately.

Occurrence: Widespread, but good ores primarily of Bi are restricted to a few areas. Main supplies come from Mexico, Peru, the U.S.A., Korea and Canada. Bolivia, Argentine, Germany and China have also produced large quantities.

Ore dressing: By standard techniques on lean ores. Rich native ore may be liquated like Sb ore (see page 359). A volatilizing roast at a carefully controlled temperature can be used to remove As and Sb as oxides.

Extraction: Similar to Sb. Oxidation may precede smelting, or sulphide may be smelted with charcoal and Fe, the latter to take up the sulphur into a matte. Soda ash $(Na_2 CO_3)$ is the main flux. Recovery from anodic slimes from electrolytic lead refining (page 375) is a complicated process depending in detail on what values are present. A controlled oxidation of the fused slimes eliminates most of the Pb, Sb and As by drossing or fuming. Further oxidation at a higher temperature would oxidize Cu and Bi to a slag leaving precious metals intact. The slag can be reduced to impure metal which can be purified electrolytically. The large number of steps can be justified by the fact that other and more valuable metals are also being won.

Cadmium

Ores: The one important mineral, greenockite (CdS), occurs only as a minor constituent in ores of Zn, Pb or Cu. Cd is always recovered as a by-product of the extraction of one of these metals, from flue dusts or fumes taken from bag-houses or electrostatic precipitators, or from leach solutions or distillates.

Reserves: Supply is limited to about 0.3% of the output of Zn.

Occurrence: Widespread and like Zn. The U.S.A. is by far the largest producer with Mexico second. South-West Africa exports a large amount as concentrates. Canada and Australia also produce a few hundred tons per year out of a total of about 5000.

Extraction 1: *From flue dust* − Dust may be·recycled through a lead blast furnace or zinc sinter plant until the concentration of Cd warrants treatment. Dust is leached with sulphuric acid under conditions designed to keep Pb out of solution as $PbSO_4$. Solution is purified of As, Sn and Cu in a series of chemical treatments and the Cd deposited either by electrolysis or by displacement from solution by Zn. In this latter case it forms a sponge which must be re-melted and may require further refinement from Zn, e.g. by electrolysis.

Extraction 2: *From leach solutions* − Cd in liquors from the leaching of Zn concentrates is precipitated with Zn dust and the sludge obtained contains Zn, Cu, Pb, etc. It must be brought back into solution from which unwanted metals are precipitated and removed. The Cd can then be recovered as above.

Extraction 3: *From distillates* − Cd concentrates in the first vapour to come off in the retorting of Zn and could be so concentrated in the old horizontal retort process. Otherwise it may appear as an impurity in the Zn "spelter". In either case it would be recovered by fractional distillation (see page 329). Cd may also be present in the distillates from vertical retorts and Zn blast furnaces, and is recovered from these in the same way.

Chromium

Ores: The spinel chromite ($FeO \cdot Cr_2O_3$) is the only ore mineral for Cr and that is always diluted with other spinels. Thirty per cent Cr would be a rich ore. Good ones of suitable composition go to refractories manufacture.

Reserves: High-grade ores are still available.

Occurrence: The major producers are Turkey, the U.S.S.R., the Union of South Africa, Rhodesia, Philippines, Cuba and New Caledonia.

Ore dressing: Usually little or none beyond handpicking except where unusually lean ore is being handled.

Extraction 1: *For ferro-alloy* − Finely ground chromite ore is smelted in an electric arc furnace with anthracite rather more than sufficient to reduce Cr and Fe. The Cr content of the alloy is about 60% depending on the grade of the ore. The carbon content would be in the range 1–8% depending on the excess used, a lower carbon being penalized by a lower yield of metal. Low carbon ferro-chrome can be made using ferro-silicon as reducing agent instead of anthracite or by oxidizing high carbon ferro-alloy with chromite ore in a second smelting operation (cf. steelmaking).

Extraction 2: *For pure Cr* – Extensive pre-refining is necessary. Sodium chromate is formed in a sintering roast with Na_2CO_3, leached, and the solution purified with respect to Cr. Pure $Na_2Cr_2O_7$ is obtained by crystallization and converted to pure Cr_2O_3 by reaction with sulphur–

$$Na_2Cr_2O_7 + S = Na_2SO_4 + Cr_2O_3.$$

The oxide can be reduced with charcoal, or with Al in a thermit process. Some preheat of the charge is necessary to ensure a complete melt out. Electrolytic extraction from a chloride solution is also possible.

Cobalt

Ores: Sulphides, arsenides and arsenosulphides such as cobaltite (CoAsS) and smaltite ($CoAs_2$) are found in Cu or Ni ores or with pyrites or in Pb ores. When Co is the major metal present Ag or Au is often present too. Some deposits are oxidized and asbolite, a mixed hydrated oxide with manganese, occurs in New Caledonia as a residual of the same serpentine as gave rise to the famous nickel ore there. This once supplied the world's needs.

Reserves: Slender. Production overdependent on Zaire.

Occurrence: Major source of supply is Zaire. Also produced in Zambia, Canada, Morocco, the U.S.A., Burma.

Extraction: Very wide range of techniques. The scale of production is usually small and the method is often dictated by the major extraction process from which the Co is a by-product. Some oxidized ores in Katanga are rich enough to justify working primarily for Co. Flotation concentration to about 8% is followed by electro-smelting to an Fe–Cu–Co alloy which is shipped for further treatment by leaching and electrolytic deposition of the Co. More often Co is collected in a matte or speiss which is further concentrated by careful smelting methods to a cobalt-rich speiss which is brought into solution by leaching with acid and treated chemically or electrolytically for the maximum recovery of values. Alternatively the speiss may be roasted out to oxide or sulphate which can be leached for electrolysis. Lean oxidized ores can be roasted to sulphate for leaching and electrolytic deposition. The electrolysis of cobalt has been discussed on page 315. Cathodes require a final fire-refine to bring the metal into commercially acceptable form.

Copper

Ores: Primary and secondary sulphides — chalcocite (Cu_2S), chalcopyrite ($CuFeS_2$), bornite ($Cu_2S \cdot CuS \cdot FeS$) and others. Oxidized secondary minerals — cuprite (Cu_2O), brochantite ($CuSO_4 \cdot 3Cu(OH)_2$), azurite ($2CuCO_3 \cdot Cu(OH)_2$) and chrysocolla ($CuSiO_3 \cdot 2H_2O$). Also native. Usually associated with iron and often with Ni, Au, Ag, the Pt metals and traces of Se, Te, Zn, Mo, As, Sb, etc. Cu usually in the range 0.5–2% with quartz, granitic or limestone gangue and often highly disseminated. Lean ores usually mined from open pits.

Reserves: Despite high extraction rate the reserves appear to be good but grade is steadily falling. A vast amount of Cu could ultimately come from ocean floor "nodules".

Occurrence: Widespread in the Americas from Chile through Peru and the eastern Unites States into British Columbia and Alaska; in Zambia, Zaire and South Africa; in Australia, the Philippines and recently in Papua, New Guinea. The U.S.S.R. and China both have good supplies and there is a little also in Europe and in Japan.

Ore dressing: Essential and elaborate except when heap leaching. Flotation should bring Cu up to about 25%.

Extraction 1: About 80% of all primary Cu is produced by pyrometallurgical procedures. The standard route is based on the old Welsh process practised in Swansea when that town was producing most of the world's Cu from Cornish, Spanish and American ores until well through the nineteenth century. A partial oxidizing roast removes some of the S to prepare the concentrate for a matte smelt in which all the Cu goes to matte with some Fe while the rest of the Fe as FeO helps to flux the gangue to a low-Cu slag which is discarded. Roasting may be carried out in a multiple-hearth furnace or in a fluidized bed. Fine concentrates are smelted in reverberatory or electric smelting furnaces. In small-scale operations some lump ore and sinter is still smelted in water-jacketed blast furnaces (see page 288). There is a trend toward combining the functions of roasting and smelting in single "flash-smelting" processes. These have high production rates and are very efficient thermally but have the disadvantage that copper losses to the slag are so high that treatment for its recovery is necessary.

Matte may contain between 20 and 60% Cu or, exceptionally, higher. It is transferred in the molten state to a Pierce–Smith converter where it is blown in successive batches to "white metal" (Cu_sS) and then to "blister copper" which is cast into slabs. Converter slag is recycled through the smelter. Blister copper is about 98–99.5% pure. If there are precious metals to be recovered it is further fire-refined and cast into anodes for electrolytic refining. The precious metals collected in the anode sludge along with S and Se and some insoluble compounds of As, Sb and Bi; soluble salts of Fe, Ni, Co, As, etc. will accumulate in the electrolyte and will contaminate the cathode copper if they are not held to very low concentrations by continuously bleeding off a portion of the electrolyte for chemical purification. Cathodes are re-melted to get rid of hydrogen and adjust the oxygen to the level appropriate to the required grade, and cast. In the absence of precious metals blister copper is fire-refined to give a saleable product.

Recent advances in the pyrometallurgical extraction of Cu include several attempts to develop a continuous process in which matte produced in a flash smelting stage is blown to blister copper within the same production unit.[86] Copper losses to the slags are higher than in conventional process and impurities in the blister copper are also higher but commercial viability is being approached in both Canada and Japan. At the same time methods of recovering copper from flash smelting slags, where the losses are also high, by settling, leaching and flotation are continually under examination. The TORCO (treatment of refractory copper ores) process utilizes a sophisticated ore-dressing technique applied to some oxide and silicate ores which respond neither to flotation nor to leaching. Finely ground ore is preheated in a fluidized bed fired with pulverized coal. It is then passed into a reactor with coal and NaCl at 900°C, in which Cu_2Cl_2 is produced as a vapour and is reduced by H_2 from the coal with deposition of metallic Cu on to the surface of the coal particles. The metallized coal is separated from gangue by flotation and smelted in a reverberatory. This is a costly process which would be used only on refractory ores and then only if they contained sufficient values to justify this extra cost.[87]

Extraction 2: Only about 20% of primary copper is extracted by leaching. It is used most appropriately on oxidized ores. Sulphide ores can be leached with oxidizing reagents, but slowly. Sulphide ores may also be leached after roasting to oxide or preferably to sulphate. This course is sometimes preferred when there are no precious metals present. These can be collected only by matte smelting. Dumps of mine wastes are frequently leached over periods of years to recover copper discarded as being at uneconomically low concentration. Leach liquors may be concentrated by solvent extraction. They must be adjusted to low acidity and low Fe^{3+} content by prolonged contact with fresh ore. Chlorine may have to be reduced with cement Cu to precipitate Cu_2Cl_2. Any Mo must be removed with SO_2. The solution is then electrolysed with Pb—Te anodes. The cathodes must be re-melted and fire-refined to eliminate traces of S, Pb, Zn, etc., and adjust the oxygen content. Ni and other values must be removed from the electrolyte which is re-circulated to the leaching plant. A less pure Cu can be produced by cementation with scrap steel.

The principal grades of Cu are "tough pitch" copper, "phosphorus deoxidized" copper and "oxygen-free high conductivity" copper. Tough pitch copper is deoxidized by poling with greenwood logs until it casts into a mould with a level surface. This denotes an oxygen content of 0.015–0.04% with only a trace of hydrogen and less than 0.001% sulphur. This has a good electrical conductivity because the oxygen precipitates traces of Fe and other impurities out of solution in the copper. Tough pitch copper cannot easily be gas welded so Cu required to be welded is deoxidized with 0.01–0.05% phosphorus. This impairs the electrical conductivity. If conductivity and weldability are both required a special treatment with carbon is needed to produce the "oxygen-free" grade with less than 0.001% of that element. Only a small amount of this quality is made. Continuous measurement oxygen concentration using an oxygen-sensitive solid state electrode now replaces the level surface test (see page 266).

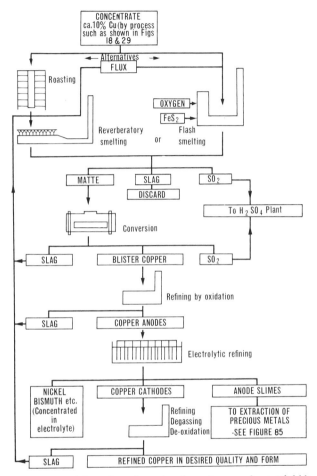

FIG. 83. Typical flow sheet for extraction of copper from sulphide concentrates by pyrometallurgical means, followed by electrolytic refining.

Gold

Ores: Gold occurs native in veins and placer deposits derived from veins. The deep Rand deposits are consolidated gravels in which gold is in the quartzite cement which binds the barren pebbles together. Some Ag is usually alloyed with Au, and pyrites (FeS_2), arsenopyrite (FeAsS), chalcopyrite ($CuFeS_2$), galena (PbS), etc., are commonly associated with Au in veins with, of course, quartz. The only compounds found are mixed tellurides with Ag. In all cases the Au is disseminated extremely finely, nuggets being rather rare occurrences.

Reserves: Apparently large relative to demand. There is plenty of Au in the ground at very low concentration but the feasibility of working it profitably is very sensitive to the price which is determined independently of production costs. When the price is high enough Au can be extracted from ore with only a few ppm present.

Occurrence: Widespread in small amounts. Main sources are South Africa (about a third of the total), the U.S.S.R., Canada, the U.S.A., Australia, Ghana, Rhodesia, Zaire, South America, Mexico and the Philippines.

Ore dressing: Panning of alluvial deposits was at once ore-dressing and extraction. On a larger scale sluicing or jigging and tabling produces a concentrate which can be amalgamated with mercury (see below). Lean deep mined ore must be crushed and finely ground below 75 μm (or "all-slimed") in (obsolescent) stamp mills or pebble mills. Larger grains of Au can be segregated by streaming pulp across ribbed corduroy stretched on a sloping table (strake).

Extraction: Au in concentrates can be amalgamated by flowing the pulp, one particle deep, across a copper plate laden with mercury or by tumbling with mercury in a drum. Au dissolves in Hg and the amalgam is scraped off periodically, pressed, and the metal finally recovered by distilling off the Hg. Finely ground lean ore pulp is thickened and leached with NaCN with agitation which ensures aeration to promote the reaction—

$$2Au + 4NaCN + \tfrac{1}{2}O_2 + H_2O = 2NaAu(CN)_2 + 2NaOH.$$

Some adjustment of pH with CaO precipitates heavy metals and assists the settlement of the solids. The solution must next be filtered absolutely clear. It is then de-aerated by vacuum and the gold precipitated with Zn dust according to

$$2NaAu(CN)_2 + Zn = Na_2Zn(CN)_4 + 2Au.$$

A slime of Au and Zn is recovered by filtration, calcined to oxidize the

Zn and smelted with a flux of borax and silica to bullion and a clean slag.

Refining: Chlorine gas bubbled through molten Au forms a slag of AgCl (in borax) and volatilizes base metal chlorides excepting the last traces of Cu. An electrolytic method is also sometimes used.

Iron

Ores: Rich haematite (Fe_2O_3) and magnetite (Fe_3O_4) ores with small amounts of siliceous gangue are the major sources of iron today. Lean limonite ($Fe_2O_3 \cdot 3H_2O$) and siderite ($FeCO_3$) ores with siliceous, argillaceous or limy gangue and high P content are still mined in the U.K. and Europe but low-cost transportation of rich ores in very large ore-carriers makes their importation by sea for smelting at coastal sites so much more economically attractive that the use of lean indigenous ores is rapidly diminishing. Fe content is usually in the range 50–70% but 20% ores have been smelted where self-fluxing burdens were possible.

Reserves: Immense — especially if the lean taconites underlying the rich haematites and from which they have been derived are included.

Occurrence: Widespread, but countries in which iron industries first grew up have now exhausted their reserves. Rich deposits in Scandinavia, North and West Africa, Brazil, Peru, Newfoundland, Labrador and northern Australia are now being exploited by industrialized nations to augment their own supplies. China, India and the U.S.S.R. are self-sufficient while Europe and the U.S.A. can satisfy a large part of their own needs. Japan and the U.K. must import almost all their requirements.

Ore dressing: Large ore is crushed to about 30 mm and the −10-mm fraction removed for sintering. Magnetic and heavy medium separation are occasionally practised on ores in which iron is distributed as nodules or otherwise as discrete particles. Magnetic separation may follow magnetic roasting (see pages 87 and 281). Blending of ore mixes in bedding plants is now commonly practised either before charging direct or prior to sintering or pelletizing. This has the purpose of providing chemically and physically uniform feed to the blast furnace in order to improve the uniformity of operation and product. All fines, all soft friable ores and some ores which are difficult to reduce are converted to sinter or pellets, depending on their nature (see page 95), so as to improve their reducibility and flow characteristics through the furnace. These expensive steps become more necessary as the size of the blast furnaces increases.

Extraction: Ore is smelted in large blast furnaces with hard coke and limestone flux to produce an impure "pig" iron containing about 4% C and a slag with a CaO/SiO_2 ratio between 1.2 and 1.4, about 5% MgO and 12% Al_2O_3. At a hearth temperature of 1550°C and an oxygen potential determined by an excess of carbon ($p_{O_2} \sim 10^{-16}$ atm) all of the P_2O_5 and most of the MnO are reduced into the metal. Some SiO_2 is also reduced but fortunately the kinetics of reduction are slow and the reaction does not approach completion. Typically, iron contains 0.5% P and 1% Mn which depend on what was charged, and about 1% Si which is determined by slag composition (a_{SiO_2}), hearth temperature and contact time between slag and metal. High phosphorus ores are smelted to iron in which P is about 2%. Low S ($\sim 0.03\%$) is desirable but cannot always be achieved without raising the Si because of the kinetics of the reactions involved (see page 204). It is generally considered to be better to keep Si steady and to control S by external desulphurization in the launder or ladle after casting. This is effected by treatment with soda ash or with metallic Mg incorporated into a preparation known as magnesian coke or even injected in the vapour state. In all cases the resultant slag must be carefully removed.

The efficiency of the process is usually measured in terms of the carbon rate (kg carbon per kg Fe produced) which depends on the richness of the ore, the reducibility of the oxide in the form in which it is charged, on the reactivity of the coke, the driving rate of the furnace and the efficiency with which the counter-current streams of solids and gases interact with each other. Ideally the solids should flow smoothly down the stack with little difference in rate at different points across its section and should offer only low uniform resistance to the upward flow of furnace gases. Every particle at any level should have reached the same temperature and should be reduced to the same degree. In practice this is not possible. The gases form race-ways above each tuyère and the solids flow is directed toward these areas where the coke is gasified and solids are melted fastest. Higher in the stack size segregation in the burden offers preferential routes for the gases through coarser material where the burden is heated more efficiently and reduced more rapidly than average — leaving other material colder and badly reduced. Changing flow patterns can lead to cooling down of partially molten burden which forms accretions which cannot then flow properly and may adhere to the wall and ultimately form "scaffolds". Periodic collapse of scaffolds brings cold unreduced ore into the bosh and hearth where it must then be reduced and smelted without the benefit of counter-current processing, that is, in a thermodynamically irreversible and therefore thermally inefficient manner. The blast-furnace process is, in fact, very efficient compared with alternatives but still capable of improvement. More elaborate burden preparation in recent years is intended to improve the efficiency of interaction of gases and solids and the use of high top pressure serves the same end. The top gas is taken off at about 1 atm over atmospheric pressure so that the pressure at all levels has been raised by about the same amount against conventional practice. Thus a higher mass flow rate can be used without increasing the volume flow rate or velocity of the gases. Recent developments include the injection of some fuel gas or oil through the tuyères for economy and for greater flexibility and control over the heat balance. Steam and oxygen may also be injected through the tuyères for several reasons of which the control over gas–solid temperature differences and hence of temperature distribution inside the furnace ought to be the most important.

The most obvious development in blast furnaces is in their size. New furnaces have a capacity of 10,000 tons per day. These should produce cheaper iron than smaller furnaces if working to capacity but they can only be less flexible in operation and re-lining will be difficult to organize. A less obvious development is a gradual increase in the blast temperature, which has risen, over the course of 20 years, from about $700°C$ to about $1200°C$, thus necessitating the re-designing of the Cowper stoves.

Pig iron is now referred to as "hot metal" or "blast-furnace metal". It contains only about 93% Fe. It is an intermediate product and is passed directly or via a desulphurizing station and still molten, for conversion into steel.

Refining 1: Steelmaking is the removal of C, Mn, Si and P by oxidation followed by deoxidation and adjustment of the composition to that required of the steel being made. The steel is usually cast to ingots for rolling or forging. Most steel is made from pig iron by oxidizing with gaseous oxygen in top-blown converters by the L–D or BOS (basic oxygen steelmaking) process or by some modification of it. The old open-hearth process persists but no more open-hearth furnaces will be built. A small amount is being made by the new Q–BOP process which shows great promise and may replace the BOS in time. The charge to the BOS converter is about 70% desulphurized iron and 30% scrap with burnt limestone as flux. Oxygen is blown on to the metal until the carbon and particularly the phosphorus are sufficiently low in the iron. This is achieved using a slag made from the oxides of all the Si and Mn charged, along with some iron oxide and lime sufficient not only to balance the P_2O_5 formed but also to saturate the slag. In contact with such a slag and at the high oxygen pressure imposed, the phosphorus can be brought down toward 0.01% while the carbon will fall to 0.1% or an even lower value. Only traces of silicon manganese remain; sulphur rises a little; and oxygen in solution in the melt rises to about 0.05%. When high P iron is being refined, a modified two-slag process is necessary. The first slag removes most of the P while the second has to be made to remove it to a low concentration. This second slag is retained in the vessel for use in the next heat. To hasten the removal of P, lime may be injected with the oxygen in what is called the LD/AC process.

When oxidation has been completed the temperature must be adjusted and the carbon brought up to the required value, if necessary with additions of pig iron or ferro-alloys. This effects some deoxidation. Further deoxidation is carried out using ferro-manganese and ferro-silicon in the ladle and possibly using aluminium in the ladle or in the moulds. Practice varies widely depending on the type of steel required (see page 318). Molten steel may be degassed; it may be cast to rimming or killed ingots; or it may be cast by the continuous technique (see pages 320 and 352).

Refining 2: A high proportion of steel is made from scrap only. Some of this is still made in open-hearth furnaces but these are being replaced for the purpose by large electric arc furnaces on account of their higher production rates and capability for producing higher-quality steel. The scrap charge requires only limited oxidation but some pig iron is added to generate a carbon boil – an evolution of CO bubbles which agitates the bath and helps to speed up reactions. Oxidation is by oxygen lance or ore additions and is continued until P is low enough. The slag is limy and can be more basic in the electric arc furnace than in other processes because the viscosity can be kept low at the higher temperature which can be attained. An advantage of the electric process is that the oxidizing slag can be removed and replaced with a reducing slag made up with CaO, CaF_2 and carbon which forms CaC_2. Under this slag, steel can be desulphurized very effectively and it can be deoxidized and alloy additions made without losses to the slag being incurred even of such easily oxidized elements as Cr, Mo and V. In general it is easier to make very high-quality steel this way than by the BOS process but the costs are higher.

Residuals: A number of impurities cannot be oxidized out of iron. These include Cu, Ni, Sn, As and Sb. These may be present in ore, in which case they will report in the pig iron and again in the steel. When the steel is recycled as scrap they will report again, repeatedly and with cumulative effect. Scrap may also carry in elements which have been added as alloys or as coatings (Ni, Sn) or metals associated with the scrap in use like brass or bronze fitments. The accumulation of these elements in modern steel is becoming increasingly worrying. Scrap should always be carefully sorted out but this is obviously an expensive process. Ultimately Cu and Sn may be removed by vacuum treatment but removal of Ni will be even more difficult. These are problems for the not too distant future.

Alternative processes: Many suggestions have been made of ways of improving the manufacture of steel. Some of these aim at making it a continuous rather than a batchwise process. Of these, probably "spray steelmaking" looked most promising – a process in which blast-furnace iron was poured in a narrow stream through a ring of oxygen jets which oxidized it and dispersed it into a rain of droplets which fell through a layer of slag to collect in a ladle. Lime was blown along with the oxygen and the process showed promise with low P metal but experimentation has been discontinued. Other suggestions are directed at finding an alternative to coke either for countries with no coking coals or for use when these coals are exhausted. Unfortunately the alternative sources of energy are not likely to be available much longer than coke – except for other coals which can, of course, be gasified. Most of these processes are "direct" reduction processes in which ore is reduced to iron without going through the pig-iron stage. If this is done at low temperatures the product is likely to be pyrophoric but at high temperatures it tends to sinter together making separation of iron difficult. The only really promising application at present is to the production of pellets of reduced ore for use as feed to electric arc steelmaking furnaces as a partial replacement for scrap steel. This can only be economic if the pellets can be made to be cheaper than scrap. It will be most successful if the ore from which the pellets are made is of high purity so that the un-separated gangue which must be fed to the steel furnace will not add too great a metallurgical load to the process. Pellets with about 90% Fe can be produced by conventional pelletizing followed by induration and reduction in rotating or shaft kilns in a reducing atmosphere. Similar pellets have been suggested as feed for blast furnaces where a lower Fe content would be acceptable.

Lead

Ores: Primary galena (PbS); secondary cerussite ($PbCO_3$) and anglesite ($PbSO_4$) usually in upper workings. Galena is commonest, associated with FeS_2, ZnS, Ag_2S and many other sulphides in minor proportions.
Reserves: Satisfactory.
Occurrence: Widespread. Biggest producers are the U.S.A., Mexico, Australia, Canada and the U.S.S.R. The U.K. production very small.
Ore dressing: Extensive. Differential flotation from zinc blende (ZnS) should yield a concentrate containing ca. 70% Pb. This is followed by a roast to PbO – usually carried out in a Dwight–Lloyd sintering machine which also prepares feed suitable for blast furnace.
Extraction: Normal large-scale reduction is in blast furnace 7 m high, 7 m long and 1.5 m across at the tuyères, much of its stack and end plates being of water-cooled steel construction. Charge is sinter, flux, and coke which is smelted to Pb "bullion", and a slag of FeO, CaO, SiO_2 with 15–20% ZnO, 1% PbO. The fuel requirement is low and about half the carbon charged reacts with PbO to form CO_2 rather than CO, so that an atmosphere oxidizing to Fe and Zn is maintained (see Fig. 45). The top gas $CO/CO_2 \simeq 1$. Pb is tapped continuously by

syphon tap hole. Slag, matte and speiss are tapped at a higher level and are all worked for their values. Reduction of sulphide in the Newnam mechanically rabbled ore hearth is practically obsolete. See also zinc blast furnace (page 392).

Refining: *Drossing* – hot bullion is cooled to just above freezing point $(327°C)$ and held with some agitation while impurities separate which are less soluble at the lower temperature. As, Sb, Sn, and Fe (with a lot of Pb) form an oxide dross. Sulphur additions dross off Cu here.

Softening – at 750°C, in a wide, open-hearth furnace or in the open saucer-shaped lead "kettle" which presents a very large surface for oxidation, Sb, Sn and As are slowly oxidized out with air or litharge, forming a series of slags of "skims" of lead antimonate, stannate and arsenate. An alternative process uses $NaNO_3$ as the oxidizing agent to make a slag of sodium oxy-salts at a lower temperature. Bismuth is not eliminated.

Desilverizing (Parkes process) – Ag·is partitioned between Pb and Zn at 480°C. Zn stirred into bullion collects Au and Ag as $Ag_2 Zn_3$ into a crust which is skimmed off as the temperature is dropped to 420°C. A second scavenging treatment brings Ag down to 0.0001% and provides the first charge of zinc for the next batch. A continuous process has been devised.

Dezincing – 0.5% Zn must then be removed, by oxidation as in softening; by conversion to $ZnCl_2$ by pumping metal through a chamber containing chlorine gas; or by vacuum distillation (which recovers Zn for re-use).

Alternative refinement: (Bett's process) When Bi is present softened bullion is cast to anodes and electrolytically refined using acid lead fluosilicate solution as electrolyte. Impurities adhere as slime to the anode, and these include Bi. Bi can also be removed by stirring a Ca–Mg alloy into molten Pb. Bi alloys with Ca, forming a crust which can be skimmed off.

The working-up of the many drosses, slags and slimes for their values is an important and most interesting part of the operations in a lead refinery.

Magnesium

Ores: Minerals brucite $(Mg(OH)_2)$, magnesite $(MgCO_3)$ and dolomite $((CaMg)CO_3)$ are all rich in Mg and widely available but a large part of

Mg production comes from sea water (Mg = 0.13%) Magnesite is of greater interest for making refractories.

Reserves: Unlimited.

Occurrence: Good refractory magnesites are restricted to Greece, Austria, the U.S.S.R., the U.S.A. and Canada. Dolomite is widespread.

Ore dressing: $Mg(OH)_2$ is precipitated from sea water by $Ca(OH)_2$ or preferably calcined dolomite by the reaction $MgCl_2 + Ca(OH)_2 = Mg(OH)_2 + CaCl_2$. The precipitate is thickened and filtered off.

Extraction 1: $MgCl_2$ must be formed either by solution in HCl and dehydration or by a chloridizing roast under reducing conditions. A solution of anhydrous $MgCl_2$ in a mixture of $CaCl_2$ and NaCl can be electrolysed in a special cell designed to segregate the Mg from the anodic chlorine (which is recycled).

Extraction 2: Methods for reducing MgO with C or Si have been used. They depend upon rapid condensation of Mg vapour before it can re-oxidize. The Pidgeon process using Si in a closed Ni–Cr alloy retort with an internal condenser is the simplest to operate. Reduction is at $1150°C$ under low pressure (see Fig. 77b).

Refining: Either cathodes or condensate must be re-melted. Iron may be reduced in the melt by adding small amounts of Mn and Zr which form insoluble phases which sink to the bottom of the melt. Great care has to be taken to avoid contamination by the chloride and fluoride fluxes used to protect Mg from the atmosphere. SO_2 gas is often preferred because it cannot become entrained in the metal.

Manganese

Ores: All minerals of interest are oxide type, pyrolusite (MnO_2) being the most important. Very rich ores go to the manufacture of dry batteries. The leanest ores are smelted with iron ores to supplement the Mn content of the blast furnace charge. Intermediate ores are used to make ferro-manganese (70–80% Mn), spiegeleisen with only about 20% Mn, or silico-manganese (70% Mn–20% Si–10% Fe).

Reserves: These seem high but it is a "strategic" metal, a high continuous rate of supply being essential to all industrialized nations. It is a major constituent of ocean floor nodules.

Occurrence: Widespread, but a few countries supply most of the world's needs — the U.S.S.R., India, Ghana, Morocco, the Union of South Africa, Brazil, Egypt and the U.S.A.

Ore dressing: Some up-grading of leaner Mn ores is obviously desirable especially in a country like the U.S.A. where most indigenous ores are of low grade. Treatment should be as simple as possible but flotation followed by agglomeration has sometimes been used.

Extraction: Reduction in blast furnaces like iron smelting is commonly used for spiegeleisen and ordinary ferro-manganese. Furnaces are smaller and are run hotter than iron blast furnaces and with a more limy slag to ensure a high yield of Mn. The product has about 6% carbon. Low carbon (1% or less) ferro-manganese can be made by electro-smelting in which process losses of Mn by volatilization are substantially lower.

Refining: Pure manganese is prepared electrolytically from a sulphuric acid leaching liquor after extensive wet purification. A diaphragm cell is necessary. A thermit process can also be used but this leaves a little Al in the metal.

Mercury

Ores: Cinnabar (HgS) is the only useful mineral, usually found in a highly disseminated state in sandstone or limestone gangue, and usually with less than 1% Hg.

Reserves: These appear to be quite large but restricted to rather a small number of places. A greatly expanded production rate could probably not be sustained for long.

Occurrence: Major sources of supply are Italy, Spain, the U.S.A., Mexico, China and Japan. The U.S.S.R., Yugoslavia and Algeria also produce some.

Ore dressing: Because extraction is a low-temperature process involving no fusion or fluxing, concentration is seldom practised. Because the ore is usually friable crushing is seldom necessary but the separation of coarse and fine fractions for treatment in different furnaces is common.

Extraction: Ore is calcined in air at 600°C to effect the decomposition of the sulphide. The Hg is recovered from the flue gases in large water-cooled condensers. The kilns may be simple shaft furnaces for coarse ore, or shaft furnaces of special design to facilitate the flow of fine ore — and the counter-flow of combustion gases. Small multiple-hearth kilns of the

Wedge type have been used, and also rotating kilns. Rich ore or concentrate is sometimes retorted with lime which takes up the sulphur according to $HgS + 4CaO = 4Hg + 3CaS + CaSO_4$. This process makes condensation much easier but cannot be applied economically to large volumes of lean ore.

Refining: Condensers contain liquid Hg which can be filtered through cloth and marketed, and "soot" which must be liquated or retorted with lime to recover the metal. Further refinement before use would be double or triple distillation under reduced pressure, only a third of any batch going forward as fully refined product.

Molybdenum

Ores: Molybdenite (MoS_2) is the chief ore mineral. Ores are very lean — less than 1% Mo. Otherwise production has been mainly as a by-product of the extraction of Cu from its porphyry ores.

Occurrence: U.S.A. produces about 90% of the world supply and most of that comes from one mine. Elsewhere production has been small and spasmodic.

Ore dressing: Molybdenite is very easily concentrated by flotation, having physical properties rather like graphite.

Extraction 1: *For ferro-alloy.* Molybdenite is roasted to oxide which is reduced in an electric arc furnace with carbon in the presence of scrap iron. The alloy produced contains about 2% C. If low carbon is required reduction may be effected by Al in a thermit smelt. Some of the Fe would be reduced from its oxide by the Al as well as the Mo, in order that the heat of the reaction would be high enough to melt out the charge.

Extraction 2: *For pure Mo.* MoO_3 can be purified by sublimation and then reduced with H_2 to produce Mo powder which can be consolidated by vacuum arc melting or by pressing and sintering.

Nickel

Ores: Various sulphides and arsenides but particularly nickeliferous pyrrhotite or pentlandite (($NiFe$)S) associated with chalcopyrite ($CuFeS_2$) and small amounts of Co, Ag and Pt metals as sulphides and arsenides with siliceous gangue. Garnierite is a hydrated mixed silicate of

Mg and Ni found as a residual deposit from weathering of nickel-bearing serpentine.

Reserves: Distribution widespread in small deposits, but few really large deposits are available. Principal untapped reserves may be in the West Indies and Australia.

Occurrence: Major source Sudbury, Ontario, and neighbouring mines (sulphides). Garnierite is worked in New Caledonia, and some sulphide ore in Finland but it is difficult to compete with Sudbury. The U.S.S.R. production is second to that of Canada but newly discovered deposits in Australia are now being developed.

Ore dressing: (Sulphides) Ore with 1–2% of Ni and Cu is concentrated by flotation to a Cu-rich and a Ni-rich fraction, which latter is partially roasted in a Wedge furnace.

Extraction 1: (Sulphide) (See flow-sheet, Fig. 84) Calcine is matte-smelted in a reverberatory (cf. Cu) and the matte is partially converted to burn out Fe and leaves a 50% Ni–25% Cu matte. At this point a choice of method is available. (a) The obsolete Orford "tops and bottoms" process was to smelt matte with Na_2SO_4 in a cupola. The cast product separated into a solution of CuS_2 in $NaSO_4$ in the "tops" and Ni_3S_2 settled in the "bottoms". Double treatment reduced Cu in the "bottoms" to under 1%. (b) Matte is cooled slowly through the solidification range (875–593°C) to produce coarse primary crystals of CuS_2 and a degraded eutectic with Ni_3S_2. During further slow cooling to 510°C and during a transformation from β to α-Ni_3S_2 more CuS_2 is rejected from the nickel-rich phase and diffuses to the copper-rich primary crystals. The structure should be coarse enough to permit separation of the phases by crushing, grinding and flotation. Flotation is preceded by a magnetic separation of a magnetic metallic nickel alloy containing precious metals (which forms during conversion). Rich nickel matte may be sintered to oxide, smelted to crude nickel in a hearth furnace with coke and cast to anodes to be refined electrolytically with further separation of precious metals as slimes. (c) Alternatively, nickel matte may be cast to anodes and electrolysed directly as indicated on page 314. In either case the electrodes are in separate compartments, the catholyte being chemically purified anolyte and the two being separated by a fabric partition (diaphragm) through which the solution is allowed to flow only from the cathode side, to prevent contamination of the purified solution. Cathodes require only to be melted and cast to a suitable form. (d) The other important extraction process, the carbonyl process, produces fairly pure metal. Matte is crushed, ground and roasted to oxide which is then reduction roasted to metal in a rotating kiln with producer gas at 400°C. The Ni volatilized as $Ni(CO)_4$ by exposure to CO at 50°C, and this gas is decomposed by passing it through a tower of growing nickel pellets at 180°C. The pellets grow very slowly. They circulate continuously and they are screened, and removed when they reach about 10 mm diameter. Copper and precious metals are left as a residue in the volatilizer. Nickel carbonyl can also be decomposed to produce nickel in powder form and if there is iron in the matte at the start of the process it can also be recovered as a by-product, also as a powder. Iron carbonyl is produced along with the nickel carbonyl and can be separated from it by distillation.

Extraction 2: Garnierite is concentrated and reduction roasted in rotating kilns at 850°C to a Ni—Fe sponge which is separated from associated gangue by electro-smelting. The product is a crude ferro-nickel which is refined to several grades by desulphurization and by blowing out C and Si with air in a side-blown converter. A part of the New Caledonia production that cannot be sold as ferro-nickel is sulphidized in a most unusual process by simultaneously blowing air and injecting liquid sulphur into the molten alloy in a Pierce—Smith type converter. The Fe is oxidized and fluxed to slag while the Ni is converted to Ni_3S_2. The matte is given further refinement in another converter and shipped to Europe.

Extraction 3: Particularly when Co and Cu are present in appreciable proportions pressure leaching with air and ammonia dissolves Ni, Co and Cu as ammines. This can be followed by a completely hydrometallurgical separation of these three metals with ammonium sulphate as an important by-product. The extraction of Ni is completed by hydrogen reduction (see page 301).

Nickel is one metal which need not be reduced to the metallic state. A proportion of production is required as nickel salts for the plating industry.

Niobium and Tantalum

Ores: These elements are usually found together. The most important minerals are the mixed iron—manganese niobotantalates, niobite and tantalite, which share the formula (FeMn) $(NbTa)_5O_6$. Niobite may contain up to 34% Ta_2O_5 and tantalite up to 27% of Nb_2O_5. Most other minerals are equally mixed and complex, the next in importance being a series between pyrochlore $((NaCa)Nb_2O_6F)$ and microlite $((NaCa)Ta_2O_6(O,OH,F))$. Distribution is widespread in mineralized regions but very sparse. Niobite is often found with Sn both in placers — in Nigeria and Malaya — and in pegmatite veins in small deposits in many places mostly in Africa. Tantalite also is found in sands and gravels but the major supplies are from pegmatitic deposits in Zaire and Brazil associated with tin and beryl respectively. Nigeria produces about three-quarters of the niobite and Zaire and Brazil the bulk of the tantalite.

Ore dressing: May be extensive to separate minerals from other values. Some grades of niobite are magnetic and magnetic methods are appropriate for the alluvial deposits. Gravity methods and especially tabling should suit these deposits too. Flotation has been used but rather on massive ore which has had to be crushed fine.

Extraction 1: *For ferro-alloy*. Ferro-niobium is made from Nb-rich concentrates by reduction with Al in a thermit process in which some pre-heat is necessary. The ferro-alloy contains about 50% Nb and perhaps 20% Ta. It can also be made in arc furnaces with reduction by silicon.

Extraction 2: *For pure metal*. Concentrates must be opened with HF or alkali fusion. Alkali fusion renders SiO_2, Al_2O_3, Sn, W and V soluble in water; and Fe, Ti, Ca, etc., soluble in HCl leaving Nb and Ta undissolved and well concentrated. These are dissolved in HF and may be separated by repeated preferential crystallization of K_2TaF_7 from the more soluble K_2NbF_7, or by liquid–liquid extraction. Several methods are possible. One which has been used employs methyl isobutyl ketone which absorbs Ta efficiently from weakly (HF) acid solutions with hardly any Nb, and will then abstract the Nb in a second operation from a more strongly acid solution. The ketone is readily stripped of Ta and Nb with pure water in turn. Of many possible reduction methods carbon reduction seems to be commonest. Carbides and oxides are prepared and mixed to give equal atomic proportions of C and O. Compacts are heated *in vacuo* ultimately to over 2000°C to eliminate CO and the product is later vacuum refined by electron beam melting. Electrolytic extraction is also possible.

The Platinum Group

Ores: The metals osmium, iridium, platinum, ruthenium, rhodium and palladium are strongly associated. They occur native as an alloy which is predominantly platinum in most cases. Alluvial deposits in the Urals, and the northern Rocky Mountains, were the main source of platinum 50 years ago but half of today's supplies comes from the nickel ores at Sudbury in Canada where Pt occurs as sperrylite ($PtAs_2$) and most of the remainder is from South Africa where native metal and the native alloy of Os and Ir are associated with Au ores and there are also large lode (vein) deposits of sperrylite. Pt can be recovered economically from ore with less than 10 ppm of precious metals along with Cu, Ni and

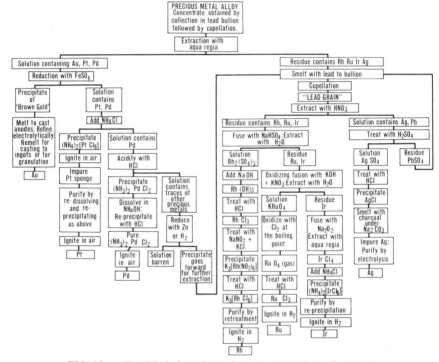

FIG. 85.   Simplified flow-sheet for the extraction of precious metals from concentrates of these obtained from operations outlined, for example, in Figs. 83 and 84.

other metals in greater concentration but of much lower value. **Extraction:** Native platinum metals can be differentiated from gold by their failure to amalgamate. From nickel ores they are collected in anode slimes or (see page 314) in a magnetic phase produced in the converter. These can be extracted by electrolytic solution of less noble constituents. The major part of the operation is the separation of the six metals which must be very thoroughly carried out. Platinum is the most plentiful – and most useful – the others accounting for only a few per cent amongst them. Most base metals are easily eliminated by melting the crude metal with lead to make a bullion, which can be cupelled to leave behind only noble metals. These are separated by the reactions of classical inorganic chemistry and Fig. 85 is a simplified flow-sheet on which the major steps are indicated. In each case the object is to produce a chloride which can be thermally decomposed to produce the metal. Rh, Ru, Ir and Os are rather difficult to chlorinate, and must be protected by hydrogen in the final stages of production because they all tend to oxidize at elevated temperatures. Os, if present, would be oxidized by chlorine along with Ru to a volatile oxide OsO_4 which is not fixed with HCl but can be collected in alkaline solution. The oxide can be precipitated with alcohol and sulphuric acid and ignited under hydrogen to the metal. In recent years solvent extraction methods have been applied to the separation of Rh, Ru and Ir but details are not yet generally available.

Silicon

Ores: Sandstone, quartzite or any convenient form of fairly pure SiO_2.
Reserves and occurrence: Unlimited and ubiquitous.
Ore dressing: Crushing to a convenient size may be necessary.
Extraction 1: *To ferro-silicon.* Calculated proportions of stone, anthracite and steel turnings are melted together in an electric arc furnace with the proportions of a small cupola. The carbon reduces the SiO_2 and the Fe dissolves the Si. The carbon in the alloy need not exceed 0.3%. The high temperature obtainable in the arc furnace is necessary to ensure rapid reduction. With a suitable charge silico-manganese (70% Mn–20% Si–10% Fe) can be made in the same way.
Extraction 2: *For pure Si.* Pure silica is reduced with charcoal in a

similar way and cast from the furnace like a metal. It melts at 1440°C and is required for alloying with Cu, Al, etc.

Silver

Ores: Principal ores contain native Ag or argentite (Ag_2S). Cerargerite (AgCl) occurs with oxidized ores and appears in gossan. A number of less common mixed sulphides, arsenides and antimonides are important mainly because they can be difficult to process. Most virgin Ag is obtained as a by-product from Pb, Cu, Zn or Au ores.

Reserves: Apparently adequate.

Occurrence: Widespread, but production is predominantly in the Americas, particularly Mexico, the U.S.A., Canada, Peru and Bolivia. In Europe, Sweden and Germany are the biggest producers, in Asia, Japan and in Africa, Zaire and the Union of South Africa. Australia produces more than any country outside America.

Ore dressing: Fine grinding in ball mills is necessary prior to cyaniding. Rich ores may need less grinding if amalgamation is practicable. A chloridizing roast may make ore more easily cyanided.

Extraction: (a) The cyanide process is so similar to the process for Au that it will not be described again (see page 368). (b) Amalgamation also follows a similar procedure to that process for Au. (c) The Au–Zn–Pb crust from the Parkes process for desilvering Pb (see page 375) is retorted to distil off the Zn, and the Pb–Ag (see page 316) alloy that is left is cupelled for Ag. The Ag may then be parted of its Au by an electrolytic process as described below. (d) Anode slimes are roasted and leached to remove base metals as far as possible. The residue is melted to an alloy of Ag and Au with other precious metals (Doré metal) which can be electrolysed in nitrate electrolyte, the Ag being transferred to the cathode where it is deposited as loose crystals, while Au and Pt group metals are retained at the anode in a fabric envelope or on a specially positioned tray.

Refining: Cathodic deposit is melted under charcoal and cast to ingots of very fine silver.

Tantalum

See Niobium.

Thorium

Ores: Thorium is obtained from monazite, a mixed phosphate of Th and rare earths which is fairly widely distributed in the U.S.A., the U.S.S.R., Asia, Africa, Europe and Australia. It is associated with placer and beach deposits of gold, tin and titanium.

Ore dressing: Magnetic and gravity methods are used as with other minerals found in sands.

Extraction: Leaching with hot concentrated H_2SO_4 is followed by multi-stage elimination by wet methods of other metals present. The reduction may be by the reaction of Ca on the oxide or a halide in a sealed "bomb" or by electrolysis from a solution in fused halides. These must be followed by vacuum arc refining.

Tin

Ores: Cassiterite (SnO_2) is the only important mineral. It occurs in lode deposits but is mined mainly from placer deposits.

Reserves: A greatly increased demand could not be met quickly but there is probably a large reserve in unattractive deep lodes up country from the placers. Recent price increases have led to the re-opening of old deep mines, notably in Cornwall which is now producing as much as Nigeria.

Occurrence: Geographic distribution is very uneven. Large supplies are available only in Malay, Thailand and Indonesia, Bolivia (deep mined), the U.S.S.R., Nigeria and the U.K. China, Burma and Zaire can produce a little. The few other sources are relatively unimportant. About two-thirds of all tin is mined by dredging. Placers can have as little as 0.02% Sn and still be workable because the SnO_2 grains are so much denser than the gangue. Shallow lode deposits are often won by hydraulic mining (using high-pressure water jets). Deep mining is relatively very costly and unattractive unless other values are present.

Ore dressing: Gravels are screened, all the Sn residing in the fine fraction. Further classification and jigging or tabling for the denser part of the fines easily produces a 50% concentrate. Treatment of deep mined ore is more complex, separate steps often being directed at single impurities — e.g. magnetic separation of wolframite (($FeMn)WO_4$); flotation to remove sulphides; chloridizing roast followed by leaching to remove Fe, Cu, Zn, etc. SnO_2 is very friable and must not be over-ground. It is also chemically very unreactive so that it usually remains behind in the residue after each treatment. The flotation of cassiterite is now possible. This can eliminate the chloridizing stage in some cases.

Extraction: A first smelt with coal in a reverberatory produces impure tin and a tin-rich slag (e.g. 30%). A temperature of $1250°C$ is needed to ensure reduction of Sn from this slag whether it is acid or basic. Slag is re-smelted with coal and limestone at a higher temperature to produce "hardhead" — 95% Sn, 5% Fe — and a lean slag (which may yet need further treatment). Hardhead is smelted with fresh concentrate and siliceous flux for the reaction $2Fe + SnO_2 = Sn + 2FeO$, the FeO being fluxed with SiO_2. Because of the value of tin, painstaking scavenging of the slags, and also of flue dusts and fumes, is a feature of the processes. This is "pyrometallurgy" at its most typical. The scale of these operations is frequently small.

Refining: Tin from alluvial concentrates requires only a little refining, mainly to remove Fe first by liquation and, finally, by blowing air or steam through it. Liquation is melting on a gently sloping hearth which allows the pure tin to run free at a temperature only just above its melting point. Lower grades can be run at higher temperatures. Tin from more complex ores is also liquated but impurities other than Fe have to be removed in other ways. Drossing takes out Fe and Zn by oxidation; Pb can be drossed off by bubbling Cl_2 while sulphur will take out Cu. If Bi is present electrolytic refining is necessary. This will also purify the metal of Pb, Fe, Sb, As and Cu. The process used depends very much on the details of the problem presented by the analysis of each concentrate.

Titanium

Ores: The principal minerals are ilmenite ($FeO·TiO_2$) and rutile (TiO_2) which are found in rock formations, or concentrated in beach sands with other heavy minerals (monazite sands, black sands).

Reserves: The amount of sand available is very large and adequate to meet foreseeable requirements without recourse to leaner alternative sources.

Occurrence: Sands occur in India, Australia and Brazil. Other ores are plentiful in the U.S.A., the U.S.S.R., Canada and Norway.

Ore dressing: Magnetic separation is useful particularly on the sands. It is often desired to produce a low Fe fraction from which to extract Ti and a low Ti fraction for use as an Fe ore, avoiding a middling. Very fine grinding may be necessary to succeed in this.

Extraction 1: *For ferro-alloy.* Concentrate may be reduced with carbon in an arc furnace to produce a high carbon ferro-titanium, or with aluminium by a modified thermit process to make a low carbon alloy. Silicon will not reduce TiO_2.

Extraction 2: *For pure Ti.* Some pre-refining to pure TiO_2 or $TiCl_4$ must be carried out. Concentrates can be reduced to carbide in an arc furnace and then subjected to a chloridizing roast to produce $TiCl_4$ which is liquid at ordinary temperatures. In the Kroll reduction process this chloride· is made to react with molten Mg at $750°C$ in a reaction vessel previously purged of all reactive gases. The $MgCl_2$ and excess Mg are leached and distilled away and the sponge of Ti is consolidated by vacuum arc melting. A number of other methods are possible including the van Arkell process (page 394). Whichever is adopted the possibility of the metal reacting with O_2, N_2, H_2 or C must be strictly avoided.

Tungsten

Ores: The important minerals are wolframite ($FeWO_4$) and scheelite ($CaWO_4$), often occurring with tin ores.

Reserves: There do not appear to be large reserves of W available. Few large deposits are known and most operations are on a small scale or work the metal as a by-product.

Occurrence: China probably has most of the world's W. The major suppliers are the U.S.A., Burma, Portugal, Bolivia and Brazil.

Ore dressing: As the W content of most ores is only about 1%, concentration is necessary. The high density of W minerals makes separations depending on gravity attractive. Cassiterite (SnO_2) is similarly dense, however, but a magnetic separation is practicable from wolframite. Scheelite is non-magnetic but can be treated successfully by flotation. Ore-dressing sequences for W ores are usually very complex and may involve intermediate roasts and leaches.

Extraction 1: *For ferro-alloy.* The process is very like that for ferro-chrome. Concentrate is smelted with a calculated charge of coal, flux and possibly some iron turnings, to produce the ferro-alloy and slagged gangue. The melting point is high and the metal must usually be cooled in the hearth and dug out when cold. The ordinary ferro-tungsten may contain about 70% W and carry 3% C which may be reduced to a lower value by re-melting under an oxidizing slag.

Extraction 2: *For pure W.* Tungsten is extracted by fusion with alkali from concentrate, purified to H_2WO_4 by "wet" methods, and reduced to W powder with H_2. The metal is consolidated by vacuum arc melting or the sintering technique of powder metallurgy (cf. Mo).

Uranium

Ores: Uranium ores are complex mixtures of minerals. In the most important ore, pitchblende, it is present as U_3O_8 in association with many other minerals, mainly sulphides of base metals and silver.

Reserves: There are only a few really large deposits known but very many small ore bodies have been discovered since uranium became an important metal and reserves may well expand to match demand, especially if the value of energy relative to that of other commodities continues to rise. If nuclear energy replaces fossil fuels as the principal source of energy in the early twenty-first century, the U ores that we now know about will not satisfy world demand for many decades.

Occurrence: The biggest deposits are in Canada and Zaire. The U.S.S.R., the U.S.A., Czechoslovakia, Australia and South Africa are among other leading producers. In South Africa the tailings from the cyaniding treatment of gold ore, accumulated over many years, has proved a workable deposit of great value and these tailings continue to be produced from very lean gold ores at a high rate.

Ore dressing: The high density of U minerals makes classification, jigging and tabling appropriate methods of concentrating them. In complex ores the dressing must also produce concentrates of other minerals if possible. Uranium minerals are friable and must not be over-ground.

Extraction: The concentrate is leached with either H_2SO_4 or Na_2CO_3 and pre-refined by "wet" methods. Concentration of the uranium in the leach liquor is effected with ion exchange resins or by solvent extraction (see page 300). This would normally be expected to follow thickening and clarification but methods have been reported for performing these exchange operations on pulp. The uranium is precipitated at this stage with ammonia, filtered and calcined to UO_3. It is then re-dissolved in HNO_3, as uranyl nitrate, which is purified by ether extraction before being finally precipitated as UO_3 which is reduced to UO_2 with hydrogen at $400°C$. For the final reduction to the metal this oxide is treated with anhydrous HF to produce UF_4 which is reduced with Mg by a thermit-type reaction in a steel "bomb" heated to about $600°C$ (see page 293). (There are a number of variations of this process possible.)

Vanadium

Ores: Vanadium is thinly spread. The richest ore, in Peru, contains patronite (V_2S_5). In the U.S.A. the main mineral is carnotite $(K_2O \cdot 2UO_3 \cdot V_2O_5 \cdot 3H_2O)$ associated with a sandstone. In Zambia and South-West Africa it is associated with Pb and Zn ores in the form of complex vanadates. In Germany vanadium appears in iron ores, is collected in the pig iron and then in converter slag which can be recycled through the iron blast furnace to bring V up to 15%. It is now mined also in the Republic of South Africa.

Extraction: Pre-refinement is employed. Concentrate is sintered with Na_2CO_3 to produce sodium vanadate which can be leached out with H_2SO_4 and purified by "wet" methods to produce V_2O_5. This can be reduced along with Fe by carbon or silicon to ferro-vanadium in an electric arc furnace or by Al in a thermit process. Pure metal may be made by a similar process but using Ca as reducing agent in a sealed-off bomb and in an inert atmosphere. Vanadium can also be pre-refined and separated from other metals by halide methods.

Zinc

Ores: Sphalerite, zinc blend (ZnS) and marmatite ((ZnFe)S) are the principal primary minerals being mined today. Secondary, oxidized zincite (ZnO) and calamine or smithsonite $(ZnCO_3)$ are of diminishing importance. More than half of all Zn comes from mixed Zn−Pb ores in which Ag and occasionally Au are associated with the Pb. Straight Zn ores usually have some Cd present. Some have Zn and Cu in similar proportions.

Reserves: Satisfactory and widespread.

Occurrence: Almost half the world's supply comes from the Americas, with Canada producing 24% and the U.S.A. rather less. The U.S.S.R. and Australia are next, with Mexico, Peru, Poland, Germany, Italy, Japan and Zaire all substantial producers. These have recently been joined by Eire and Greenland.

Ore dressing: Extensive comminution followed by flotation is directed at separating Zn from Pb, Cu and as far as possible from Fe. Roasting to ZnO is essential for all purposes. This is carried out in Wedge-type roasters, by suspension roasting, in fluidized beds or by sintering. Calcine which is to be leached for electro-winning will be partially sulphated. If it is to be reduced by carbon a full conversion to ZnO is required. Calcine may be briquetted with coke for retorting.

Extraction 1: Calcine is leached with H_2SO_4 to dissolve all Zn. Roasting conditions should have minimized formation of $ZnO \cdot Fe_2O_3$ to facilitate the solution of ZnO. Solution is neutralized with ZnO to precipitate Fe, Al, Sb, etc., which are removed in thickeners. Cu and Cd are precipitated by cementation with Zn dust. Co, if present, is precipitated with nitroso β-naphthol. The clarified solution is then electrolysed between Pb−0.25% Ag anodes (which are less soluble than Pb) and Al cathodes. The Zn need only be stripped, melted and cast. If the electrolyte has been efficiently purified this should be 99.9% pure. The purity of the electrolyte is most essential because the Standard Electrode Potential of Zinc is so negative (−0.76 V) that most common metals would be co-deposited with the Zn if present in the solution.

Extraction 2: ZnO calcine is reduced in horizontal or vertical retorts with carbon (see Figs. 77 and 78). Despite its being an old labour-intensive batch process, the horizontal retort, Champion or Belgian process persists in use. Retorts are charged with a mixture of calcine or sinter and low volatile coal or coke breeze. This is heated to 1100−1200°C. The Zn vapour produced is collected as liquid in air-cooled condensers called "aludels" and, partly oxidized, as dust called "blue powder" in steel prolongs. Residues remain in the retort and must be removed by hand. CO produced is burned at the mouth of the prolong. In Fig. 45 the intersection of the C–CO and Zn–ZnO lines indicates that with Zn and CO, both at 1 atm pressure, the reduction of ZnO by carbon cannot take place except above the boiling point of Zn (907°C) so that the net reaction can be written—

$$ZnO + C = Zn_{(g)} + CO; \Delta G_T^\circ = 338.9 - 0.28T \text{ kJ mol}^{-1}. \quad (13.2)$$

$\Delta G_T^\circ = 0$ when $T = 1210$ K or 937°C so that this temperature must be exceeded if the reaction is to proceed when the reactants and products are all in their standard states. In the practical case, however, the total pressure is about 1 atm so that $p_{Zn} = p_{CO} = 1$ is not possible. The real situation is very nearly that $p_{Zn} = p_{CO} = 0.5$ atm. If the appropriate adjustments are made the reaction becomes—

$$ZnO + C = Zn_{(0.5 \text{ atm})} + CO_{(0.5 \text{ atm})}; \Delta G_T^\circ = 338.9 - 0.29T \text{ kJ mol}^{-1}. \quad (13.2a)$$

Under these conditions, $\Delta G_T^\circ = 0$ when $T = 1162$ K or 889°C. The reduced boiling point when the partial pressure of Zn is 0.5 atm is only 842°C so that reduction can only produce Zn in the gaseous state and this is true for any value of $p_{Zn} < 1.0$ atm. (The b.p. can be calculated from the vapour pressure equation $- \log p_{Zn}$ (atm) $= -6670/T + 9.12 - 1.26 \log T$.) The mechanism of the reaction must involve the gaseous phase and proceeds in two steps—

$$ZnO + CO = Zn_{(g)} + CO_2 \quad (13.3)$$

followed by

$$CO_2 + C = 2CO. \quad (13.4)$$

The atmosphere in the retort will be maintained predominantly CO if reaction (13.4) is fast enough to consume the CO_2 from reaction (13.3) as soon as it is produced. This requires that the temperature is about 1050–1100°C and that an excess of carbon is available so that even when the ZnO is nearly exhausted the CO/CO_2 ratio in the retort gas will continue to be that in equilibrium with carbon. From consideration of reaction (13.3) it can be shown that a CO/CO_2 ratio of at least 200 is needed to afford protection from oxidation down to the dew point at 842°C. This can be provided by the gas mixture in equilibrium with carbon at 1100°C but full protection through the condenser and prolong would require a much higher CO/CO_2 ratio than is available so that in practice some oxidation to "blue powder" occurs. This is deposited in the prolong and must be recycled.

In vertical retorts the chemistry is similar. Retorts are charged with hot, partially carbonized briquettes made from calcine and coal. The temperature is maintained at $1200°C$ to ensure that the CO/CO_2 ratio in the gas is high. The products are displaced toward the condenser with some of the flue gas so that the total heat to be dissipated by the condenser is greater than in the case of the horizontal retort process. A more elaborate condenser is required such as a zinc splash type working at $500°C$. This collects about 95% of the Zn. Some is deposited as "blue powder" and the last $1-2\%$ is caught in a water spray washer which is designed to serve also as an ejector for drawing the gases through the system. Spent briquettes are discharged in solid form. Iron oxide in the charge is unavoidably reduced with equivalent consumption of carbon. Neither of these retort processes is particularly efficient in its use of energy.

Extraction 3: The zinc blast furnace has a short brick stack and a water-jacketed bosh similar to the lead blast furnace. It is charged with hot sintered concentrate of mixed Zn–Pb ore, preheated coke at $800°C$ and a basic flux. Air at $500°C$ is blown through water-cooled tuyères. All of the Pb is reduced and is collected in the hearth with a limy slag which contains about 6% Zn. Most of the Zn is reduced from unfluxed sinter in the stack to Zn vapour which leaves the furnace in the top gas. It is condensed in a lead splash condenser very quickly, before it has time to suffer appreciable oxidation.

The condenser is a chamber which contains a bath of Pb at $550°C$ with impellers which throw curtains of molten Pb across the path of the Zn-laden gases. Zn dissolves in the Pb which is circulated continuously through water-cooled launders where it is cooled to $450°C$ at which temperature Zn separates out and floats on top of the Pb which is returned to the condenser. Pb dissolves 2.0% Zn at $450°C$ and 2.25% at $550°C$.

The chemistry is similar to that described above but the conditions are much more critical in all parts of the system. The CO/CO_2 ratio in the bosh and hearth must be high enough to reduce PbO and ZnO without reducing FeO. PbO is easily reduced (see Fig. 45) but if the CO/CO_2 ratio rises above about 10/1 in the hearth where the temperature is about $1300°C$, FeO in the slag will be reduced to solid Fe which, not being soluble in Pb, is difficult to dispose of in the hearth. In practice the activity of FeO is kept low (≈ 0.5) to inhibit its reduction, by using a very limy slag. The slag contains about 6% ZnO. Most of the ZnO is reduced in the solid state but the low CO/CO_2 ratio in the furnace gas limits the reaction rate that is possible. A porous, easily reducible sinter is obviously a necessity. The gases are taken off at about $1000°C$. They must obviously remain reducing toward ZnO, otherwise the Zn vapour will be oxidized to blue powder. The relevant equation for the reduction reaction is—

$$ZnO_{(s)} + CO = Zn_{(g)} + CO_2 ; \Delta G_T^\circ = 178.3 - 0.111T \text{ kJ}. \quad (13.3)$$

At $1000°C$

$$G_{1273}^\circ = 37.0 \text{ kJ}.$$

Hence

$$K_{1273} = p_{Zn} \frac{p_{CO_2}}{p_{CO}} = 0.03.$$

Now, in the top gases there is about 5% Zn so that $p_{Zn} = 0.05$. The equilibrium CO/CO_2 ratio is therefore $0.05/0.03 = 1.6$ which is the minimum value permissible if the gas is to be incapable of oxidizing the Zn vapour. The top gas typically has about 20% CO and 10% CO_2 which leaves a small margin for error. If the gas is cooled to $900°C$, however, the critical CO/CO_2 ratio rises to 6.7 and at the depressed dew point corresponding to $p_{Zn} = 0.05$ atm, namely $670°C$, the critical CO/CO_2 ratio is nearly 600. It is obvious that protection by such an atmosphere cannot be provided and it was for this reason that the lead splash condenser had to be developed. If the top gas is insufficiently reducing it is sometimes possible to return to satisfactory conditions by injecting oxygen into the top gas to burn some of the CO to raise the temperature. At $1100°C$ the critical CO/CO_2 ratio is only $0.54/1$, so about a quarter or a third of the CO might be converted to CO_2 for this purpose but the balance is a fine one.

The Zn produced contains about 1.2% of Pb with a little Fe and As and all of any Cd in the sinter. The process is flexible and is used to recover Zn from scrap and residues from other processes. Matte and speiss can be collected from the hearth if there is Cu or As to be removed.

Refining: Electrolytic Zn needs no refining. Other Zn is likely to contain Pb, Cd and Fe. Pb and Fe can be reduced sufficiently for some purposes by a liquation process but the Pb cannot be reduced below 1.2% by this method. Higher purity and the removal of Cd can be effected only by fractional distillation (see page 329).

Zirconium

Ores: Zircon ($ZrO_2 \cdot SiO_2$) is a constituent of the same beach sands as rutile (TiO_2) from which it is separated by magnetic or electrostatic methods. Baddeleyite (ZrO_2) is a less important mineral found in gravels in Brazil. Zirconium is always associated with about 1% of hafnium from which separation is difficult.

Occurrence: Australian beach sands are principal source. Otherwise, as Ti.

Ore dressing: The narrowly sized sands are amenable to separation according to density by a Humphreys spiral (see page 68) as well as by magnetic and electrostatic methods.

Extraction: Zircon is roasted to carbide in an electric arc furnace along with graphite. Most of the SiO_2 is volatilized as SiO. The carbide is then chlorinated in a shaft furnace under reducing conditions and the $ZrCl_4$ condensed separately from the more volatile $SiCl_4$ and $TiCl_4$. The difficult separation of hafnium may be effected by several tedious processes such as fractional sublimation of the chlorides or fractional crystallization of K_2ZrF_6 and K_2HfF_6 – or see page 333. Solvent extraction has also been suggested. This separation is necessary only when the metal is to be used in nuclear reactors where the high absorption cross-section of Hf for thermal neutrons makes its presence unacceptable. Purified $ZrCl_4$ is sublimed into contact with molten Mg at $825°C$ in a retort which has been carefully purged of reactive gases (Kroll process). After reduction, the zirconium sponge is freed of $MgCl_2$ by leaching and of excess Mg by distillation under vacuum (see Titanium). Further purification may be effected by reacting on sponge with iodine vapour and decomposing the ZrI_4 formed on a hot filament of pure Zr a short distance away in the same reaction vessel (van Arkell process). Sponge or refined metal must be consolidated by vacuum arc or vacuum induction melting. In the latter case a carbon crucible is used and a pick-up of about 0.2% carbon is unavoidable (though considering the temperature, 1900°C, this is remarkably small).

Other Metals

The few metals which have been omitted in this chapter are of little economic importance (though most of them have their uses) and do not involve any novel procedures. In most cases intricate "wet" methods are used to purify intermediate products of other processes. The alkali metals are unusual in that their ores are salts, and indeed except for Na the occurrence of these in reasonable concentration is very rare, the principal source being the Stassfurt deposits in Germany. Na and K are in demand to provide liquid metal cooling fluid for heat exchangers in the nuclear energy industry. Among the alkaline earths Ca is required as a reducing agent in a number of extraction processes. All three are prepared by the electrolysis of pure fused chlorides or hydroxide (Na). Rare earths are obtained from monazite sands. The concentrate is extracted with H_2SO_4 and the individual elements separated as far as is necessary by "wet" methods. Reduction is by electrolysis of fused chlorides.

FOOTNOTE

It cannot be too highly stressed that the choice of procedure to be adopted in any case depends jointly upon technological and economic considerations. It is of prime importance to the success of the venture that the sum of all the production costs per ton of metal should not exceed the price per ton on the market – a figure almost entirely beyond the control of the individual operator. By his skill in running his plant from day to day the metallurgist has always had the duty to keep production costs low but he also shares with others the responsibility for developing, selecting and introducing new procedures when appropriate. In this the metallurgist must collaborate particularly with the accountants by providing them with the best possible estimates of requirements of power, materials and labour and of production rate possibilities of alternative processes so that they can calculate with reasonable precision[88] comparative costs on the basis of which major policy decisions can be made. In this way the part played by guesswork and intuition can be minimized if not eliminated.

WORKED EXAMPLES OF CALCULATIONS EMPLOYING THERMODYNAMICS

THESE examples are intended to give guidance to the student in the application of thermodynamics to the solution of metallurgical problems. They are supplementary to those examples which have appeared in the text to help to explain the chemistry of various processes. Here, the main purpose is to show how data obtained from reference books can be combined and adjusted into forms suitable for direct application to the solution of problems.

Most of the information presented under the headings DATA have been obtained from books in the bibliography numbered 20, 31, 50, 52. In most cases the data have been converted from c.g.s. to SI units with some rounding off and the calculations have been carried out in SI units. This usually leads to calculated values which are rather different and occasionally substantially different from values calculated in c.g.s. units and converted as a final step. Most practising thermodynamists would probably adopt the latter course if only to keep the numbers smaller and minimize the work involved. Differences between values obtained by the two approaches must be within the limits of uncertainty of the data used. The accuracy of the data has not been taken into account. It is seldom quoted in the more accessible compilations but estimates are usually given in original papers. Errors in data can be substantial and can accumulate rapidly in the course of a calculation especially when the difference calculated between two values of only moderately high accuracy is small. Then the sum of the errors on the original values becomes the error on the difference.

Some of the calculations may seem to be unduly long and tedious. This is typical of what must be carried out as a routine if thermodynamics is to be used properly. On the other hand, the student will notice that small corrections are often mutually cancelling so that a tiresome exercise in arithmetic brings one back almost to the starting point. With experience one can sometimes foresee such circumstances.

Some of the examples have relevance to topics which have been dealt with in the text and forward reference will have been made to them. It will be appreciated that there is a considerable degree of simplification and approximation in many of these calculations. There is something of an art in knowing what one can "get away with". This can be learned only with practice and experience.

Example 1. Enthalpy and Kirchhoff's equation

Calculate the increase in the enthalpy of 1 mole of H_2O as ice at $0°C$ is melted, heated to $100°C$, evaporated and heated to $200°C$ at atmospheric pressure.

DATA: C_p for H_2O (liquid) (273–373 K) = 75.4 J mol^{-1} K^{-1},
C_p for H_2O (gas) (> 373 K) = 30.0 + 10.7 x $10^{-3}T$
$+ 0.33$ x $10^5 T^{-2}$ J mol^{-1} K^{-1},
L_f for H_2O at 273 K = 6013 J mol^{-1},
L_e for H_2O at 373 K = 40,700 J mol^{-1}.

Let the enthalpy of ice at 273 K = $H_{1(273)}$.
Then the enthalpy of liquid water at 273 K = $H_{2(273)} = H_{1(273)} + L_f$

$$(6.19a)$$

$$= H_{1(273)} + 6013 \text{ J}.$$

The enthalpy of liquid H_2O at any temperature T K between 273 and 373 K is given by

$$H_{2(T)} = H_{2(273)} + \int_{273}^{T} C_p \, dT \qquad (6.19a)$$

$$= H_{1(273)} + 6013 + \int_{273}^{T} 74.5 \, dT$$

$$= H_{2(273)} + 6013 + 74.5 \, T \Big|_{273}^{T}$$

so that when $T = 373$ K,

$$H_{2(373)} = H_{1(373)} + 6013 + 7450 \text{ J}$$
$$= H_{1(373)} + 13{,}463 \text{ J}.$$

When the H_2O is vaporized at 373 K the enthalpy becomes

$$H_{3(373)} = H_{2(373)} + L_e$$
$$= H_{1(273)} + 13{,}463 + 40{,}700$$
$$= H_{1(273)} + 54{,}163 \text{ J}.$$

Further heating of the vapour to T K where $T > 373$ entails increasing the enthalpy to

$$H_{3(T)} = H_{3(373)} + \int_{373}^{T} C_p \, dT$$

$$= H_{3(373)} + \int_{373}^{T} (30.0 + 10.7 \times 10^{-3} T + 0.33 \times 10^{5} T^{-2}).$$

When $T = 473$ K

$$H_{3(473)} = H_{1(273)} + 54{,}163 + (30.0T + 5.35 \times 10^{-3} T^2$$
$$- 0.33 \times 10^{5} T^{-1}) \Big|_{373}^{473}$$

$$= H_{1(273)} + 54{,}163 + 3470$$
$$= H_{1(273)} + 57{,}633 \text{ J}.$$

To facilitate plotting Enthalpy vs. Temperature in this range of temperature where the relationship is not linear, $H_{3(T)}$ should be evaluated also at at least two other temperatures. It can be shown that

$$H_{3(400)} = H_{1(273)} + 55{,}091 \text{ J} \quad \text{and} \quad H_{3(450)} = H_{1(273)} + 56{,}821 \text{ J}.$$

The results of this calculation are plotted in Fig. A1.

Note that all temperatures entered in the calculation must be in degrees Kelvin. In thermodynamic calculations this is the rule.

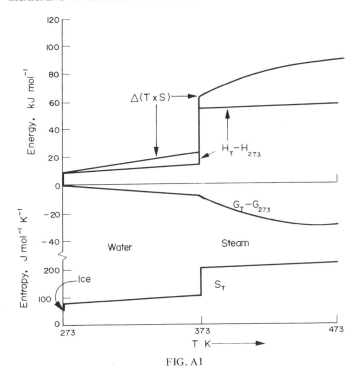

FIG. A1

Example 2. Entropy and its change with temperature

Calculate the entropy of 1 mole H_2O at various stages as ice at $0°C$ is melted, heated to $100°C$, evaporated and heated to $200°C$ at atmospheric pressure.

DATA: As in Example 1.

$$S_{(298)} = 80.49 \text{ J mol}^{-1} \text{ K}^{-1}.$$

Obtain the entropy of liquid H_2O at 273 K from

$$S_{2(273)} = S_{2(298)} + \int_{298}^{273} \frac{C_p}{T} \, dT \qquad \text{(equation (6.3))}$$

$$\text{i.e. } S_{2(273)} = S_{2(298)} + \int_{298}^{273} \frac{75.4}{T} \, dT$$

$$= 80.49 + 75.4 \ln T \Big|_{298}^{273}$$

$$= 80.49 - 6.60$$

$$= 73.89 \text{ J mol}^{-1} \text{ K}^{-1}.$$

If the water is solidified at the same temperature the entropy will be reduced by L_f/T, i.e. by $6013/273 = 22.03$ J K^{-1} so that the entropy of ice at 273 K is $S_{1(273)} = 51.86$ J mol^{-1} K^{-1}.

By exactly the same argument as was used above,

$$S_{2(323)} = 80.49 + 75.4 \ln T \Big|_{298}^{323} = 86.56 \text{ J mol}^{-1} \text{ K}^{-1},$$

$$S_{2(348)} = 80.49 + 75.4 \ln T \Big|_{298}^{348} = 92.18 \text{ J mol}^{-1} \text{ K}^{-1},$$

and $\quad S_{2(373)} = 80.49 + 75.4 \ln T \Big|_{298}^{373} = 97.42 \text{ J mol}^{-1} \text{ K}^{-1}.$

It is best to calculate from 298 K each time to avoid building up errors. When the water is vaporized there is another increment of entropy equal to L_e/T, i.e. $40,700/373 = 109.11$ so that the entropy of steam at 373 K is $S_{3(373)} = 206.53$ J mol^{-1} K^{-1}.

When the steam is superheated to T K the entropy is obtained from

$$S_{3(T)} = 206.53 + \int_{373}^{T} \frac{C_p}{T} \, dT$$

$$= 206.53 + \int_{373}^{T} \left(\frac{30.0}{T} + 10.7 \times 10^{-3} + 0.33 \times 10^5 \, T^{-3} \right) dT$$

$$= 206.53 + (30.0 \ln T + 10.7 \times 10^{-3} \, T - 1.65 \times 10^4 \, T^{-2}) \Big|_{373}^{T}$$

whence

$$S_{3(400)} = 208.93 \text{ J mol}^{-1} \text{ K}^{-1},$$

$$S_{3(450)} = 213.02 \text{ J mol}^{-1} \text{ K}^{-1},$$

and

$$S_{2(473)} = 214.24 \text{ J mol}^{-1} \text{ K}^{-1}.$$

Entropy values from 273 to 473 K are plotted in Fig. Al.
The greatest increment of entropy is clearly that associated with vaporization. Note that entropy can be given an absolute value whereas enthalpy and Gibbs free energy cannot.

Example 3. Gibbs free energy and its change with temperature

Calculate the change in the free energy of 1 mole of H_2O as ice at $0°C$ is melted, heated to $100°C$, evaporated and heated to $200°C$ at atmospheric pressure.

DATA: As in Example 1 and using the results of Examples 1 and 2.

Heat supplied to a closed system must be accountable either as vibrational energy or as structural energy, i.e. as $T \times S$ or as G. This can be expressed—

$$\Delta H = \Delta G + \Delta(T \times S).$$

In Example 1 the increase in enthalpy has been calculated at each stage, and in Example 2 the entropy at each stage has been calculated from which the vibrational energy $T \times S$ and hence the change in vibrational energy $\Delta(T \times S)$ can also be calculated.
In the initial state, ice at $0°C$, $H = H_{1(273)}$ and from Example 2 $S_{1(273)} = 51.86 \text{ J K}^{-1}$ so that $(T \times S) = 14{,}158 \text{ J}$.
After melting, at 273 K, the enthalpy of the water is $H_{2(273)} = H_{1(273)} + 6013 \text{ J}$ and $S_{2(273)} = 73.89 \text{ J K}^{-1}$ so that $(T \times S) = 20{,}172 \text{ J}$. The change in enthalpy is, therefore, 6013 J and the change in vibrational energy is 6014 J. Hence

$$\Delta G = \Delta H - \Delta(T \times S)$$
$$= 6013 - 6014$$
$$= -1 \text{ J}.$$

Since ice and water are in equilibrium at 273 K the correct value of ΔG is zero. Strictly the temperature should be 273.16 K. This and other small errors in the data can in general be expected to bring about small discrepancies in calculated values.

Repeating the calculation for the final state, water vapour at 473 K, $H_{3(473)} = H_{1(273)} + 57,633$ J and $S_{3(473)} = 214.24$ J K^{-1} so that $(T \times S) = 101,335$ J and $\Delta(T \times S) = 87,177$ J.

Hence, $\Delta G_{473} = 57,633 - 87,177$

$$= -29,544 \text{ J}.$$

ΔG is plotted in Fig. A1 along with ΔH and also $\Delta(T \times S)$.

It will be noticed that the increase in vibrational energy is consistently greater than the increase in enthalpy and that the values of G decrease as the temperature is raised. G is a continuous function of temperature whereas both H and S are discontinuous. The change in G reflects the reduction in bond energy as interatomic spacings increase with temperature.

Example 4. Enthalpy and Kirchhoff's equation

Calculate the increase in enthalpy of 1 g atom of · iron as its temperature is raised from 25°C to 2000°C.

DATA:

C_p for Fe$_\alpha$(300–1033 K) = 14.1 + 29.7 $\times$ 10^{-3} T + 1.8 $\times$ 10^5 T^{-2} J K^{-1},

C_p for Fe$_\beta$(1033–1183 K) = 43.5 J K^{-1},

C_p for Fe$_\gamma$(1183–1674 K) = 20.3 + 12.55 $\times$ 10^{-3} T J K^{-1},

C_p for Fe$_\delta$(1674–1808 K) = 43.1 J K^{-1},

C_p for Fe$_l$(1808–3000 K) = 41.8 J K^{-1},

L_t ($\alpha \rightarrow \beta$) = 1720 J g atom^{-1} at 1033 K,

$L_t(\beta \rightarrow \gamma)$ = 910 J g atom^{-1} at 1185 K,

$L_t(\gamma \rightarrow \delta)$ = 630 J g atom^{-1} at 1674 K,

$L_t(\delta \rightarrow l)$ = 16,160 J g atom^{-1} at 1808 K.

(These figures have been taken from reference 89 and converted to SI. Corresponding figures from other sources may appear to be significantly different, especially in the higher temperature range. This reflects the inherent difficulty of making the measurements and it is probable that further changes in the preferred values will continue to be made.)

The calculation is carried out as in Example 1 applying equation (6.19a). It is not practicable to apply that equation at low temperatures down toward 0 K because the C_p data are not valid at very low temperatures. Any extrapolation to evaluate H at 0 K produces only a fictitious value for H_0.

Let the enthalpy of 1 g atom of Fe at $25° = H_{(298)}$.

Then $H_{\alpha(400)} = H_{\alpha(293)} + \int_{298}^{400} (C_p \text{ for Fe}_\alpha) \, dT = H_{\alpha(298)} + 2614 \text{ J}$,

$H_{\alpha(600)} = H_{\alpha(298)} + \int_{298}^{600} (C_p \text{ for Fe}_\alpha) \, dT = H_{\alpha(298)} + 8454 \text{ J}$,

$H_{\alpha(800)} = H_{\alpha(298)} + \int_{298}^{800} (C_p \text{ for Fe}_\alpha) \, dT = H_{\alpha(298)} + 15{,}367 \text{ J}$,

$H_{\alpha(1033)} = H_{\alpha(298)} + \int_{298}^{1033} (C_p \text{ for Fe}_\alpha) \, dT = H_{\alpha(298)} + 24{,}832 \text{ J}$.

At the first transition temperature,

$$H_{\beta(1033)} = H_{\alpha(1033)} + L_{t(\alpha-\beta)} = H_{\alpha(298)} + 24{,}832 + 1720$$
$$= H_{\alpha(298)} + 26{,}552 \text{ J}.$$

Hence

$$H_{\beta(1183)} = H_{\beta(1033)} + \int_{1033}^{1183} (C_p \text{ for Fe}_\beta) \, dT = H_{\alpha(298)} + 26{,}552 + 6825$$

$$= H_{\alpha(298)} + 33{,}377 \text{ J}.$$

Since the relationship in that temperature range is obviously linear there is no need to calculate intermediate values. The rest of the calculation follows exactly the same pattern and only the results will be presented here. They will provide the reader with an opportunity for carrying out the calculations on his own.

$$H_{\gamma(1183)} = H_{\beta(1183)} + L_{t(\beta-\gamma)} = H_{\alpha(298)} + 34{,}287 \text{ J},$$

$$H_{\gamma(1400)} = H_{\gamma(1183)} + \int_{1183}^{1400} (C_p \text{ for } Fe_\gamma)\, dT = H_{\alpha(298)} + 42{,}209 \text{ J,}$$

$$H_{\gamma(1674)} = H_{\gamma(1183)} + \int_{1183}^{1674} (C_p \text{ for } Fe_\gamma)\, dT = H_{\alpha(298)} + 53{,}056 \text{ J,}$$

$$H_{\delta(1674)} = H_{\gamma(1674)} + L_{t(\gamma\to\delta)} = H_{\alpha(298)} + 53{,}687 \text{ J,}$$

$$H_{\delta(1808)} = H_{\delta(1674)} + \int_{1674}^{1808} (C_p \text{ for } Fe_\gamma)\, dT = H_{\alpha(298)} + 59{,}462 \text{ J,}$$

$$H_{l(1808)} = H_{\delta(1808)} + L_f = H_{\alpha(298)} + 75{,}622 \text{ J,}$$

$$H_{l(2273)} = H_{l(1808)} + \int_{1808}^{2273} (C_p \text{ for Fe liquid}) = H_{\alpha(296)} + 95{,}059 \text{ J.}$$

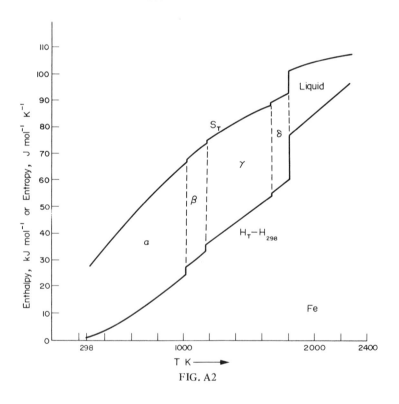

FIG. A2

Calculations of this kind are inevitably tedious. There are no short cuts but the use of electronic calculators has, in recent years, made the work rather less time consuming. It is usual to refer back to the value at 298 K – or rather at 25°C or 298.16 K – rather than to 0°C as was used in Example 1. The enthalpy of iron is plotted against temperature in Fig. A2.

Example 5. Entropy and its change with temperature

Calculate the change in entropy of 1 g atom of iron as its temperature is raised from 25°C to 2000°C.

DATA: As in Example 4.

$$S_{Fe(\alpha)(298)} = 27.16 \text{ J K}^{-1} \text{ g atom}^{-1}.$$

The calculation is carried out exactly as in Example 2 but using as many steps as in Example 4. In this case the entropy of the iron in its initial state is known. The data cannot be extrapolated down toward 0 K.

Applying equation (6.3)–

$$S_{\alpha(400)} = S_{\alpha(298)} + \int_{298}^{400} \left(\frac{C_p \text{ for Fe}_\alpha}{T} \right) dT = 27.61 + 6.73 = 34.34 \text{ J K}^{-1}.$$

Similarly,
$$S_{\alpha(600)} = 27.61 + 18.07 = 45.68 \text{ J K}^{-1},$$
$$S_{\alpha(800)} = 27.61 + 27.96 = 55.57 \text{ J K}^{-1},$$
$$S_{\alpha(1033)} = 27.61 + 38.43 = 66.04 \text{ J K}^{-1}.$$

At the first transition,
$$S_{\beta(1033)} = S_{\alpha(1033)} + \frac{L_{t(\alpha \to \beta)}}{T}$$
$$= 66.04 + 1.67 = 67.71 \text{ J K}^{-1},$$
$$S_{\beta(1183)} = 67.71 + 5.90 = 73.61 \text{ J K}^{-1}.$$

At the second transition,
$$S_{\gamma(1183)} = S_{\beta(1183)} + \frac{L_{t(\beta \to \gamma)}}{T}$$
$$= 73.61 + 0.77 = 74.38 \text{ J K}^{-1},$$

$$S_{\gamma(1400)} = 74.38 + 6.14 = 80.52 \text{ J K}^{-1},$$
$$S_{\gamma(1674)} = 74.38 + 13.20 = 87.58 \text{ J K}^{-1}.$$

At the third transition, $\quad S_{\delta(1674)} = 87.58 + 0.38 = 87.96 \text{ J K}^{-1},$

$$S_{\delta(1808)} = 87.96 + 3.32 = 91.28 \text{ J K}^{-1}.$$

At the melting point, $\quad S_{l(1808)} = 91.28 + 8.93 = 100.21 \text{ J K}^{-1},$

$$S_{l(1900)} = 100.21 + 2.07 = 102.28 \text{ J K}^{-1},$$

and finally, $\quad S_{l(2273)} = 102.28 + 4.22 = 106.50 \text{ J K}^{-1}.$

The results are plotted in Fig. A2. The increase in entropy at the first transition, which is the Curie point, indicates that some degree of disordering is involved even though this is not detectable crystallographically. The degree of disordering is, of course, much less than that at the melting point.

Example 6. Heat of transformation and its change with temperature

Calculate the heat of transformation $Fe_\beta - Fe_\gamma$ at $1401°C$.

DATA: As in Example 4.

It is frequently necessary to calculate the value of a thermodynamic function at a temperature other than that for which a value is already available, either to find a value at the temperature at which a reaction is likely to occur from published values which have been calculated for 298 K, or to find the value at another temperature at which one or more of phases present may be either superheated or supercooled.

It is necessary in this case to use Kirchhoff's law in the form—

$$\Delta H_{T_2} = \Delta H_{T_1} + \int_{T_1}^{T_2} \Delta C_p \, dT. \qquad \text{(cf. equation (6.19b))}$$

In this case $\Delta H_{T_1} = L_{t(\beta-\gamma)(1183)} = 910 \text{ J g atom}^{-1},$

$\Delta C_p = (C_p \text{ for Fe}_\gamma) - (C_p \text{ for Fe}_\beta)$ (always product minus reactant).

$$\Delta H_{T_2} = L_{t(\beta-\gamma)(1674)},$$

$$\text{i.e.}\ \Delta H_{T_2} = 910 + \int_{1183}^{1674} (-23.2 + 12.55 \times 10^{-3}\ T)\,dT$$

$$= 910 - 2588$$

$$= -1678\ \text{J g-atom}^{-1}.$$

It is known that Fe_β and Fe_δ are the same phase so that $\Delta H_{(\beta - \gamma)}$ and $\Delta H_{(\gamma - \delta)}$ should be numerically equal but of opposite sign at any particular temperature. Thus at 1674 K the expected value is -630 J. This shows that the data used in the calculation are not fully compatible, but it is not possible to deduce which item is in error without examining all the original sources in great detail. Data taken from other compilations show similar inconsistencies and it must be accepted that extrapolations beyond the stated temperature range of validity of the C_p values involved can be expected to yield only approximate values.

Example 7. Changes in values of ΔS and ΔG with temperature

Calculate the entropy change and the Gibbs free energy change when 1 g-atom of Fe_β transforms to Fe_γ at $1401°C$.

DATA: As in Example 4 and the results of Examples 5 and 6.

$$\Delta S_{t(\beta - \gamma)(1183)} = 0.77 \quad \text{(See Example 5)}.$$

$$\Delta S_{t(\beta - \gamma)(1674)} = 0.77 + \int_{1183}^{1674} \left(\frac{C_p \text{ for } \gamma - C_p \text{ for } \beta}{T} \right) dT$$

$$= 0.77 + \int_{1183}^{1674} \left(\frac{-23.2 + 12.55 \times 10^{-3}\ T}{T} \right) dT$$

$$= 0.77 + (-1.89)$$

$$= -1.12\ \text{J K}^{-1}.$$

From this it follows that—

$$\Delta G_{t(\beta - \gamma)(1674)} = \Delta H_{t(\beta - \gamma)(1674)} - 1674 \Delta S_{t(\beta - \gamma)(1674)}$$

$$= -1678 - 1674 \times (-1.12)$$

$$\text{i.e.} = +191\ \text{J g-atom}^{-1}$$

The expected value for $S_{t(\beta-\gamma)(1674)}$ is -0.38 J K^{-1}, that is equal in value but of opposite sign to $S_{t(\gamma-\delta)(1674)}$. The expected value for $\Delta G_{t(\beta-\gamma)(1674)}$ is zero because the β- and δ-phases are equally in equilibrium with the γ-phase at that temperature.

The reasons for the discrepancies are the same as were noted in the previous example. These examples demonstrate how the values of the thermodynamic functions can be adjusted for change of temperature. They also show that a high degree of precision is not always obtainable. Values obtained by extrapolation in such a manner should be used with caution.

Example 8. Change in ΔH with temperature. Kirchhoff's law. Hess's law

Calculate ΔH at $1600°$C for the reaction—

$$Fe_{(liquid)} + \tfrac{1}{2}S_2 = FeS_{(liquid)}.$$

DATA: As in Example 4.
For $Fe_\alpha + \tfrac{1}{2}S_2 = FeS_{(solid)}$; $H_{298} = -97,250$ J.

For FeS, $L_f = 20,900$ J mol^{-1} and ΔC_p on melting is about 2.0 J K^{-1}.

For the required reaction ΔC_p is about 6.0 J K^{-1}. These last two figures are estimates (see page 155 and ref. 30). Any inaccuracy incurred affects the result but does not invalidate the method of calculation.

Kirchhoff's law must be used to reduce available relevant ΔH values to their 298 K values. Then Hess's law is applied to obtain the ΔH_{298} value for the required reaction. That value is again adjusted using Kirchhoff's law to find the 1873 K ($1600°$C) value which is required.

$$Fe_\alpha + \tfrac{1}{2}S_2 = FeS_{(solid)}; \quad \Delta H_{298} \text{ (given)} \qquad\qquad = -97,250$$

$$Fe_\beta \quad = Fe_\alpha; \quad \Delta H_{298} = -1720 + \int_{1033}^{298} (C_{p(\alpha)} - C_{p(\beta)})\, dT$$
$$= \quad 4933$$

$$Fe_\gamma \quad = Fe_\beta; \quad \Delta H_{298} = -910 + \Big|_{1183}^{298} (C_{p(\beta)} - C_{p(\gamma)})$$
$$= -14,810$$

$$Fe_\delta \quad = Fe_\gamma; \quad \Delta H_{298} = -630 + \int_{1674}^{298} (C_{p(\gamma)} - C_{p(\delta)})\, dT$$
$$= 13{,}715$$

$$Fe_{(liquid)} = Fe_\delta; \quad \Delta H_{298} = -16{,}160 + \int_{1808}^{298} (C_{p(\delta)} - C_{p(l)})\, dT$$
$$= -18{,}123$$

$$FeS_{(solid)} = FeS_{(liquid)}; \quad \Delta H_{298} = 20{,}900 + \int_{1473}^{298} (2.0)\, dT \quad = 18{,}503$$

$$Fe_{(liquid)} + \tfrac{1}{2} S_2 = FeS_{(liquid)}, \text{ by Hess's law; } \Delta H_{298} \qquad = -92{,}985$$

Hence
$$\Delta H_{1873} = -92{,}985 + \int_{298}^{1873} (6.0)\, dT$$
$$= -83{,}535 \text{ J}.$$

The units are J per mole of FeS, or per mole of Fe, or per half-mole of S_2. It is sometimes convenient to refer to the heat of a reaction in terms of J per g-formula weight. In that case if the equation were written—

$$2Fe + S_2 = 2FeS$$

the heat of the reaction would be $-167{,}070$ J per g-formula weight.

Example 9. Activity and change of standard state

TABLE A.1

1	2	3	4	5	6	7	8	9
wt% C	$\dfrac{p^2_{CO}}{p_{CO_2}}$	$sat^a C$	x_C	$h^a C$	$\%^a C$	$sat^\gamma C$	$h^\gamma C$	$\%^\gamma C$
0.216	93	0.006	0.010	0.01	0.216	0.6	1.0	1.0
0.425	191	0.0125	0.0195	0.022	0.44	0.64	1.1	1.05
0.64	292	0.019	0.029	0.032	0.68	0.65	1.1	1.07
0.85	400	0.026	0.0385	0.043	0.93	0.68	1.1	1.1
1.28	670	0.044	0.057	0.072	1.55	0.78	1.25	1.23
1.68	1030	0.067	0.073	0.111	2.39	0.92	1.52	1.43
2.10	1510	0.099	0.0908	0.162	3.50	1.07	1.78	1.67
2.92	2930	0.19	0.123	0.424	6.80	1.55	3.45	2.32
4.12	7200	0.47	0.172	0.775	16.7	2.25	4.50	4.0
5.20	15300	1.00	0.203	1.64	35.5	4.9	8.1	6.8

The activity of carbon in solution in molten iron at $1540°C$ has been investigated[90] by equilibrating iron carbon alloys with various mixtures of CO and CO_2. The experimental observations are given in the first two columns of Table A.1. Calculate the activity of C at each composition, referring to the saturated solution (pure C), to the Henrian standard state and to the 1% solution as standard state. The 5.20% solution is saturated with carbon.

Consider the reaction—

$$\underline{C}_{(in\ Fe)} + CO_2 = 2CO.$$

The equilibrium constant is—

$$K = \frac{p_{CO}^2}{p_{CO_2}} \cdot \frac{1}{a_C}.$$

Rearranging, $a_C = K' \dfrac{p_{CO}^2}{p_{CO_2}}$ where $K' = K^{-1}$.

The activity of the carbon is therefore proportional to the ratio in the second column of the table. This is true for any selected standard state. If the chosen standard state is pure carbon or the saturated solution, then at 5.2% C, $_{sat}a_C = 1.0$, so that—

$$1.\overset{\prime}{0} = K' \times 15,300,$$

$$\therefore\ K' = 1/15,300.$$

At other competitions,

$$_{sat}a_C = \frac{1}{15,300} \cdot \frac{p_{CO}^2}{p_{CO_2}}.$$

Values of $_{sat}a_C$ calculated from this relationship have been entered into column 3 of Table A.1.

Before the Henrian activity values can be calculated it is necessary to convert the values of wt% C to mole fractions. This is obtained from—

$$\text{Mole fraction } x_c = \frac{\%\ C/12}{\%\ C/12 + (100 - \%\ C)/56} = \frac{56.0\%\ C}{44.0\%\ C + 1200}$$

the molecular weights of C and Fe being 12 and 56 respectively. It is not appropriate to use the simplifying assumption that % C is negligibly small over the range of compositions used here. Values of x_C so calculated have been entered in column 4 of the table.

At this stage a graph should be drawn to determine the range of validity of Henry's law, p^2_{CO}/p_{CO_2} being plotted aginst x_C. In fact such a graph shows that only the first point could lie on a Henry's law line. In the absence of supporting evidence the best that can be done is to put $_ha_C = x_C$ for the alloy with 0.216% C. Since $_ha_C$ is then equal to 0.01,

$$0.01 = K'' \times 93$$

so that $$K'' = 1.075 \times 10^{-4}$$

and at other compositions,

$$_ha_C = 1.075 \times 10^{-4} \frac{p^2_{CO}}{p_{CO_2}}.$$

Values of $_ha_C$ calculated from this equation have been entered into column 5 of Table A.1.

Continuing to assume that Henry's law is valid up to 0.216% C, we can write—

$$_{\%}a_C = \% \, C = 0.216 = K''' \times 93.$$

Hence, $$K''' = 2.32 \times 10^{-3}$$

so that $$_{\%}a_C = 2.32 \times 10^{-3} \frac{p^2_{CO}}{p_{CO_2}}.$$

Values of $_{\%}a_C$ calculated from this equation are entered into column 6 of Table A.1.

Values of $$_{sat}\gamma_C = \frac{_{sat}a_C}{x_C}, \quad _h\gamma_C = \frac{_ha_C}{x_C} \quad \text{and} \quad _{\%}\gamma_C = \frac{_{\%}a_C}{\% \, C}$$

have also been included in the last three columns of Table A.1.

It will be appreciated that it would be more satisfactory if additional measurements were available at lower concentrations to enable a graphical determination of γ° to be made. Reliance on the accuracy of one observation is not satisfactory.

It is interesting to note the wide range of numbers which can be used to describe the same solution. It is obvious that an activity value is of no value unless the standard state to which it is referred is known.

Example 10. Chemical potential and change of standard state

Calculate the chemical potentials of carbon relative to pure gaseous carbon at 1 atm pressure, of pure solid carbon, of carbon in a saturated solution in iron, and of carbon in the Henrian and wt % standard states – all values to refer to $1540°C$.

DATA : As in Example 9.

The vapour pressure of solid carbon, p_C atm, at T K, is given by—

$$\log_{10} p_C = -\frac{47,000}{T} - 0.75 \log_{10} T + 11.99.$$

At $1540°C$ or 1813 K, at which the data of Example 9 are valid, $p_C = 4.19 \times 10^{-17}$ atm.

Let the standard state of carbon be the pure gas at 1 atm pressure. That this state cannot be realized is no obstacle in thermodynamics.

Then the chemical potentials of carbon in various states, referred to this standard state and always at 1813 K, are as follows—

At the standard state, 1 atm pressure, $RT \ln$ in $1 =$ 0
At 10^{-1} atm pressure, $RT \ln 10^{-1}$ (from equation (6.44c)) = $-34,712$ J
At 10^{-10} atm pressure, $RT \ln 10^{-10} =$ -347.120 J
At 4.19×10^{-17} atm pressure, $RT \ln (4.19 \times 10^{-17}) =$ $-556,194$ J
As pure solid carbon (same as vapour in equilibrium), $-556,194$ J
As carbon in a saturated solution in $Fe_{(sat}a_C = 1)$(same as C), $-556,194$ J
As carbon in a 2.1% solution in $Fe_{(sat}a_C = 0.099)$,
   $(-556,194 + RT \ln 0.099) =$ $-591,057$ J
As carbon in a 0.64% solution in $Fe_{(sat}a_C = 0.029)$,
   $(-556,194 + RT \ln 0.029) =$ $-609,567$ J
As carbon in a 0.216% solution in $Fe_{(sat}a_C = 0.006)$,
   $(-556,194 + RT \ln 0.006) =$ $-633,318$ J
As carbon in the unreal Henrian standard state
   in which $_{sat}a_C = 0.60$, $(-556,194 + RT \ln 0.60) =$ $-563,895$ J

As carbon in a 1% solution in Fe in which $_{sat}a_C = 0.027$,

$(-556{,}194 + RT \ln 0.027) =$ $-610{,}644$ J

Notes: The value of $_{sat}a_C$ for the Henrian standard state is obtained by extrapolating the Henry's law line to intersect the ordinate through $x_C = 1.0$ (unreal) in $_{sat}a_C = 0.60$. The value of $_{sat}a_C$ for the wt % standard state is obtained by extrapolating the Henry's law line as far as $x_C = 0.045$ which (see *Footnote below*) is the mole fraction when % C = 1. At this point, which corresponds to an unreal condition,

$$_{sat}a_C = \frac{0.006}{0.01} \times 0.045 = 0.027.$$

It will be appreciated that chemical potentials of carbon which are referred to pure solid carbon or to the saturated solution in iron as the standard state are less negative than those listed by 556,194 J, and those referred to the Henrian standard state are less negative by 563,895 J. Consequently chemical potentials of carbon which are referred to the Henrian standard state are more negative than those referred to the saturated solution by $563{,}895 - 556{,}194 = 7701$ J while those referred to the wt % standard state are more negative than if referred to the Henrian standard state, by $610{,}644 - 563{,}895 = 46{,}749$ J. This may help to clarify the arguments presented on page 127.

Footnote: Consider 100 g alloy with 1% C and 99% Fe. There are 1/12 moles C and 99/56 moles Fe so that the mole fraction $x_C = (1/12 \div (1/12 + 99/56))$, i.e. 0.045).

Example 11. Interaction coefficients

Calculate the $_{\%}a_C$ value for the carbon in a molten steel at $1600°C$ which contains 0.216% C, 1.2% Mn, 0.3% Si, 0.03% S, 0.03% P, 0.5% Mo, and 0.01% O.

DATA : As in Example 9.

The following interaction coefficients have been taken from

reference (50) in the Bibliography:

$$e_C^C = 0.19\text{--}0.225;\ e_C^{Mn} = -0.001;\ e_C^{Si} = 0.1;\ e_C^S = 0.5;$$

$$e_C^P = 0.05;\ e_C^{Mo} = -0.009;\ \text{and}\ e_C^O = -0.35.$$

All values are for $1600°C$.

The range of values for e_C^C from 0.19 to 0.225 is not untypical of the level of uncertainty in data currently available. Using the value 0.19 to calculate $\%a_C$ in an Fe–C binary alloy containing 0.216% C, we obtain by equation (6.58),

$$\log \%\gamma_C = e_C^C \times wt\% \ C$$

$$= 0.19 \times 0.216$$

$$= 0.0410.$$

Therefore

$$\%\gamma_C = 1.099$$

and

$$\%a_C = 0.237.$$

If the highest value of e_C^C is used, i.e. 0.225, the value of $\%a_C$ is calculated to be 0.242. Such a small difference would seldom be significant, so that it does not matter which value of e is used. In general where there is a choice of values offered one in the middle of the range has been selected and in the case of carbon, 0.21 will be used below.

It will be noticed that the $\%a_C$ is appreciably greater than the wt % C value 0.216 which was obtained in Example 9. It would appear that the assumption made in the working of Example 9 that Henry's law is valid up to 0.216% has not been substantiated by more recent investigators.

Putting all of the data into equation (6.58),

$$\log \%\gamma_C^* = e_C^C \cdot wt\% \ C + e_C^{Mn} \cdot wt\% \ Mn + e_C^{Si} \cdot wt\% \ Si + e_C^S \cdot wt\% \ S$$

$$+ e_C^P \cdot wt\% \ P + e_C^{Mo} \cdot wt\% \ Mo + e_C^O \cdot wt\% \ O.$$

$$= (0.21 \times 0.216) - (0.001 \times 1.2) + (0.1 \times 0.3) + (0.05 \times 0.03)$$

$$+ (0.05 \times 0.03) - (0.009 \times 0.5) - (0.35 \times 0.01)$$

i.e. $\log \%\gamma_C^* = 0.0454 - 0.0012 + 0.03 + 0.0015 + 0.0015 - 0.0045 - 0.0035 = 0.0692.$

Therefore

$$\%\gamma_C = 1.173$$

and

$$\%a_C = 0.253 \quad \text{when} \quad C = 0.216\%$$

This value is about 5% higher than that obtained above where the interaction of the other elements with carbon was ignored but it is 17% above the value 0.216 obtained by assuming that activity can be put equal to wt% C. It will be seen that apart from the effect of carbon upon itself, silicon makes by far the largest contribution to the change in the activity coefficient because the e values for other elements are all small. The coefficients for the interactions of H and N on C are greater than most others at about 0.7 and 0.1, respectively, but in both cases the concentration in steel would always be very small and their total effect would also be small.

To demonstrate the possible importance of using these coefficients to obtain more accurate values of activities, Example 12 shows the effect their use has on the examination of the equilibrium between carbon and oxygen in molten steel.

Example 12. Interaction coefficients and $\Delta G° = -RT \ln K$

Calculate the concentration of oxygen which would be in equilibrium with 0.216% C in molten steel of the composition given in Example 11. Compare this with the concentration calculated when the interactions of the solute elements are ignored.

DATA : As in Example 11.
Interaction coefficients for oxygen from Bibliography (50) $-e_O^O = -0.2$; $e_O^C = -0.45$; $e_O^{Mn} = -0.025$; $e_O^{Si} = -0.13$; $e_O^S = -0.09$; $e_O^P = 0.06$; and $e_O^{Mo} = 0.0035$.

For the reaction $C_{(in\ Fe)} + O_{(in\ Fe)} = CO$, $m = 1/K = \%a_C \times \%a_O$

$= 0.0020$, at $1600°C$ (see Example 13).

The activity coefficient of oxygen, calculated as for carbon in Example 11, is $\%\gamma_O^* = 0.68$. Example 11 shows that $\%a_C = 0.253$, so that—

$$\%a_O = \frac{0.0020}{\%a_C} = \frac{0.0020}{0.253} = 0.0079$$

and

$$wt\%\ O = \frac{\%a_O}{\%\gamma_O} = \frac{0.0079}{0.68} = 0.0116.$$

If the activities of C and O are simply put equal to their concentrations in per cent by weight, as is done on page 317, the concentration of oxygen in equilibrium with 0.216% carbon would be calculated to be $0.0020/0.216 = 0.0093\%$. This would be an underestimate of the concentration of dissolved oxygen in equilibrium with the carbon, by about 20%. This could be important if, for example, the figure was being used to calculate the necessary addition of aluminium to complete the deoxidation. (Note that the effect of silicon as a deoxidizing agent has been ignored in this example.)

Example 13. $\Delta G° = -RT \ln K$, *interaction coefficients and temperature change*

Calculate the change in the oxygen concentration in the molten steel of Example 9 which would be in equilibrium with the 0.216% of carbon if the temperature were reduced to $1550°C$.

DATA :　As in Example 9.

Results of Example 12.

For reaction $C_{(1\%\ in\ Fe)} + O_{(1\%\ in\ Fe)} = CO$; $\Delta G_T° = -22,390 - 39.7T$ J.

The change in temperature alters $\Delta G_T°$ and hence K_C and m (see previous example). The values of γ_C^* and γ_O^* are also modified.

At $1600°C$, $\Delta G_{1873}^{\circ} = -96,748$, $K_C = 499$ and $m = \%a_C \times \%a_O = 0.0020$.

At $1550°C$, $\Delta G_{1823}^{\circ} = -94,763$, $K_C = 518$ and $m = 0.00193$.

The change in the activity coefficients can be estimated only by assuming that the solutions of C and O in Fe behave as regular or at least like semi-regular solutions. Regularity can be tested by using equation (6.69) from which the condition for regularity is that the ratio $\log {}_h a_i/(1-x_i)^2$ is a constant. The results in Example 9 can be used to show that the Fe–C solutions are certainly not regular and they provide no positive evidence of semi-regular behaviour when the solutions are very dilute. Indeed Henrian behaviour, which has already been assumed, and regularity are not compatible. Even in the absence of regularity, however, the activity coefficients may be adjusted for temperature by assuming that they are at least *approximately* inversely proportionate to the absolute temperature. Thus, at $1550°C$,

$$\%\gamma_C^* = 1.173 \times \frac{1873}{1823} = 1.21 \quad \text{and} \quad \%\gamma_O^* = 0.68 \times \frac{1873}{1823} = 0.699$$

so that

$$\%a_O = \frac{0.00193}{1.21 \times 0.216} = 0.00738$$

and

$$\text{wt}\% \text{ O} = \frac{0.00738}{0.699} = 0.0106.$$

Clearly, in view of the inadequate justifications of assumptions made, this must be regarded as no more than an estimate of the change in the value of the oxygen concentration in equilibrium with the carbon in the steel. Where precision is not possible such an estimate, used with caution, is better than nothing.

Example 14. $\Delta G° = RT \ln K$ and calculation of excess reagent

Examine the efficiency with which hydrogen can be utilized for the reduction of molybdic oxide at 1000 K.

DATA: $MoO_3 + H_2 = MoO_2 + H_2O$; $\Delta G^{\circ}_{1\,T} = -175{,}750 + 144.9T$ J,
$MoO_2 + 2H_2 = Mo + 2H_2O$; $\Delta G^{\circ}_{2\,T} = 53{,}570 - 31.1T$ J.

For reaction (1) at 1000 K, $\Delta G^{\circ}_1 = -30{,}850$ so that $K_1 = \dfrac{H_2O}{H_2}$

$$= 40.9.$$

For reaction (2) at 1000 K, $\Delta G^{\circ}_2 = 22{,}470$ so that $K_2^{\frac{1}{2}} = \dfrac{H_2O}{H_2} = 0.26.$

The first stage of the reduction can be continued until the atmosphere in the reaction vessel retains only about 2.5% H_2 but the second stage cannot proceed after the hydrogen concentration has fallen below about 80%. If the critical value of the H_2/H_2O ratio is rounded off to 4.0, the effective reaction can be written as

$$MoO_3 + 15H_2 = Mo + 12H_2 + 3H_2O.$$

This level of utilization cannot be improved upon at 1000 K but the critical H_2/H_2O ratio is somewhat lower at higher temperatures (e.g. 2.25 at 1200 K). It would be necessary to devise means of removing the water and recycling the hydrogen.

Some excess of reagent is always required. See also Example 15.

Example 15. $\Delta G^{\circ} = -RT \ln K$ *and calculation of excess reagent*

How much aluminium must be added, per tonne, to a heat of silicon deoxidized steel at 1600°C to reduce its dissolved oxygen concentration from 0.01% to 0.0001%?

DATA: $2\underline{Al}_{(1\% \text{ in } Fe)} + 3\underline{O} = Al_2O_3$; $\Delta G^{\circ}_T = -1{,}225{,}000 + 392T$ J.

At 1600°C, $\Delta G^{\circ}_{1873} = -490{,}784$, so that $K = 4.85 \times 10^{13}$.

Since the product is pure alumina at unity activity, and both %$\underline{Al}$ and %$\underline{O}$ are very small,

$$\frac{1}{K} = (\%\underline{Al})^2 \times (\%\underline{O})^3 = 2.06 \times 10^{-14}.$$

When % $\underline{O} = 10^{-4}$, %$\underline{Al} = 0.1435.$

This required an addition of 1.435 kg Al per tonne of steel. This must be retained in solution. In addition Al must be added, equivalent to the 0.099 kg of oxygen which is precipitated as Al_2O_3 from each tonne of steel, i.e. 0.099 x $\frac{54}{48}$ = 0.111 kg. The total addition of Al to the steel is, therefore, 1.55 kg per tonne, most of which goes into solution in the steel. In practice the addition would be rather smaller to avoid undesirable effects of such a high concentration of Al in the finished steel. The reaction might not go to completion but it would approach an equilibrium with a $1/K$ value lower than 2.06×10^{-14} as the metal cooled toward its solidification temperature. At 1530, for example, $1/K = 9.72 \times 10^{-16}$ and only 0.03% Al is required to be in equilibrium with 0.0001% of $\underline{O}$. Allowing for failure to reach equilibrium, an addition of about 1 kg per tonne would probably bring the dissolved oxygen to a sufficiently low level.

Example 16. Thermochemistry and the charge to a converter

A BOS converter is normally charged with 200 t of blast furnace iron and 100 t of steel scrap. The iron is charged at $1300°C$ and scrap and flux at $25°C$. The heat is finished at $1600°C$. The CaO/SiO_2 ratio in the slag is 3 (molar). Calculate how much additional scrap must be added to prevent the temperature rising above $1600°C$ if the silicon content of the iron is raised by 0.1% with no other significant changes made.

DATA: $Si_{(s)} + O_2 = SiO_{2 \text{ (glass)}}$; $\Delta H_{f(298)} = -857.9$ kJ.
Other information is provided below as required.

The additional Si amounts to 0.2 t or 7143 moles.
The calculation will be based on 1 mole of Si and energies expressed in kJ mol^{-1}. The heat available from the oxidation of the additional Si is equal to $\Delta H_{f(298)}$ plus the heat in the Si in its state of solution in Fe at $1300°C$, minus the heat required to bring the SiO_2 produced up to $1600°C$ and dissolve it into the slag. Allowances should be made for heating up to $1600°C$ enough CaO to flux the additional SiO_2 and for oxidizing enough Fe to FeO to maintain the slag analysis unchanged.

Heat in Si at $1300°C$ = Heat content of solid Si at $1300°C$, $H_{Si(1573)}$
+ Latent heat of fusion of Si, $L_{f(Si)}$ at $1300°C$
− Heat of solution of Si in Fe at 1% level at $1300°C$.

DATA: $C_{p(Si, liq)} = 31.0 \times 10^{-3}$ kJ K^{-1},

$C_{p(Si, s)} = 23.9 \times 10^{-3} + 4.27 \times 10^{-6} T - 4.44 \times 10^2 T^{-2}$ kJ K^{-1},

$L_{f(Si)} = 46.5$ kJ at 1683 K,

$Si_{(liq)} = Si_{(1\% \text{ in } Fe)}$; $\Delta H_m = -128$ kJ mol^{-1} approx. at 1873 K.[91]

From these data—

$$L_{f(1573)} = L_{f(1683)} + \int_{1683}^{1573} (C_{p(Si, liq)} - C_{p(Si, s)}) \, dT$$

$$= 46.5 - 1.36$$
$$= 45.14 \text{ kJ.}$$

$$H_{Si(1573)} = \int_{298}^{1573} C_{p(Si, s)} \, dT$$

$$= 23.49 \text{ kJ,}$$

$$\Delta H_{m(1573)} = -128 \text{ kJ (approximately, as at 1873)}$$

Hence, for the change—

$$Si_{(in Fe at 1573)} - Si_{(solid, 298)}; \Delta H_{Si} = -L_{f(1573)} - H_{Si(1573)}$$
$$+ H_{m(1573)}$$
$$= -45.14 - 23.49 + 128.4$$
$$= +59.8 \text{ kJ.}$$

The heat required to bring the SiO_2 up to $1600°C$ is calculated in three parts: heating α-quartz to 848 K, heating β-quartz to 1873 K and melting the β-quartz at 1873.

DATA: $C_{p(\alpha)} = 4.68 \times 10^{-2} + 3.4 \times 10^{-5} T - 1.13 \times 10^3 T^{-2}$ kJ K^{-1},

$C_{p(\beta)} = 6.03 \times 10^{-2} + 8.1 \times 10^{-6} T$ kJ K^{-1},

$L_{t(\alpha - \beta)} = 0.6$ kJ,

$L_{f(SiO_2)} = 8.54$ kJ (not altered much by temperature changes).

The heat required to bring 1 mole of SiO_2 from 298 K to 1873 K is given by—

$$\Delta H_{SiO_2} = \int_{298}^{848} C_{p(\alpha)} dT + L_{t(\alpha-\beta)} + \int_{848}^{1873} C_{p(\beta)} dT + L_{f(SiO_2)}$$

$$= 34.0 + 0.6 + 73.1 + 8.54$$

$$= 116.2 \text{ kJ.}$$

The SiO_2 required 3 moles of CaO to be heated up to 1873 K and combined with it to make slag.

DATA: $C_{p(CaO)} = 4.18 \times 10^{-2} + 1.02 \times 10^{-5} T - 4.52 \times 10^2 T^{-2}$

$$\text{kJ K}^{-1},$$

$L_{f(CaO)} = 75 \text{ kJ mol}^{-1}$ (estimated),

$\Delta H_{f(3CaO.SiO_2)} = -125 \text{ kJ mol}^{-1}$ in the crystalline state. It is not likely to be very different when formed as a liquid from liquids.

Heat required for CaO $= \Delta H_{CaO} = 3 \int_{298}^{1873} C_p(CaO) \, dT + 3L_{f(CaO)}$

$$= 246.0 + 225 = 471 \text{ kJ per mole Si.}$$

Heat obtained from slag formation = 125 kJ per mole Si.

If the FeO in the slag is to be maintained at the original concentration it is necessary to oxidize about 1 mole of Fe for each addition mole of Si.

The heat of formation of FeO, $\Delta H_{f(FeO)}$ is about -250 kJ mol^{-1}.

The total extra energy available per mole of additional Si is obtained from—

$$\Sigma \Delta H = \Delta H_{f(SiO_2, 298)} + \Delta H_{Si} - \Delta H_{SiO_2} - \Delta H_{CaO} + \Delta H_{f(slag)}$$

$$+ \Delta H_{f(FeO)}$$

$$= -857.9 + 59.9 + 116.2 + 471 - 125 - 250 = -585.9 \text{ kJ.}$$

The total extra heat available in the 300-t heat is therefore $585.9 \times 7143 = 4.18 \times 10^6 \text{ kJ.}$

If this heat is not used to heat additional scrap the steel will rise in temperature above $1600°C$. The heat capacity of molten Fe at $1600°C$ is 4.42 J g atom^{-1} K^{-1}. This is equivalent to 790 kJ t^{-1} K^{-1} so that 4.18×10^6 kJ will raise the temperature of 300 t of steel by $17.6°C$. This ignores the heat capacity of the vessel and that of the slag. If it is required to keep the temperature at $1600°C$ more scrap must be added.

DATA: The heat content of liquid Fe at $1600°C$ $(H_{1873} - H_{298})$ is listed as $81,306$ J g atom^{-1}, equivalent to 1.456×10^6 kJ t^{-1}.

The required amount of scrap is $\dfrac{4.18 \times 10^6}{1.456 \times 10^6} = 2.9$ t.

It is possible that a heat may have a high silicon content, unknown to the operator because of errors in the analytical procedure. In that case the finishing temperature would be high and the slag analysis would not be as intended.

This calculation is, of course, only a small part of the computation needed to control a BOS process (see page 272).

Example 17. Electrolysis, the Nernst equation and activities in aqueous solutions

When Zn is deposited electrolytically from a sulphate solution, the net reaction is—

$$ZnSO_4 + H_2O = Zn + H_2SO_4 + \tfrac{1}{2}O_2$$

for which $E_{298}^{\circ} = 2.0$ V. What would be the minimum e.m.f. required to deposit Zn from a solution containing 100 g l^{-1} Zn and 25 g l^{-1} H$_2$SO$_4$?

When reactants or products are in non-standard states the value of the reversible e.m.f. is obtained from the Nernst equation (6.73a)—

$$E = E^{\circ} + \frac{RT}{nF} \ln \frac{a_{Zn} \times a_{H_2SO_4} \times p_{O_2}^{\frac{1}{2}}}{a_{ZnSO_4} \times a_{H_2O}}.$$

Zinc is usually deposited as very pure metal so that $a_{Zn} = 1.0$.

The standard state for H_2SO_4 is implicitly the normal solution because $E^°$ is measured against the standard hydrogen electrode in which the electrolyte is 1 N HCl or more precisely 1.2 N HCl, the difference being slight.

Normal H_2SO_4 is the 0.5 molal solution (very nearly) for which the activity coefficient is $_{aq}\gamma_{H_2SO_4} = 0.15$ so that $_{aq}a^*_{H_2SO_4} = 0.075$. The concentration of H_2SO_4 in the cell is given as 25 gl^{-1}. This is very close to 0.25 molal at which concentration $_{aq}\gamma_{H_2SO_4} = 0.197$ so that $_{aq}a_{H_2SO_4} = 0.049$ and the activity relative to that of the standard state must be entered as

$$\frac{0.049}{0.075} = 0.66.$$

In a similar, but not identical, manner the standard state for $ZnSO_4$ is a solution containing 1 g ion l^{-1} of Zn^{2+} which is very nearly the 1 molal solution of $ZnSO_4$ (see page 143). In this solution $_{aq}\gamma_{ZnSO_4} = 0.0435$ so that $_{aq}a^*_{ZnSO_4} = 0.0435$. The electrolyte containing 100 g Zn per litre has a molality of about 1.7 at which concentration $_{aq}\gamma_{ZnSO_4} = 0.036$ so that $_{aq}a_{ZnSO_4} = 0.61$. The activity relative to the standard state at which the $E^°$ value is measured is therefore

$$\frac{0.061}{0.0435} = 1.40.$$

The activity coefficients were taken from reference (31) in the Bibliography.

The activity of water is usually entered as 1.0 but would probably be reduced to about 0.95 in this electrolyte.

It is now possible to evaluate the reversible e.m.f. under the prescribed conditions—

$$E = 1.99 + 0.0295 \ln \frac{1.0 \times 0.66 \times 1.0}{1.40 \times 0.95}$$

$$= 2.00 - 0.02$$

$$= 1.98 \text{ V}.$$

This is obviously so near to the E° value that in practice there is seldom any advantage to be gained from calculating the difference, unless concentrations very far from molal are being considered. The difference in E° due to the operating temperature being about $70^\circ C$ would also be very small. In cells used for electrowinning Zn the potential drop per cell is about 3.3 V. The effects of small changes in the composition of the electrolyte are, obviously, not very important.

Example 18. Electrolysis and energy requirements

Compare the production rate and the energy required per kg of Zn produced in an electrolytic extraction process when the cell voltage has the values 2.5, 2.75, 3.0, 3.25 and 3.5. Consider the effect of production rate on production costs. Assume a current efficiency of 85%.

DATA: As in Example 17.
Electrode dimensions—1 m x 1 m, spaced 8 cm apart.
Specific resistivity of electrolyte—6Ω cm^{-3} (ohms per cm cube).
Resistance due to polarization of anode—$0.003\ \Omega$; and at cathode—$0.001\ \Omega$.
Resistance in connections and bus-bars—$0.0005\ \Omega$.

The resistance of the electrolyte between two electrodes is obtained from equation (12.3) —

$$R = r \cdot \frac{1}{s} = 6 \times \frac{8}{10,000} = 0.0048\ \Omega.$$

The total resistance of the cell = 0.0048 + 0.003 + 0.001 + 0.0005

$$= 0.0093\ \Omega.$$

If the applied voltage across the cell is 2.50 and if this is opposed by $E_R = 2.0$ (see Example 17) the net potential drop across the combined resistance is 0.5 V.

The cell current is, therefore,

$$I = \frac{0.5}{0.0093} = 53.76 \text{ A}.$$

The production rate per cell $= \dfrac{53.76 \times 0.85 \times 3600}{96,500} \times \dfrac{65.38}{2} \times \dfrac{1}{1000}$

$$= 0.056 \text{ kg Zn per hour}.$$

The energy requirement $= 2.5 \times 53.76 \times 0.001$

$$= 0.134 \text{ kWh for } 0.056 \text{ kg Zn}.$$

Energy per kg Zn $\quad = \dfrac{0.134}{0.056} = 2.39 \text{ kWh}.$

Repeating for the other applied voltages and tabulating the results (Table A.2) shows clearly that the energy requirement per kg of Zn rises steadily as the production rate is raised.

It is obvious that if the energy cost per unit of production rises as the production rate is increased and if, as is usual, other charges like the cost of capital, labour and overheads accumulate at constant rate, independent of production rate, then the portion of these standing charges attributable to each unit of production falls as the production rate rises, so that the cost per unit of production may pass through a minimum value at some particular production rate. That this may be so is demonstrated in

TABLE A.2

Applied potential (V)	Production rate (kg h^{-1})	Power (kW)	Energy per kg Zn (kWh kg^{-1})	Energy cost per kg*	Standing charges per kg*	Total cost per kg
2.5	0.056	0.134	2.39	2.39	1.79	4.18
2.75	0.084	0.221	2.63	2.63	1.19	3.82
3.0	0.112	0.322	2.88	2.88	0.89	3.77
3.25	0.140	0.435	3.11	3.11	0.71	3.82
3.5	0.168	0.564	3.35	3.35	0.60	3.95

*Energy at 1 arbitrary monetary unit per kWh and standing charges at one-tenth of a monetary unit per hour.

Table A.2 but it should be made clear that this need not occur at a production rate which is practicable. This depends entirely on the relative values of the price of energy and of the standing charge. If in the above example the standing charge is raised to 0.5 monetary units per hour the minimum cost per kg Zn occurs only at a production rate so high that an applied voltage in excess of 3.5 is required, at which hydrogen is evolved vigorously.

In practice the polarization potentials would increase as the cell current was raised so that an additional energy cost would be incurred. The magnitude of this increase is not readily estimated.

Example 19. Reaction rates and the Arrhenius equation

A reaction occurring at an interface has an activation energy of 100 kJ mol^{-1}. Would the rate of this reaction be increased more by an increase in temperature from $800°C$ to $850°C$ or by an extension of its effective interfacial area by a factor of 2?

1. Doubling the area would increase the mass of matter transferring in any time interval by a factor of 2.0.
2. An increase in temperature from $800°C$ to $850°C$ will increase the rate of reaction from

$$A \exp - \left(\frac{100,000}{8.315 \times 1073} \right) \quad \text{to} \quad A \exp - \left(\frac{100,000}{8.315 \times 1123} \right)$$

where A is not known.

The ratio of these rates can be expressed as

$$\frac{\text{Rate at } 850°}{\text{Rate at } 800°} = \exp - \left(\frac{100,000}{8.135} \cdot \left(\frac{1073 - 1123}{1073 \times 1123} \right) \right)$$

$$= 1.65.$$

As this is less than 2, the increase in temperature has less effect on the reaction rate than doubling the interfacial area.

Example 20. Reaction rates and the Arrhenius equation

The sulphur content of the steel in an electric arc furnace fell from 0.035% to 0.030% in 60 minutes when the temperature was $1580°C$ and in

50 minutes when it was 1600°C, with the conditions otherwise unchanged. Calculate the time required for the same reduction in sulphur content at 1630°C.

This is another application of the Arrhenius equation (8.6).

At 1580°C the rate of sulphur removal can be expressed as—

$$r_{1853} = \frac{0.005}{60} = A \exp\left(\frac{-Q}{8.315 \times 1853}\right) \quad \% \text{ S min}^{-1} \tag{1}$$

and the rate at 1600°C as—

$$r_{1873} = \frac{0.005}{50} = A \exp\left(\frac{-Q}{8.315 \times 1873}\right) \quad \% \text{ S min}^{-1} \tag{2}$$

where A and Q are both unknown.

To eliminate A, divide (1) by (2)—

$$\frac{50}{60} = \exp - \left[\frac{Q}{8.315}\left(\frac{1}{1853} - \frac{1}{1873}\right)\right],$$

$$\therefore 0.833 = \exp - (6.93 \times 10^{-7} Q),$$

$$\therefore \ln 0.833 = -6.93 \times 10^{-7} Q,$$

$$\therefore Q = 263,667 \text{ J}.$$

To find A enter Q in equation (1)—

$$\frac{0.005}{50} = A \exp - \left(\frac{263,667}{8.315 \times 1853}\right),$$

$$\therefore 10^{-4} = A \, 44.4 \times 10^{-8},$$

$$\therefore A = 2252.$$

At 1630°C the rate of sulphur removal is obtained from—

$$r_{1903} = \frac{0.005}{t} = 2252 \exp - \left(\frac{263,667}{8.315 \times 1903}\right)$$

where t is the time in minutes required to reduce the sulphur from 0.035% to 0.030%. It follows that $t = 38$ minutes.

It is assumed that the removal of sulphur is controlled by a process which involves the formation of an activated complex. There is no evidence in the question that this is the case. An activation energy as high as that calculated might be associated with control by diffusion in the slag layer.

Example 21. Reaction rates, evaporation and diffusion

Calculate the rate at which Mn evaporates from the clean surface of a molten alloy of Fe with 1% Mn under a virtually perfect vacuum in an induction furnace at 1600°C. Calculate also the concentration of Mn remaining in the alloy after 10 minutes if the melt is 20 cm deep. Assume that the metal is well stirred inductively and that the evaporation rate is controlled by diffusion across a boundary layer of thickness 0.003 cm.

DATA: Fe–Mn solutions behave ideally; $\gamma_{Mn}^{\circ} = 1.0$.

Vapour pressure of pure Mn, p° atm is obtained from—

$\log p_{Mn}^{\circ} = -13{,}900 T^{-1} - 2.52 \log T + 14.39$.

M.W. of Mn = 54.94.

Density of Fe = 7160 kg m^{-1} = 7.16 cm^{-3}.

Diffusion coefficient for Mn in Fe, $D_{Mn} = 10^{-4}$ cm^2 s^{-1}.

Let the mole fraction of Mn at the surface of the alloy be x_s. Since the molecular weights of Fe and Mn are almost the same, use the approximation that the weight per cent of Mn at the surface is $100 x_s$.

Use equation (12.9a) to obtain the rate of evaporation of Mn in terms of x_s.

$$w_{Mn} = 44.33 \cdot \gamma_{Mn}^{\circ} x_s \cdot p_{Mn}^{\circ} \left(\frac{M.W._{Mn}}{T} \right)^{1/2}$$

At 1600°C, $p_{Mn(1873)}^{\circ} = 0.053$ atm.

$$w_{Mn} = 44.33 \times 1.0 \times x_s \times 0.053 \times \left(\frac{54.94}{1873} \right)^{1/2}$$

$$= 0.402 x_s \text{ g cm}^{-2} \text{ s}^{-1}.$$

This must be equal to the rate at which Mn diffuses across the boundary layer as given by equation (8.16)—

$$w_{Mn} = D_{Mn} \cdot \frac{a_b - a_s}{\delta} \text{ g cm}^{-2} \text{ s}^{-1}.$$

The activities in this example are equal to the mole fractions and can be replaced by concentrations but these must be entered in units compatible with those of D and δ, i.e. in g cm^{-3}. The appropriate values are

$$c_b = 1\% \text{ Mn} = 0.0716 \text{ g cm}^{-3} \quad \text{and} \quad c_s = 100x_s = 7.16x_s \text{ g cm}^{-3}.$$

Hence, $$w_{Mn} = \frac{10^{-4}}{0.003} \cdot (0.0716 - 7.16x_s) = 0.402x_s, \qquad (A.21.1)$$

i.e. $$0.00238 - 0.238x_s = 0.402x_s$$

so that $$x_s = 0.00372$$

and $$w_{Mn} = 0.0015 \text{ g cm}^{-2} \text{ s}^{-1}.$$

The rate at which the concentration falls depends on the mass of alloy in the furnace and on the area exposed to the vacuum. The rate must obviously fall as the residual concentration of Mn is reduced.

In general, equation (A.21.1) can be written—

$$w_{Mn} = \frac{10^{-4}}{0.003} \cdot 7.16(x_b - x_s) = 0.402x_s$$

so that $$(x_b - x_s) = 1.684x_s$$

and $$x_b = 2.684x_s$$

or $$x_s = 0.37x_b$$

$$= 0.0037 \ (\% \text{ Mn}).$$

At any concentration of Mn, $w_{Mn} = 0.402x_s = 0.0015 \ (\% \text{ Mn})$ g cm^{-2} s^{-1}.

If the exposed area is A cm^2 the rate of evaporation of Mn can be expressed as

$$\frac{dw_{Mn}}{dt} = 0.0015A \ (\% \text{ Mn}) \text{ g s}^{-1}$$

and if the volume of alloy is V cm^3, the rate of reduction of the concentration expressed as % Mn is obtained from—

$$-\frac{d(\% \text{ Mn})}{dt} = 0.0015A \ (\% \text{ Mn}) \cdot \frac{100}{\rho V}.$$

But $V/A =$ the depth of the melt, d (assuming the pot to be cylindrical), so that

$$-\frac{d(\% \text{ Mn})}{dt} = \frac{0.15}{\rho \cdot d}(\% \text{ Mn}) = \frac{0.15}{7.16 \times 20}(\% \text{ Mn})$$

$$= 1.04 \times 10^{-3} \ (\% \text{ Mn}).$$

Separating the variables and integrating from $t = 0$ to t,

$$-\frac{d(\% \text{ Mn})}{(\% \text{ Mn})} = 1.04 \times 10^{-3} \ dt,$$

i.e.

$$\ln\left(\frac{\% \text{ Mn}_{t=0}}{\% \text{ Mn}_{t=t}}\right) = 1.04 \times 10^{-3} \ t.$$

If the value of $\% \text{ Mn} = 1.0$ when $t = 0$, then after 10 minutes when $t = 600$ s,

$$\ln\left(\frac{1}{\% \text{ Mn}_{t=600}}\right) = 1.04 \times 10^{-3} \times 600 = 0.624$$

so that $\% \text{ Mn}_{t=600} = 0.536.$

The working of this example has been made exceptionally easier than usual by the fact that the molecular weights of the two metals involved are very similar. The result obtained is close to what might be found in practice but it will be appreciated that the calculated rates of evaporation are very dependent on the values used for the diffusion coefficient and the thickness of the boundary layer. Diffusion coefficients are extremely difficult to measure with accuracy. The values for the thicknesses of surface diffusion zones are even more uncertain and are in any case subject to variations in the physical conditions obtaining in particular circumstances.

CONVERSION FACTORS AND VALUES OF PHYSICAL CONSTANTS

Physical quantity	SI units	Units used in other systems and their equivalents in SI		Notes
Length	km	1 mile	= 1.609 km	
	m	1 foot (ft)	= 0.3048 m	
	mm	1 inch (in.)	= 25.4 mm	
	μm	1 cm	= 10 mm	For mesh sizes, see Appendix 3
Area	m^2	1 cm^2	= 10^{-4} m^2	
Volume	m^3	1 cm^3	= 10^{-6} m^3	
		1 litre (l)	= 10^{-3} m^3	
Mass	(tonne) t	1 t	= 10^3 kg or 10^6 g	
	kg	1 lb	= 0.4539 kg	
	g	1 ton	= 1.016 t	(long ton = 2340 lb)
		1 short ton	= 0.907 t	(2000 lb)
Velocity	m s^{-1}	1 cm s^{-1}	= 0.01 m s^{-1}	
Density	kg m^{-3}	1 g cm^{-3}	= 10^3 kg m^{-3}	
				(Specific gravity, S.G., is numerically equal to density in g cm^{-3} but is dimensionless)
Force	(Newton) N	1 dyne	= 10^{-5} N	
Energy	(Joule) J	1 cal	= 4.186 J	(calorie or g calorie)
	kJ	1 k cal	= 4.186 kJ	(kilogramme calorie)
	MJ	1 kW h	= 3.6 MJ	(1 J = 1 watt-sec)
Surface tension	N m^{-1}	1 dyne cm^{-1}	= 10^{-3} N m^{-1}	
		1 erg cm^{-2}	= 10^{-3} N m^{-1}	
Viscosity	N s m^{-2}	1 poise	= 10^{-1} N s m^{-1}	
Pressure	N m^{-2}	1 atm	= 1.01325×10^6 dynes cm^{-2}	(normal atmosphere = 760 mm Hg or torr)
		1 torr	= 133.322 N m^{-2}	
		1 bar	= 10^5 N m^{-2}	

Physical quantity	SI units	Units used in other systems and their equivalents in SI	Notes

The molar gas constant R = 8.3143 J mol^{-1} K^{-1}
= 1.986 cal deg^{-1} mol^{-1}
= 0.08205 litre-atm mol^{-1}
= 2.27 x 10^{19} erg mol^{-1}
The normal molar volume of an ideal gas at 0°C = 273.16 K and
1 normal atmosphere pressure
= 0.022413 m^3
= 22.413 litre
Planck's constant h = 6.626 x 10^{-34} J s
= 6.626 x 10^{-27} erg s
Boltzmann's constant k = 1.38 x 10^{-23} J K^{-1} $(=R/N)$
Avogadro's number N = 6.023 x 10^{23} mol^{-1}

SCREEN AND SIEVE APERTURE SIZES

THE nominal aperture sizes of the members of each of the principal series are listed in inches where appropriate and in millimetres down to 1 mm and in micrometres below 1000 μm. The Mesh number is also given where appropriate but the trend is toward designation by nominal aperture in mm or μm. The British and American Mesh numbers indicate the number of wires in the cloth per linear inch; the French number has been more arbitrarily selected. The B.S. and A.S.T.M. series are now very similar with respect to the nominal apertures in use but the Mesh numbers do not always correspond. In British and American systems the ratio of adjacent sizes is approximately $\sqrt{2}$ (or $\sqrt[4]{2}$ at the top of the range) except in Tyler where it is exactly $\sqrt{2}$ or $\sqrt[4]{2}$. The German and French systems use a ratio of 1.2 ($= \sqrt[10]{10}$). These are identical except that the DIN series extends to larger sizes. The various specifications cover the gauge of wire which must be used (which determines the screening area available) and the tolerances permitted on wire gauge and on apertures. They also refer to methods of constructing screens and to the use of perforated plate with either square or round holes.

British Standard Test Sieves–B.S. 410, Old Series 1943: Inches or Mesh no.	Old Series 1943: Equivalent mm or μm	New Series 1970 mm or μm	U.S. Standard A.S.T.M. E/11/1970: Inches or Mesh no.	U.S. Standard: mm or μm	Tyler Standard screens: Mesh no.	Tyler: Inches	Tyler: mm or μm	IMM Standard screens: Mesh no.	IMM: Inches	IMM: mm or μm	German Standard DIN 4188: mm or μm	French Standard screens: Mesh no.	French: mm or μm
5	127	125	5	125									
	106	106	4.24	106									
4	101.6		4	100									
$3\frac{1}{2}$	88.9	90	$3\frac{1}{2}$	90									
3	76.2	75	3	75									
$2\frac{3}{4}$	69.9												
$2\frac{1}{2}$	63.5	63	$2\frac{1}{2}$	63									
$2\frac{1}{4}$	57.2												
		53	2.12	53									
2	50.8		2	50									
$1\frac{7}{8}$	47.6												
$1\frac{3}{4}$	44.5	45	$1\frac{3}{4}$	45									
$1\frac{3}{8}$	41.3												
$1\frac{1}{2}$	38.1	37.5	$1\frac{1}{2}$	37.5									
$1\frac{3}{8}$	34.9												
$1\frac{1}{4}$	31.7	31.5	$1\frac{1}{4}$	31.5									
$1\frac{1}{8}$	28.6												
		26.5	1.06	26.5		1.050	26.67						
1	25.4		1	25.0							25		
$\frac{7}{8}$	22.2		$\frac{7}{8}$	22.4							20		
$\frac{3}{4}$	19.1	19.0	$\frac{3}{4}$	19.0		0.742	18.85				18		
$\frac{5}{8}$	15.9	16.0	$\frac{5}{8}$	16.0							16		
		13.2	0.530	13.2		0.525	13.33						
$\frac{1}{2}$	12.7		$\frac{1}{2}$	12.5									
		11.2	$\frac{7}{16}$	11.2							12.5		

B.S.S. mesh	B.S.S. aperture (µm)	B.S. mm	A.S.T.M.	B.S. mm	Tyler mesh	Tyler in.	Tyler mm	I.M.M. mesh	I.M.M. in.	I.M.M. mm	Metric No.	Metric mm	Metric mm
												10	
3/8	9.5	9.5	3/8	9.5		0.371	9.423						
5/16	7.9	8.0	5/16	8.0								8	
		6.7	0.265	6.7	3	0.263	6.680						
1/4	6.35		1/4 (No. 3)	6.3								6.3	
		5.6	No. 3½	5.6									
											38	5.0	5.0
3/16	4.75	4.75	No. 4	4.75	4	0.185	4.699						
											37	4.0	4.0
		4.00	No. 5	4.00									
											36	3.15	3.15
5	3.353	3.35	No. 6	3.35	6	0.131	3.327						
1/8	3.17												
6 m	2.812	2.80	No. 7	2.80									
								5	0.1	2.540			
											35	2.5	2.5
7 m	2.411												
		2.36	No. 8	2.36	8	0.093	2.362						
8 m	2.057	2.00	No. 10	2.00							34	2.0	2.0
10 m	1.676	1.70	No. 12	1.70	10	0.065	1.651						
											33	1.6	1.6
12 m	1.58							8	0.062	1.574			
14 m	1.405	1.40	No. 14	1.40									
								10	0.05	1.270	32	1.25	1.25
16 m	1.204	1.18	No. 16	1.18	14	0.046	1.168						
								12	0.0416	1.056			
18 m	1.003	1.00	No. 18	1.00							31	1.0	1.0
22 m	853	850	No. 20	850	20	0.0328	833						
								16	0.0312	792	30	800	800
25 m	699	710	No. 25	710									
								20	0.025	635	29	630	630
30 m	599	600	No. 30	600	28	0.0232	589						
36 m	500	500	No. 35	500							28	500	500
44 m	422	425	No. 40	425	35	0.0164	417	30	0.0166	421			
											27	400	400
52 m	353	355	No. 45	355									
								40	0.0125	317	26	315	315
60 m	295	300	No. 50	300	48	0.0116	295						
72 m	251	250	No. 60	250				50	0.010	254	25	250	250
85 m	211	212	No. 70	212	65	0.0082	208	60	0.0083	211			
											24	200	200
100 m	178	180	No. 80	180				70	0.0071	180			
								80	0.0062	157	23	160	160
120 m	152	150	No. 100	150	100	0.0058	147	90	0.0055	139			
150 m	124	125	No. 120	125				100	0.005	127	22	125	125
170 m	104	106	No. 140	106	150	0.0041	104	120	0.0042	107			
											21	100	100
200 m	89	90	No. 170	90									
								150	0.0033	84	20	80	80
240 m	76	75	No. 200	75	200	0.0029	74						
300 m	66	63	No. 230	63				200	0.0025	63	19	63	63
350 m	53	53	No. 270	53									
											18	50	50
400 m		45	No. 325	45									
											17	40	40
		38	No. 400	38									

REFERENCES

1. NIGGLI, P. *Ore Deposits of Magmatic Origin, Their Genesis and Natural Classification.* Thos. Murby, London, 1929, p. 3.
2. GAUDIN, A. M. See Bibliography 9, p. 71.
3. CAREY, W. F. and STAIRMAND, C. J. See Bibliography, 12, p. 117.
4. HUKKI, R. T. See Bibliography, 13, p. 85.
5. MYERS, J. F. See Bibliography, 12, p. 137.
6. CAVANAGH, W. J. and ROGERS, D. J. See Bibliography, 16, p. 91.
7. BOND, F. C. *Ttrans. A.I.M.E.* **193**, 484 (1952).
8. HOLMES, J. A. *Trans. Inst. Chem. Eng.* **35**, 125 (1957).
9. CHAMPION, F. C. and DAVY, N. *The Properties of Matter.* Blackie, Glasgow, 1936, p. 253.
10. HEYWOOD, H. See Bibliography, 12, p. 31.
11. See Bibliography, 10, p. 706.
12. RINELLI, G. and MARABINI, A. M. See Bibliography, 16, p. 493.
13. MELOY, T. P. *Froth Flotation.* A.I.M.E., New York, 1962, p. 91.
14. MALINOWSKII, V. A. *et al.* See Bibliography, 16, p. 717.
15. READ, A. D. and WHITEHEAD. See Bibliography, 16, p. 949.
16. KRUKIEWICZ, R. and LASKOWSKI, J. See Bibliography, 16, p. 391.
17. FISHER, R. A. *J. Agric. Soc.* **16**, 492 (1926).
18. RUMPF, H. See Bibliography, 17, p. 379.
19. STIRLING, H. T. See Bibliography, 17, p. 177.
20. See Bibliography, 21, p. 51.
21. MORRIS, J. C. and BUEHL, R. C. *Trans. A.I.M.E.* **188**, 137 (1950)
22. SHERMAN, C. W. and CHIPMAN, J. *Trans. A.I.M.E.* **194**, 597 (1952).
23. CHIPMAN, J. See Bibliography, 22, p. 23.
24. BODSWORTH, C. *Physical Chemistry of Iron and Steel Manufacture.* 1st ed., Longmans, London, p. 47.

25. TAYLOR, C. R. and CHIPMAN, J. *Trans. A.I.M.E. Tech. Pub.* 1499 (1942).
26. DERGE, G. See Bibliography, 25, p. 15.
27. CHIPMAN, J. See Bibliography, 22, p. 23.
28. CHIPMAN, J. *Metals Handbook.* A.S.M. Cleveland, 1948, p. 1215.
29. SRIDMAR, R. and JEFFES, J. H. E. *Trans. Inst. Min. Metall.* **76**, C44 (1967).
30. See Bibliography, 20, p. 111.
31. SMITHELLS, C. J. *Metals Reference Book*, Vol. 2, 3rd ed., Butterworth, London, 1962, pp. 618–661.
32. See Bibliography, 31, Appendices.
33. RICHARDSON, F. D. and JEFFES, J. H. E. *J.I.S.I.* **160**, 261 (1948).
34. RICHARDSON, F. D., JEFFES, J. H. E. and WITHERS, G. *J.I.S.I.* **166**, 213 (1950).
35. RICHARDSON, F. D. and JEFFES, J. H. E. *J.I.S.I.* **171**, 165 (1952).
36. RICHARDSON, F. D. *J.I.S.I.* **135**, 53 (1953).
37. PELKE, R. D. See Bibliography, pp. 297–301.
38. See Bibliography, 50, p. 485.
39. ELLINGHAM, H. T. *J. Soc. Chem. Ind.* **63**, 125 (1944).
40. TURKDOGAN, E. T. and PEARSON, J. *J.I.S.I.* **175**, 398 (1953).
41. BOOKEY, J. D. *J.I.S.I.* **172**, 61 (1952).
42. See Bibliography, 49, pp. 11 and 124.
43. See Bibliography, 24, p. 12.
44. GOODEVE, C. See Bibliography, 22, p. 9.
45. SAUNDERS, H. C. and WILD, R. *J.I.S.I.* **165**, 198 (1950).
46. DARKEN, L. S. See Bibliography, 22, p. 101.
47. MEADOWCROFT, T. R. and RICHARDSON, F. D. *Trans. Farad. Soc.* **61**, 54 (1965) or see Bibliography, 28, p. 87.
48. RICHARDSON, F. D. *J. Soc. Chem. Ind.* **71**, 50 (1952).
49. BOCKRIS, J. O'M., MACKENZIE, J. D. and KITCHENER, J. A. *Trans. Farad. Soc.* **51**, 1734 (1955).
50. BAAK, T. See Bibliography, 25, p. 84.
51. PAULING, L. *The Nature of the Chemical Bond.* Cornell University Press, Ithaca, 1960.
52. WELLS, A. F. *Structural Inorganic Chemistry.* Clarendon Press, Oxford, 1950.
53. RICHARDSON, F. D. See Bibliography, 22, p. 248.

54. BOWEN, N. L., SCHAIRER, J. F. and POSNJAK, E. *Amer. J. Sci.* 5th ser. **26**, 193 (1933).

55. KOZAKEVITCH, P. *Rev. Met.* **46**, 505 and 572 (1949).

56. OSBORN, E. F. and MUAN, A. See Bibliography, 29, p. 243.

57. DARKEN, L. S. and GURRY, R. W. *Metals Handbook.* A.S.M. Cleveland, 1948, p. 1212.

58. OSBORN, E. F. and MUAN, A. See Bibliography, 29, p. 234.

59. HAY, R. *J. West of Scotland I.S.I.* **49**, 89 (1941–2).

60. TURKDOGAN, E. T. and PEARSON, J. *J.I.S.I.* **173**, 217 (1953).

61. ABRAHAM, K. P., DAVIES, M. W. and RICHARDSON, F. D. *J.I.S.I.* **196**, 82 (1960).

62. REIN, R. H. and CHIPMAN, J. *Trans. Metall. Soc. A.I.M.E.* **233**, 415 (1965).

63. See Bibliography, 28, Vol. 1, p. 285.

64. KUBASCHEWSKI, O. *et al. Metallurgical Chemistry* (Symposium). N.P.L–H.M.S.O., London, 1972, pp. 629–650.

65. See Bibliography, 27, p. 19.

66. KRIVSKY, W. A. and SCHUHMANN, R. *J. Metals* **9**, 981 (1957).

67. See Bibliography, 28, p. 324.

68. See Bibliography, 31, Appendices.

69. REULEAUX, O. *Metall. u. Erz*, **99** (1927) or see Bibliography, 27, p. 21.

70. SHERIDAN, A. T. *Ironmaking Steelmaking* **2**, 262 (1975).

71. WORTHINGTON, P. M. *J.I.S.I.* **200**, 894 (1962).

72. INGRAHAM. T. R. in *Applications of Fundamental Thermodynamics to Metallurgical Processes.* Gordon & Breach, New York, 1967, p. 190.

73. INGRAHAM, T. R. *Trans A.I.M.E.* **233**, p. 359.

74. See Bibliography, 45, p. 68.

75. HOPKINS, D. W. See Bibliography, 19, p. 91.

76. VILLA, H. *J. Soc. Chem. Ind. (Suppl.)*, p. 9 (1950).

77. SMITHELLS, C. *Metals Reference Book.* 3rd ed., Vol. 2. Butterworth, London, 1962, p. 1027.

78. WEIR, D. R., EVANS, D. J. I. and MACKIW, V. N. See Bibliography, 37, p. 3.

79. LINDSTROM, R. and WALLDEN, S. See Bibliography, 38, p. 111.

80. BROTZMANN, K., LANKFORD, W. T. and BRISSE, A. H. *Ironmaking Steelmaking* **3**, 259 (1976).

81. BELL, H. B. *J. Met. Club, Roy. Coll. Sci. and Tech.,* Glasgow **9**, 23 (1957).
82. OLETTE, M. See Bibliography, 23 part II, p. 1065.
83. SMITHELLS, C. *Metals Reference Book.* 3rd ed., Vol. 2. Butterworth, London, 1962, p. 638.
84. LANGMUIR, I. *Phys. Rev.* **2**, 329 (1913).
85. BARRACLOUGH, K. C. *Ironmaking Steelmaking* **4**, 92 (1977).
86. See Bibliography, 45, Chapter 11.
87. See Bibliography, 45, p. 110.
88. BRISBY, M. D. J., WORTHINGTON, P. M. and ANDERSON, R. J. *J.I.S.I.* **202**, 721 (1964).
89. *Handbook of Chemistry and Physics.* 55th ed. C.R.C. Press, Cleveland, Ohio, 1974, p. D56.
90. MARSHALL, S. and CHIPMAN, J. *Trans. Amer. Soc. Metals* **34**, 695 (1942) or see Bibliography, 22, p. 42.
91. CHART, T. G. *High Temp. High Press.* **2**, 461 (1970).

BIBLIOGRAPHY

Chapter 1
1. ATCHISON, L. *A History of Metals* (2 vols.). Macdonald & Evans, London, 1960.
2. TYLECOTE, R. F. *A History of Metallurgy*. Metals Society, London, 1976.

Chapter 2
3. BATEMAN, A. M. *The Formation of Mineral Deposits.* Wiley, New York, 1951.
4. GARRELS, R. M. and CHRIST, C. L. *Solutions, Minerals and Equilibria.* Harper & Row, New York, 1965.
5. EDWARDS, A. B. *Textures of Ore Minerals*, Australian Institute of Mining and Metallurgy, Melbourne, 1960.
6. READ, H. H. *Rutley's Elements of Mineralogy*, 26th ed. Thos. Murby, London, 1970.
7. PRYOR, E. J. *Economics for Mineral Engineers.* Pergamon Press, Oxford, 1958.

Chapters 3 and 4
8. PRYOR, E. J. *Mineral Processing*, 3rd ed. Elsevier, Amsterdam, 1965.
9. GAUDIN, A. M. *Principles of Mineral Dressing.* McGraw-Hill, New York, 1939.
10. COULSON, J. H. and RICHARDSON, J. F. *Chemical Engineering*, Vol. 2. *Unit Operations.* Pergamon Press, London, 1955.
11. CREMER, H. W. and DAVIES, T. (Eds.) *Chemical Engineering Practice*, Vol. 4, *Solid Systems.* Butterworth, London, 1957.
12. *Recent Developments in Mineral Dressing* (Symposium). Institute of Mining and Metallurgy, London, 1953.
13. *Progress in Mineral Dressing* (Transactions of an International Conference at Stockholm). Almqvist & Winskell, Stockholm, 1958.
14. ROBERTS, A. (Ed.) *Mineral Processing* (Proceedings of a Conference at Cannes in 1963). Pergamon Press, London, 1969.
15. JONES, M. J. (Ed.) *Mineral Processing in Extractive Metallurgy* (Vol. 3 of *Proceedings of 9th Commonwealth Mining and Metallurgical Congress, 1969*). Institute of Mining and Metallurgy, London, 1970.
16. JONES, M. J. (Ed.) *10th International Mineral Dressing Congress, 1973.* Institute of Mining and Metallurgy, London, 1974.

Chapter 5
17. KNEPPER, W. A. (Ed.) *Agglomeration* (Symposium). Interscience, New York, 1962.

18. BALL, D. F. *et al.* (Eds.) *The Agglomeration of Iron Ores.* Heinemann, London, 1973.

Chapters 6, 7, 8 and 9

19. HOPKINS, D. W. *Physical Chemistry and Metal Extraction.* J. Garnet Miller, London, 1954.
20. KUBASCHEWSKI, O. and ALCOCK, C. B. *Metallurgical Thermochemistry,* 5th ed. Pergamon Press, Oxford, 1979.
21. WAGNER, C. *The Thermodynamics of Alloys.* Addison Wesley, New York, 1952.
22. *The Physical Chemistry of Process Metallurgy* (Discussion of the Faraday Society No. 4). Gurney & Jackson, London, 1948.
23. *The Physical Chemistry of Process Metallurgy,* Parts I and II (Symposium). Interscience, New York, 1961.
24. MOELWYN-HUGHES, E. A. *The Kinetics of Reactions in Solution.* Clarendon Press, Oxford, 1933.
25. ELLIOT, J. F. (Ed.) *The Physical Chemistry of Steelmaking* (Symposium). Massachusetts Institute of Technology and Wiley, New York, 1958.
26. *The Physical Chemistry of Melts* (Symposium). Institute of Mining and Metallurgy, London, 1953.
27. RUDDLE, R. W. *The Physical Chemistry of Copper Smelting.* Institute of Mining and Metallurgy, London, 1953.
28. RICHARDSON, F. D. *The Physical Chemistry of Melts in Metallurgy* (2 vols.). Academic Press, London, 1974.
29. SIMS, C. E. (Ed.) *Electric Furnace Steelmaking* (Vol. 2). *Theory and Fundamentals.* Interscience, New York, 1963.
30. ELLIOT, J. F. and GLIESER, M. *Thermochemistry for Steelmaking* (2 vols.). Addison Wesley, Reading, Mass., 1963.
31. ROBINSON, R. A. and STOKES, R. H. *Electrolyte Solutions.* Butterworth, London 1959.

See also relevant chapters in 49, 50 and 51 below.

Chapters 10 and 11

32. GILCHRIST, J. D. *Fuels, Furnaces and Refractories.* Pergamon Press, Oxford, 1977.
33. JONES, E. B. *Instrument Technology* 3rd ed. (2 vols.) Butterworth, London, 1974.
34. DIEFENDERFER, A. J. *Principles of Electronic Instrumentation.* Saunders, Philadelphia, 1972.

Chapter 12

35. ALCOCK, C. B. *Principles of Pyrometry.* Academic Press, London, 1976.
36. VAN ARSDALE, G. D. (Ed.) *Hydrometallurgy of the Base Metals.* McGraw-Hill, New York, 1955.
37. WADSWORTH, M. E. and DAVIS, F. T. (Ed.) *Unit Processes in Hydrometallurgy* (Symposium). Gordon & Breach, New York, 1974.
38. EVANS, D. J. I. and SHOEMAKER, R. S. (Eds.) *International Symposium on Hydrometallurgy.* A.I.M.E., New York, 1973.
39. BELK, J. A. *Vacuum Techniques in Metallurgy.* Pergamon Press, Oxford, 1963.
40. WINKLER, O. and BAKISH, R. *Vacuum Metallurgy.* Elsevier, Amsterdam, 1971.

41. DUCKWORTH, W. E. and HOYLE, E. *Electroslag Refining.* Chapman & Hall, London, 1969.
42. McBRIDE, D. C. and DANCY, T. E. (Eds.) *Continuous Casting* (Symposium). Interscience, New York, 1962.

Chapter 13
43. DENNIS, W. H. *Metallurgy of the Non-ferrous Metals.* Pitman, London, 1954.
44. GERARD, G. and STROUP, P. T. (Eds.) *The Extractive Metallurgy of Aluminium* (2 vols.) (Symposium). Interscience, New York, 1963.
45. BISWAS, A. K. and DAVENPORT, W. G. *The Extractive Metallurgy of Copper.* Pergamon Press, Oxford, 1976.
46. BOLDT, J. R. *The Winning of Nickel.* Longmans, Toronto, 1967.
47. WRIGHT, P. A. *The Extractive Metallurgy of Tin.* Elsevier, Amsterdam, 1966.
48. MATHEWSON, C. H. *Zinc.* Hafner, New York, 1970.
49. WARD, R. G. *Introduction to the Physical Chemistry of Steelmaking.* Arnold, London, 1962.
50. BODSWORTH, C. and BELL, H. B. *Physical Chemistry of Iron and Steel Manufacture,* 2nd ed. Longmans, London, 1972.

General Textbooks
51. NEWTON, J. *Extractive Metallurgy.* Wiley/Chapman & Hall, N.Y./London, 1959.
52. PEHLKE, R. D. *Unit Processes in Extractive Metallurgy.* American Elsevier, New York, 1973.
53. ROSENQVIST, T. *Principles of Extractive Metallurgy.* McGraw-Hill, New York, 1974.

This selection of books may be of assistance to readers who wish to study the subject more deeply than is possible using only one small textbook. The general textbooks cover a similar range of topics to that covered in this book but with different approaches. Monographs 44 and 45 also offer wide coverage of processes as applied in each case to a single metal but most of the monographs are as restricted as their titles suggest. Books identifiable as proceedings of symposia and conferences contain collections of papers on special topics and sometimes useful review articles which may provide the student with an easy introduction to the literature at the date of the conference. Many of the references in the text are to papers in these books. For more recent papers it is necessary to consult metallurgical abstracts.

INDEX

Place names have been omitted. Their principal appearances are in Chapter 13 where the sources of the various metals are indicated under the heading "Occurrence". Only principal references to common metals and minerals are listed. Other frequently occurring words and phrases have been treated in a similar manner where appropriate

445

Other Titles in the Series